Volker Schulze

Modern Mechanical Surface Treatment

Related Titles

Champion, Y., Fecht, H.-J. (eds.)

Nano-Architectured and Nanostructured Materials
Fabrication, Control and Properties

166 pages with 101 figures and 16 tables
2004
Hardcover
ISBN 3-527-31008-8

Herlach, D. M. (ed.)

Solidification and Crystallization

322 pages with 204 figures and 20 tables
2004
Hardcover
ISBN 3-527-31011-8

Zehetbauer, M., Valiev, R. Z. (eds.)

Nanomaterials by Severe Plastic Deformation

872 pages with 600 figures and 65 tables
2004
Hardcover
ISBN 3-527-30659-5

Kainer, K. U. (ed.)

Magnesium
Proceedings of the 6th International Conference Magnesium Alloys and Their Applications

1088 pages with 1005 figures and 83 tables
2004
Hardcover
ISBN 3-527-30975-6

Leyens, C., Peters, M. (eds.)

Titanium and Titanium Alloys
Fundamentals and Applications

532 pages with 349 figures and 56 tables
2003
Hardcover
ISBN 3-527-30534-3

Kainer, K. U. (ed.)

Metallische Verbundwerkstoffe

342 pages with 286 figures and 37 tables
2003
Hardcover
ISBN 3-527-30532-7

Wagner, L. (ed.)

Shot Peening

584 pages with 522 figures and 77 tables
2003
Hardcover
ISBN 3-527-30537-8

Schumann, H., Oettl, H.

Metallographie

976 pages with 1080 figures and 134 tables
2004
Hardcover
ISBN 3-527-30697-X

Volker Schulze

Modern Mechanical Surface Treatment

States, Stability, Effects

WILEY-VCH

WILEY-VCH Verlag GmbH & Co. KGaA

The Author

Priv.-Doz. Dr.-Ing. habil. Volker Schulze
Universität Karlsruhe (TH)
Inst. f. Werkstoffkunde I
Kaiserstr. 12
76131 Karlsruhe

Original title
Stabilität von Randschichtzuständen in
mechanisch oberflächenbehandelten
metallischen Werkstoffen und deren
Auswirkungen bei thermischen und
mechanischen Beanspruchungen
Habilitationsschrift, Fakultät für
Maschinenbau, Universität Karlsruhe
(TH), 2004

Translation
J. K. Schwing, Germany

Library of Congress Card No.: applied for

British Library Cataloguing-in-Publication Data
A catalogue record for this book is available
from the British Library.

**Bibliographic information published by
Die Deutsche Bibliothek**
Die Deutsche Bibliothek lists this publication
in the Deutsche Nationalbibliografie; detailed
bibliographic data is available in the Internet at
<http://dnb.ddb.de>.

Printed in the Federal Republic of Germany.
Printed on acid-free paper.

Composition   Kühn & Weyh, Satz und Medien,
Freiburg
Printing   Strauss GmbH, Mörlenbach
Bookbinding   J. Schäffer GmbH, Grünstadt
Cover Design   Grafik-Design Schulz, Fußgönheim

ISBN-13:   978-3-527-31371-6
ISBN-10:   3-527-31371-0

Contents

1   Introduction   *1*

2   **Procedures of Mechanical Surface Treatments**   9
2.1   Shot Peening   *9*
2.1.1   Definition and Delimitation of Procedure   *9*
2.1.2   Application Examples   *9*
2.1.3   Devices, Tools and Important Parameters   *11*
2.2   Stress Peening   *14*
2.2.1   Definition and Delimitation of Procedure   *14*
2.2.2   Application Examples   *14*
2.2.3   Devices, Tools and Important Parameters   *15*
2.3   Warm Peening   *15*
2.3.1   Definition and Delimitation of Procedure   *15*
2.3.2   Application Examples   *15*
2.3.3   Devices, Tools and Important Parameters   *16*
2.4   Stress Peening at Elevated Temperature   *16*
2.5   Deep Rolling   *16*
2.5.1   Definition and Delimitation of Procedure   *16*
2.5.2   Application Examples   *17*
2.5.3   Devices, Tools and Important Parameters   *18*
2.6   Laser Peening   *19*
2.6.1   Definition and Delimitation of Procedure   *19*
2.6.2   Application Examples   *20*
2.6.3   Devices, Tools and Important Parameters   *20*

3   **Surface Layer States after Mechanical Surface Treatments**   25
3.1   Shot Peening   *25*
3.1.1   Process Models   *25*
3.1.2   Changes in the Surface State   *44*
3.2   Stress Peening   *72*
3.2.1   Process Models   *72*
3.2.2   Changes in the Surface State   *74*
3.3   Warm Peening   *81*

Modern Mechanical Surface Treatment. Volker Schulze
Copyright © 2006 WILEY-VCH Verlag GmbH & Co. KGaA, Weinheim
ISBN: 3-527-31371-0

3.3.1   Process Models   81
3.3.2   Changes in the Surface State   84
3.4   Stress Peening at elevated Temperature   87
3.5   Deep Rolling   89
3.5.1   Process Models   89
3.5.2   Changes in the Surface State   92
3.6   Laser Peening   101
3.6.1   Process Models   101
3.6.2   Changes in the Surface State   108

4   **Changes of Surface States due to Thermal Loading**   135
4.1   Process Models   135
4.1.1   Elementary Processes   135
4.1.2   Quantitative Description of Processes   137
4.2   Experimental Results and their Descriptions   140
4.2.1   Influences on Shape and Topography   140
4.2.2   Influences on Residual Stress State   142
4.2.3   Influences on Workhardening State   157
4.2.4   Influences on Microstructure   170

5   **Changes of Surface Layer States due to Quasi-static Loading**   179
5.1   Process Models   179
5.1.1   Elementary Processes   179
5.1.2   Quantitative Description of Processes   180
5.2   Experimental Results and their Descriptions   184
5.2.1   Influences on Shape and Deformation Behavior   184
5.2.2   Influences on Residual Stress State   186
5.2.3   Influences on Workhardening State   227
5.2.4   Influences on Microstructure   243

6   **Changes of Surface States during Cyclic Loading**   247
6.1   Process Models   247
6.1.1   Elementary Processes   247
6.1.2   Quantitative Description of Processes   250
6.2   Experimental Results and their Descriptions   260
6.2.1   Influences on Residual Stress State   260
6.2.2   Influences on Worhardening State   291
6.2.3   Influences on Microstructure   298
6.3   Effects of Surface Layer Stability on Behavior during Cyclic Loading   303
6.3.1   Basic Results   303
6.3.2   Effects on Cyclic Deformation Behavior   304
6.3.3   Effects on Crack Initiation Behavior   310
6.3.4   Effects on Crack Propagation Behavior   313
6.3.5   Effects on Fatigue Behavior   319

7 **Summary** *355*

Acknowledgments *365*

Index *367*

1
Introduction

Technological practice today, particularly in the spring-manufacturing, automotive and aerospace industries, is hardly imaginable without mechanical surface treatments. The origins of these processes date back to ancient history. [1.1] states that in the city of Ur, gold helmets were hammered and thus mechanically enhanced, as early as 2700 BC. The knights of the Crusades used the same method to reinforce their swords when shaping them. The first modern-day applications, again, are to be found in military technology, but also in railroad technology. [1.1] reports that in 1789, the outer surfaces of artillery gun barrels were hammered in order to improve their strength, and by 1848, train axles and bearing bolts were evened out by rolling. Until that point, the methods had been intrinsically connected to the skill and experience of the craftspeople, who used strict confidentiality in passing on their knowledge in order to keep their competitive advantage.

It was only in the 1920s and -30s that surface treatment evolved into technical processing methods. Föppl's seminal treatises of 1929 [1.2, 1.3] establish the correlation between mechanical surface treatment and increased fatigue strength, indicating significantly higher fatigue strength in surface-rolled samples than in polished samples. Consequently, Föppl's group [1.4] extended their examinations to include notched components and found that the fatigue strength increased by 20–56 % in the case of deep-rolled thread rods. These findings were confirmed by Thum [1.5] in his systematic examination of the relation of rolling and fatigue strength, published in 1932. Thum also found that resistance to corrosion fatigue [1.6, 1.7] and fretting fatigue [1.8] increased.

An alternative to deep rolling emerged in the form of shot peening. Its precursor was developed in 1927 by Herbert [1.9], a process he termed "cloudburst", in which large quantities of steels balls are "rained" onto component surfaces from a height of 2–4 meters. Herbert observed increases of hardness, but did not give any indications regarding contingent increases of fatigue strength. In his aforementioned [1.2, 1.3] paper of 1929, Föppl showed that samples treated with a ball-shaped hammer also exhibit significantly higher fatigue life under cyclic stress than polished samples do. In 1935, Weibel [1.10] independently proved that sandblasting increases the fatigue strength of wires. This additional precursor of present-day shot peening methods builds on the British patent taken out by the American, Tilgham [1.11], in 1870, which was originally geared at drilling, engraving

Modern Mechanical Surface Treatment. Volker Schulze
Copyright © 2006 WILEY-VCH Verlag GmbH & Co. KGaA, Weinheim
ISBN: 3-527-31371-0

and matting of iron and other metals and deals with surface treatment using sand accelerated by pressurized air, steam, water or centrifugal force. In 1938, Frye and Kehl [1.12] proved the positive effect of blast cleaning treatments on fatigue strength, and in 1939 v. Manteuffel [1.13] found higher degrees of fatigue strength in sandblasted springs than in untreated springs. Crucial systematic examinations were published in the US in the early 1940s. Working at Associated Springs Co., Zimmerli [1.14] used shot peening to increase the fatigue strength of springs and analyzed the influence of peening parameters. At General Motors, Almen [1.15, 1.16] demonstrated fatigue strength improvements in engine components and achieved increased reproducibility of the peening process by introducing the Almen strips named after him. In 1948, fatigue strength improvements were proven also for shot peened components under conditions of corrosion [1.17].

The development of special methods brought an additional impetus for the technical application of mechanical surface treatment processes. Straub and May [1.18] were the first to report increases of fatigue strength in springs which were shot peened under pre-stress. While they presented models in which the state of residual stress was to be shifted toward higher compressive residual stress by means of tensile prestressing, this was not proven until 1959, when Mattson and Roberts [1.19] analyzed residual stress states after 'strain peening' combined with tensile or compressive prestrains. Today, this method is called stress peening and is predominantly used on springs [1.20–1.25], but also on piston rods [1.26, 1.27]. Supplying thermal energy simultaneous or consecutive to the actual peening process constitutes an approach for increasing the effect of the mechanical surface treatment even further. Warm peening, i.e. shot peening at high workpiece temperatures, was first suggested in a 1973 Japanese patent [1.28] to achieve increased fatigue strength in springs by using the "Cottrell effect". In the meantime, applications in the spring manufacturing industry have been examined [1.29–1.35] and fundamental research by the Vöhringer and Schulze group [1.36–1.38], in particular, has been pushing toward a deeper understanding of the processes and an optimization of warm peening. Conventional shot peening and consecutive annealing was examined more closely by the teams of Scholtes [1.39] as well as Vöhringer and Schulze [1.41] as an alternative method. These examinations show that appropriately selected annealing temperatures and times are able to achieve effects comparable to warm peening, while complexity is reduced. Wagner and Gregory [1.42–1.46] increased the density of nuclei for re-crystallization or precipitation in the surface layers of titanium and aluminum alloy workpieces which is effective during annealing after shot peening or rolling, and thus enables fine grain formation and selective or preferred surface hardening. These procedures, too, allow for considerable increases of fatigue strength at room temperature or higher temperatures. A completely new method has been developing since the 1970s in the form of laser shock treatment. However, it has attained technical relevance only gradually. Its importance has started to increase since suitable laser technologies have become available and the enhancement process has been transferred from laboratory lasers, which are irrelevant for technical applications, to industrially applicable lasers [1.47–1.52].

In the course of method development, at first the question remained which surface changes of the workpieces the observed increases in fatigue strength could be attributed to. Samples manufactured by machining were used to prove and to quantitatively record the influence of surface topography on fatigue strength. Houdremont and Mailänder [1.53] demonstrated that the difference in roughness between polished and coarsely cut surfaces leads to fatigue strength changes which become more pronounced the greater the strength of a material is. Siebel and Gaier [1.54] in 1956 stated a factor for roughness that expresses the effect on fatigue strength and decreases linearly with the logarithm of roughness. At first, an intense and controversial debate centered on whether the cause for fatigue strength increases was to be found in the effects of mechanical workhardening, as postulated by Föppl and his team [1.2, 1.3], or the effects of the induced compressive residual stress states, as Thum and his team [1.5, 1.55] assumed. Fig. 1.1 summarizes the essential approaches. Today it is commonly accepted knowledge that the inhomogeneous plastic deformations required for generating residual stresses always involve local alterations of the material state, which may affect a component's fatigue strength. However, the residual stress stability within the given operating conditions of a component determines whether the residual stresses are to be treated as loading stresses, in which case they are predominant in comparison with the effect mentioned first. Both effects may be taken into account in the so-called concept of the local fatigue limit [1.56, 1.57] and be super-

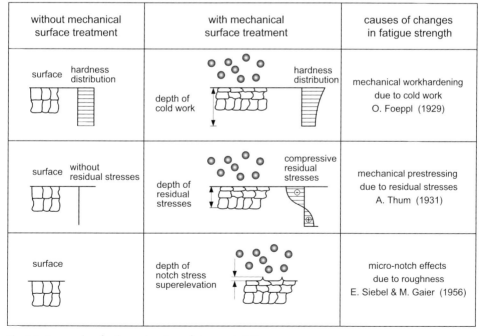

Fig. 1.1: Approaches for the explanation of changes in fatigue behaviour due to mechanical surface treatments

posed with the aforementioned roughness effects and those of additional potential phase transformations.

Mechanical surface treatment processes commonly used today may be roughly divided into cutting and non-cutting methods. The main focus of cutting methods is on shaping, while achieving optimal surface layer states for later use is only a secondary objective. Therefore, study is restricted to describing non-cutting methods which serve to enhance the surface layer state with respect to the future application. Fig. 1.2 shows a systematized compilation of these methods. The methods indicated are subdivided into those without or with relative movement between the tools and the workpiece and those with a static or an impulsive tool impact. The description of methods without relative movement is limited to impulsive impact, which has a repetitive irregular pattern in shot peening and a repetitive regular pattern in laser shock treatment. Among the methods involving relative movement, the focus is on the rolling movement of deep rolling. The aforementioned process modifications are always included in the description. As indicated earlier, it is crucial for the effects of mechanical surface treatment on component properties that the modifications imparted on the surface layer state are as stable as possible and are not reduced significantly during loading. This applies, in particular, to the residual stress states created. Therefore, the following description of the individual methods and the surface layer alterations they cause goes on to examine their stability during thermal, quasi-static and cyclic loading and combinations thereof. In addition to the experimental results and the causes, the focus is also on approaches toward a quantitative modeling of the changes of the surface layer state. In conclusion, the effects of mechanical surface treatments on cyclic loading behavior are discussed systematically and integrated into quantitative model approaches, as well.

		without relative movement	with relative movement			
			rolling		sliding	
			without slip	with slip	solid medium	liquid medium
static	singular	smooth embossing, flat embossing, size embossing				autofretting, stressing
static	repetitive regular		deep rolling, finish rolling, size rolling		spinning, smooth drawing, smooth spinning	
impulsive	singular					
impulsive	repetitive regular	hammering, laser shock treating, high pressure water peening				
impulsive	repetitive irregular	shot peening, needle peening, ultrasonic peening			brushing	

Fig. 1.2: Overview of the principal non-cutting processes of mechanical surface treatment

References

1.1 G. F. Bush, J. O. Almen, L. A. Danse, J. P. Heise: How, when and by whom was mechanical prestressing discovered, In: Sociecty of Automotive Engineers ISTC, Div. 20 Meeting, SAE, Colorado Springs, Colorado, 1962

1.2 O. Föppl: Stahl und Eisen (1929) 49, pp. 775.

1.3 O. Föppl, G. v. Heydekampf: Dauerfestigkeit und Konstruktion, Metallwirtschaft Wissenschaft und Technik 8(1929) 45, pp. 1087–1094.

1.4 H. Isemer: Die Steigerung der Schwingungsfestigkeit von Gewinden durch Oberflächendrücken, In: Mitteilungen des Wöhler-Instituts, Braunschweig, 1931.

1.5 A. Thum, H. Wiegand: Die Dauerhaltbarkeit von Schraubenverbindungen und Mittel zu ihrer Steigerung, Verein Deutscher Ingenieure Zeitschrift 39(1933), pp. 1061–1063.

1.6 E. Hottenrott: Mitteilungen des Woehler-Instituts, Braunschweig(1932) 10, pp. 1.

1.7 A. Thum, O. Ochs: Verein Deutscher Ingenieure Zeitschrift 76(1932), pp. 951.

1.8 A. Thum, F. Wunderlich: Verein Deutscher Ingenieure Zeitschrift 77(1933), pp. 851.

1.9 E. G. Herbert: The work-hardening of steel by abrasion, with an appendix on the "cloudbrust" test and superhardening, Journal of Iron and Steel 11(1927), pp. 265–282.

1.10 E. E. Weibel: The Correlation of Spring-Wire Bending and Torsion Fatigue Tests, Transactions of ASM(1935) 57, pp. 501–516.

1.11 I. Horowitz: Oberflächenbehandlung mittels Strahlmitteln – Handbuch über Strahltechnik und Strahlanlagen – Band 1: Die Grundlagen der Strahltechnik, Vulkan-Verlag, Essen, 1982.

1.12 J. H. Frye, G. L. Kehl: The fatigue resistance of steel as affected by some cleaning methods, Transactions of ASM(1938), pp. 192–218.

1.13 R. Z. v. Manteuffel: Dissertation, TH Darmstadt, 1939.

1.14 F. P. Zimmerli: Shot blasting and its effects on fatigue life, In: Surface Treatment of Metals, ASM, Metals Park, 1941, pp. 261–278.

1.15 J. O. Almen: Peened surfaces improve endurance of machine parts, Metal Program 43(1943) 2, pp. 209–315.

1.16 J. O. Almen: Shot blasting to increase fatigue resistance, SAE Transactions 51(1943) 7, pp. 248–268.

1.17 A. J. Gould, U. R. Evans: The effect of shot-peening upon the corrosion-fatigue of a high-carbon steel, Journal of the Iron and Steel Institute 10(1948), pp. 164–168.

1.18 J. C. Straub, D. May: Stress Peening, The Iron Age(1949), pp. 66–70.

1.19 R. L. Mattson, J. G. Roberts: The effect of residual stresses induced by strain peening upon fatigue strength, In: G. M. Rassweiler, W. L. Grube (eds.), Symposium internal stresses and fatigue in metals, New York, 1959, pp. 338–357.

1.20 C. G. Robinson, E. Smart: The use of specialised shot peening techniques an tapered leaf suspension springs for road vehicles, In: H. O. Fuchs (ed.), Proc. Int. Conf. Shot Peening 2, American Shot Peening Society, Paramus, 1984, pp. 79–83.

1.21 B. Kaiser: Randschichtverfestigung und Schwingfestigkeit hochfester Parabelfedern, VDI Bericht 852, VDI-Verlag, Düsseldorf, 1991, pp. 587–600.

1.22 J. M. Potter, R. A. Millard: The effect of temperature and load cycling on the relaxation of residual stresses, Advances in X-Ray Analysis 20(1976), pp. 309–320.

1.23 L. Bonus, E. Müller: Spannungsstrahlen von Fahrzeugtragfedern – Untersuchungen des Relaxationsverhaltens von spannungsgestrahlten Schraubenfedern, Draht 47(1996) 7/8, pp. 408–410.

1.24 E. Müller, L. Bonus: Lebensdauer spannungsgestrahlter Schraubenfedern unter Korrosion, Draht 48(1997) 6, pp. 30–33.

1.25 E. Müller: Der Einfluß des Plastizierens und des Kugelstrahlens auf die Ausbildung von Eigenspannungen in Blattfedern, Hoesch Berichte aus Forschung und Entwicklung,1992, pp. 23–29.

1.26 F. Engelmohr, B. Fiedler: Erhöhung der Dauerfestigkeit geschmiedeter Pleuel durch Kugelstrahlen unter Vorspannung, Materialwissenschaft und Werkstofftechnik 22(1991), pp. 211–216.

1.27 F. Engelmohr, B. Fiedler: Festigkeitsstrahlen unter Vorspannung, Auswirkungen auf Eigenspannungszustand und Schwingfestigkeit von Bauteilen, In: 1991, pp. 77–91.

1.28 I. Gokyu: Japanisches Patent 725630, 1973, Japan.

1.29 A. Tange, H. Koyama, H. Tsuji: Study on warm shot peening for suspension coil spring, SAE Technical Paper Series 1999–01–0415, International Congress and Exposition, Detroit, 1999, pp. 1–5.

1.30 A. Tange, K. Ando: Study on the shot peening processes of coil spring, In: Proc. Int. Conf. Residual Stresses 6, IOM Communications, Oxford, 2000, pp. 897–904.

1.31 L. Bonus: Versuchsbericht Warmverdichten am Funker 2000, Abteilung EF12 Hoesch-Hohenheimburg, Federnwerke, 1989.

1.32 G. Kühnelt: Der Einfluß des Kugelstrahlens auf die Dauerfestigkeit von Blatt- und Parabelfedern, In: A. Niku-Lari (eds.), Proc. Int. Conf. Shot Peening 1, Pergamon, Paris, 1981, pp. 603–611.

1.33 J. Ulbricht, H. Vondracek: Möglichkeiten zur Federwegvergrößerung bei Drehstäben, Estel-Berichte(1976) 3/76, pp. 125–132.

1.34 M. Schilling-Praetzel: Einfluß der Werkstücktemperatur beim Kugelstrahlen auf die Schwingfestigkeit von Drehstabfedern, Dissertation, RWTH Aachen, 1995.

1.35 MAN: Die Trucknology Generation – TG-A der MAN Nutzfahrzeuge, Automobiltechnische Zeitschrift 102(2000) 9, pp. 666–674.

1.36 A. Wick, V. Schulze, O. Vöhringer: Kugelstrahlen bei erhöhter Temperatur mit einer Druckluftstrahlanlage, Materialwissenschaft und Werkstofftechnik 30(1999), pp. 269–273.

1.37 A. Wick, V. Schulze, O. Vöhringer: Effects of warm peening on fatigue life and relaxation behaviour of residual stresses of AISI 4140, Materials Science and Engineering A293(2000), pp. 191–197.

1.38 R. Menig, A. Wick, V. Schulze, O. Vöhringer: Shot peening and stress peening of AISI 4140 at elevated temperatures – Effects on fatigue strength and stability of residual stress, In: K. Funatani G. E. Totten (eds.), 20th ASM Heat Treating Society Conference, ASM, Metals Park, 2000, pp. 257–264.

1.39 I. Altenberger, B. Scholtes: Improvement of fatigue behaviour of mechanically surface treated materials by annealing, Scripta Materialia 42 (1999), pp. 873–881.

1.40 I. Altenberger: Mikrostrukturelle Untersuchungen mechanisch randschichtverfestigter Bereiche schwingend beanspruchter metallischer Werkstoffe, Dissertation, Universität Gesamthochschule Kassel, 2000.

1.41 R. Menig, V. Schulze, D. Löhe, O. Vöhringer: Shot peening plus subsequent short-time annealing – A way to increase the residual stress stability and alternating bending strength of AISI 4140, Technical Paper Series 2002–01–1409, Society of American Engineers, Las Vegas, 2002, pp. 1–8.

1.42 L. Wagner, C. Müller, J. K. Gregory: In: Proc. Fatigue 93, EMAS, 1993, pp. 471.

1.43 J. K. Gregory, C. Müller, L. Wagner: Bevorzugte Randschichtaushärtung: Neue Verfahren zur Verbesserung des Dauerschwingverhaltens mechanisch belasteter Bauteile, Metall 47(1993), pp. 915–919.

1.44 L. Wagner, J. K. Gregory: Thermomechanical surface treatment of titanium alloy, In: Second European ASM Heat Treatment and Surface Engineering Conference, ASM, Metals Park, 1993, pp. 1–24.

1.45 L. Wagner, J. K. Gregory: Improve the fatigue life of titanium alloys, Part II, Advanced Materials & Processes 145(1994) 7, pp. 50HH-50JJ.

1.46 F. Bohner, J. K. Gregory: Mechanical Behavior of a Graded Aluminium Alloy, In: W. A. Kaysser (ed.), Proc. Functionally Graded Materials 1998, Trans Tech Publications, 1999, pp. 313–318.

1.47 J. J. Daly, J. R. Harrison, L. A. Hackel: New laser technology makes lasershot peening commercially affordable, In: A. Nakonieczny (ed.), Proc. Int. Conf. Shot Peening 7, Warschau, 1999, pp. 379–386.

1.48 A. H. Clauer: Laser Shock Peening for fatigue resistance, In: J. K. Gregory, H. J. Rack, D. Eylon (eds.), Proc. Symp. Surface Performance of Titanium, Cincinnati, 1997, pp. 217–230.

1.49 A. H. Clauer, D. F. Lahrman: Laser Shock Processing as a Surface Enhancement Process, In: Durable Surface Symposium, International Mechanical Engineering Congress & Exposition, Orlando, 2000.

1.50 M. Obata, Y. Sano, N. Mukai, M. Yoda, S. Shima, M. Kanno: Effect of laser peening on residual stress and stress corrosion cracking for type 304 stainless steel, In: A. Nakonieczny (ed.), Proc. Int. Conf. Shot Peening 7, Warschau, 1999, pp. 387–394.

1.51 Y. Sano, N. Mukai, K. Okazaki, M. Obata: Residual stress improvement in metal surface by underwater laser irradiation, Nuclear Instruments and Methods in Physics Research B 121(1997), pp. 432–436.

1.52 K. Eisner: Prozeßtechnologische Grundlagen zur Schockverfestigung von metallischen Werkstoffen mit einem kommerziellen Excimerlaser, Dissertation, Universität Erlangen-Nürnberg, 1998.

1.53 E. Houdremont, R. Mailänder: Archiv für das Eisenhüttenwesen(1929) 49, pp. 833.

1.54 E. Siebel, M. Gaier: Untersuchungen über den Einfluß der Oberflächenbeschaffenheit auf die Dauerschwingfestigkeit metallischer Bauteile, Verein Deutscher Ingenieure Zeitschrift 98(1956) 30, pp. 1715–1723.

1.55 H. Oschatz: Gesetzmäßigkeiten des Dauerbruches und Wege zur Steigerung der Dauerhaltbarkeit, Dissertation, Mitteilungen der Materialprüfungsanstalt an der Technischen Hochschule Darmstadt, 1933.

1.56 H. Wohlfahrt: Einfluß von Eigenspannungen, In: W. Dahl (ed.), Verhalten von Stahl bei schwingender Beanspruchung, 1978, pp. 141–164.

1.57 E. Macherauch, H. Wohlfahrt: Eigenspannungen und Ermüdung, In: D. Munz (ed.), Ermüdungsverhalten metallischer Werkstoffe, DGM-Informationsgesellschaft Verlag, Oberursel, 1985, pp. 237–283.

2
Procedures of Mechanical Surface Treatments

2.1
Shot Peening

2.1.1
Definition and Delimitation of Procedure

DIN 8200 [2.1] defines peening as mechanical surface treatment processes in which peening media with a specific shape and a sufficiently high degree of hardness (compare DIN 8201 [2.2]) are accelerated in peening devices of various kinds and interact with the surface of the treated workpiece. The methods summarized in Table 2.1 are to be distinguished depending on the objective. The creation of compressive residual stresses close to the surface is the main focus of the shot peening process, whereas in the other methods, these effects are more or less significant side effects. Accordingly, shot peening is the sole method used for increasing the load capacity of technical components. Therefore, the following report is limited to this peening process.

2.1.2
Application Examples

Due to its flexibility, shot peening may be used on components of almost any shape, particularly on those possessing a complex geometry. It is thus predestined for use on cross-sectional variations, chamfers, boreholes and bore edges. Components which are typically shot peened in technical mass production are springs, con-rods, gears, stepped or grooved shafts and axles, turbine vane and blade bases and heat-affected zones of welded joints. Due to the positive effects on resistance to stress corrosion cracking and corrosion fatigue, shot peening is also used in apparatus engineering and plant construction, in order to protect e.g. interior pipe surfaces against corrosive media.

Modern Mechanical Surface Treatment. Volker Schulze
Copyright © 2006 WILEY-VCH Verlag GmbH & Co. KGaA, Weinheim
ISBN: 3-527-31371-0

Tab. 2.1: Subdivision of peening treatments

Peening Type	Shot Peening	Blast Cleaning	Finish Peening	Abrasive Peening	Peen Forming
Main Aim	Generation of compressive residual stresses and work-hardening close to the surface	Removal of contaminants, coatings (rust, scale), laquers	Generation of a uniform rough or smooth surface structure	Machining of the workpiece with geometrically undefined cutting edge	Forming, Straightening
Side Effects	Mostly increase of roughness, cleaning	Generation of compressive residual stresses and work-hardening in usually very thin surface layers	Generation of compressive residual stresses and work-hardening in usually very thin surface layers	Generation of compressive residual stresses and work-hardening in usually very thin surface layers	Generation of compressive (convex, small depth of deformation) or tensile (concave, large depth of deformation) residual stresses, workhardening and increase of roughness
Applications	Cyclically, wear, corrosively or fretting corrosively loaded components, generation of compressive residual stresses prior to coating	Fettling, removal of tarnish at CrNi-steels, descaling, derusting, desanding, cleaning prior togalvanizing	Surface roughening prior to coating, improvement of load capacity and lubrication properties, optical effects as roughening, smoothing, polishing or matting	Deburring, removal, separation	Sheet components, integrally reinforced plankings
Peening Media	Round: rounded cut wire shot, cast steel shot (d = 0.2 to 3 mm), ceramic (d = 0.15 to 1.5 mm) or glass (d = 0.05 to 0.85 mm) beads	Chiselled: chilled cast grit, corundum, quartz sand, glass beads, polymer granulate	Round or chiselled: steel, aluminum, ceramic, glass, silicon carbide	Chiselled: chilled cast grit, corundum	Round: steel balls of high hardness (d = 2 to 10 mm)

2.1.3
Devices, Tools and Important Parameters

According to Fig. 2.1, shot peening may be applied using rotating wheel, compressed-air, injector and injector gravitational peening systems that are distinguished by their respective technique for accelerating the shot. A rotating wheel peening installation consists of one or more rotating wheel units. The rotating guide vanes receive the shot through the wheel hub and centrifugal forces thus serve to accelerate the shot. These systems are characterized by a high throughput of the shot, but also by a broad velocity distribution within the cross section of the shot stream. The drop velocity of the shot is controlled by the rotation of the rotating wheel. In compressed-air peening systems, the shot is conveyed from an unpressurized storage container to a pressurized storage container by a supply system. The shot is then passed into a jet of air of equal pressure via a shot dispensing unit. The mixture of air and shot is finally accelerated through a nozzle onto the surface of the workpiece. The essential factor which determines the velocity of the shot is the pressure of the air/shot mixture. In contrast to rotating wheel systems, the shot delivery rate and the spread of shot velocities are comparatively low. Compressed-air peening systems are primarily suited for the treatment of geometrically complex components due to the greater freedom of movement of the nozzle, and for high-intensity treatment of small areas because of the small jet diameters. Injector peening systems use the effect of suction to transfer the shot from a storage container into the air stream. The shot is transported by the air stream and accelerated through nozzles onto the workpiece surface, as in compressed-air systems. The simple construction of these systems allows for high delivery rates and medium velocity spreads of the shot, while nozzle mobility and shot intensity are reduced in comparison to compressed-air systems. In injector gravitational peening systems the storage container is located above the workspace and the supply of shot to the air stream is thus supported by the force of gravity. This permits improved control of velocity and delivery rate.

a)   b)   c)   d)

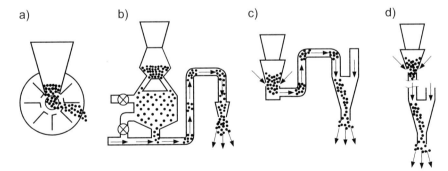

Fig. 2.1: Different devices for shot peening treatments: rotating wheel (a), compressed air (b), injector (c) and injector gravitational (d) peening devices, [acc. 2.3]

A special form of shot peening is represented by ultrasonic shot peening, in which the shot and the workpiece are placed in a chamber together and are exposed to a strong ultrasonic field at frequencies of about 20 kHz. This field accelerates the ball bearings that are used as shot. The shot interacts with the surface of the workpiece in an impact process at similar velocities as in conventional shot peening methods. Appropriate chamber design permits use of the method for large component surfaces and mobile applications [2.4]. Furthermore, the encapsulation of the shot prevents contamination by small shot and shot particles. As the shot deviates only slightly from exact spherical shape, the aim is to achieve lower degrees of surface roughness [2.4].

The shot acts as the tool of the peening process. Depending on type, the shot is delivered either as balls, chiseled grit, rounded cut wire, beads or granulate. According to Fig. 2.2, the classification distinguishes between metallic shot – such as chilled cast steel or steel in the form of rounded cut wire –, inorganic non-metallic shot – such as ceramic or glass beads –, organic shot – such as polymer granulate –, and a number of natural materials. It is characterized by its chemical composition, size, size distribution, shape, stiffness, hardness and density. While dry rotating wheel systems, for instance, generally use iron-based shot, special applications in wet shot peening systems in the nonferrous metals industry also call for the use of nonferrous metal shot whenever contamination of the workpiece by shot dust of different material is to be avoided. Among the types of inorganic non-metallic shot, glass beads and beads made of zirconium oxide, silicon oxide and corundum are particularly significant. Organic peening media, some of which are of natural origin, are also used for special applications, specifically superfinishing [2.5].

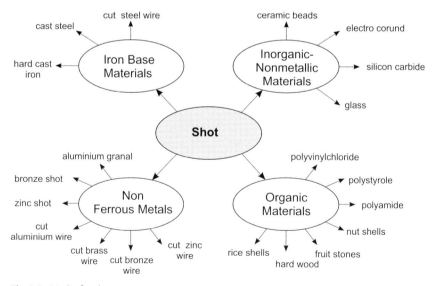

Fig. 2.2: Media for shot peening

According to the detailed discussion of the elementary processes of shot peening in Sect. 3.1.1, the hardness of the shot must, at least approximately, equal the hardness of the workpiece. Therefore, hardened or case hardened steels require the use of steel shot of at least grade 54–58 HRC instead of the common cast steel shot of grade 46–51 HRC [2.6]. Owing to their hardness, glass or ceramic beads are also an option for achieving high compressive residual stress states. It must be taken into account, however, that their lower density results in a smaller impulse upon impact on the workpiece and thus, less deep action of the peening process [2.7,2.8].

The shot size is to fit the dimensions and geometry of the components to be treated. On the one hand, the shot size must be sufficiently small to reach areas which are hard to access, such as small notches. This also serves to avoid notch effects due to impact-induced roughness. On the other hand, the dimensions and shape of thin-walled components are not to be changed inadmissibly [2.9]. Regarding the quality of a peening process, regular monitoring of the shot state is essential, as the shape and the size spread of the shot will change during application due to fragmentation and wear. Thus, the shot is conditioned [2.10–2.12] by removing dust, screening out particles that exceed an upper or lower limit and separating chiseled particles. One way of achieving the latter is to pour the shot onto a conveyor belt which is tilted sideways, thus allowing sufficiently round particles to quickly roll off [2.13].

Steel workpieces are usually treated with steel shot. Glass beads are primarily suited for low-intensity shot peening treatments and for components which would otherwise be contaminated and, for example, become more prone to corrosion. Therefore, they are used primarily on small-diameter components and on titanium or aluminum alloys sensitive to contamination by iron [2.14]. Ceramic beads combine high hardness with medium densities and are used e.g. for the shot peening of titanium alloys [2.14]. Recent research deals with shot peening using zirconium oxide and hard metal shot for the strengthening of ceramic samples made of silicon nitride and aluminum oxide [2.15–2.17].

In special cases, different shot types are used consecutively. After being treated with cast steel, case hardened gears may receive a secondary peening using glass beads to shift the maximum compressive residual stress closer to the surface [2.18,2.19]. This secondary treatment also reduces the roughness of components made of titanium alloys [e.g. see 2.20–2.23]. Shot of different size is used for peening welded joints [e.g. see 2.24,2.25]. Notches with small radii are primary treatment using fine shot, while the required intensity is achieved by using coarser shot subsequently.

Fig. 2.3 shows the essential parameters of the shot peening process. Apart from the factors defined by the peening system which influence the amount and the velocity of the shot and thus, peening intensity and coverage, the aforementioned properties of the shot are crucial. According to the detailed description in Sect. 3, the properties and the state of the workpiece, determined by its temperature and the degree of prestress, are so essential that alterations of these factors have led to a variety of methods, described separately below.

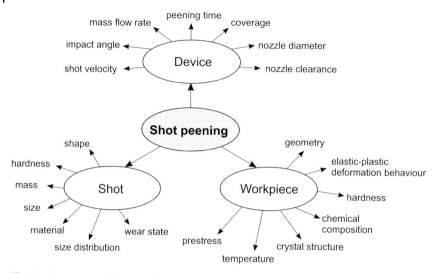

Fig. 2.3: Parameters influencing the results of shot peening treatments

2.2
Stress Peening

2.2.1
Definition and Delimitation of Procedure

Stress peening is a variant method of shot peening in which mechanical prestress of the same direction as the future operational load is applied to the workpiece during the peening treatment [e.g. see 2.26–2.35]. This prestressing is used to shift the residual stress state after the shot peening process towards higher compressive values. Since the stress peening process itself, aside from applying prestress, is identical to the shot peening process described in the preceding section, it requires the same equipment.

2.2.2
Application Examples

Stress peening is primarily used in the final treatment stage of springs, such as leaf, coil, turning-rod and brake accumulator springs [e.g. see 2.34–2.39]. There are also studies available regarding the use of stress peening on con-rod shafts [2.32,2.33]. Concerning components peened under torsional prestress, it must be taken into account that the resulting residual stress components vary significantly due to the biaxial load. In this particular case it is crucial that the future service load will not be alternating, but swelling, or should at least have a pronounced mean stress component.

2.2.3
Devices, Tools and Important Parameters

In terms of systems technology, there are no different prerequisites than for conventional shot peening. The same kind of shot is utilized for stress peening treatments as for conventional peening treatments. The only difference regarding the required installations is presented by the devices for applying pre-stresses, which are component-specific because they must permit load to be introduced in the same manner as service load on the component will be later. In the industrial area, the actual application of prestress is usually effected by screw connections [2.32], while within a laboratory environment prestress may also be applied hydraulically [2.31] or pneumatically [2.40]. The actuating variables of the conventional shot peening process, too, remain in effect without change. The amount of prestress applied is a significant additional factor. Concerning samples of simple geometry, the type and direction of prestress as determined by the future service loading of the component are important variables as well. Thus, the distinction is made between homogeneous uniaxial tensile prestressing, non-homogeneous uniaxial bending prestress and non-homogeneous biaxial torsional stress.

2.3
Warm Peening

2.3.1
Definition and Delimitation of Procedure

Warm peening is a variant of the common shot peening process used exclusively on steels, in which the workpiece exhibits an elevated temperature. Peening generally takes place at temperatures ranging from 170 °C to 350 °C [e.g. see 2.29,2.41–2.44]. Apart from the elevated temperature, the process is identical to conventional shot peening, and therefore principally requires the same equipment. An additional concern is that the effective temperature levels should only occur in the workpiece and that the shot, in particular, should not be warmed up significantly, as this would lead to decreases in hardness due to annealing effects and thus, to a reduced benefit of the peening process.

2.3.2
Application Examples

Examinations of warm peening applications in technological practice have so far focused exclusively on the surface treatment of various steel springs [e.g. see 2.29,2.44–2.49].

2.3.3
Devices, Tools and Important Parameters

As mentioned above, there are no particular requirements beyond heating the workpiece and avoiding undue warming of the shot. Industrial processes frequently utilize the residual heat of preceding heat treatments to achieve the desired workpiece temperature. However, maintaining a constant temperature can become problematic in case of even the slightest deviation from scheduled timing. Therefore, scientific research relies on a specific warming of the workpiece in compressed-air peening systems. It is achieved either by adding heated air streams to the mixture of air and shot, as indicated for the peening nozzle in Fig. 2.4 [2.40], or by inducing eddy current [2.50–2.52].

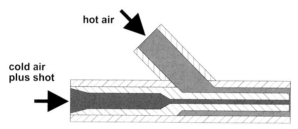

Fig. 2.4: Peening nozzle for shot peening at elevated temperatures [2.39]

2.4
Stress Peening at Elevated Temperature

To date, only a few examinations of stress peening at elevated temperatures [e.g. see 2.46,2.53,2.54] have been carried out. Applications, if economically relevant at all, are to be found in the spring manufacturing industry. The required technical systems and the actuating variables described in connection with stress peening and warm peening are to be considered jointly.

2.5
Deep Rolling

2.5.1
Definition and Delimitation of Procedure

Deep rolling is a non-cutting production method which, next to finish rolling and size rolling, is counted among the fine surface rolling methods (see Fig. 2.5) according to VDI guideline 3177 [2.55]. Finish rolling aims at creating surfaces with a plateau-like profile and especially low roughness, without significantly altering the geometric shape, in order to achieve good sliding and antifrictional qualities and to reduce wear. Size rolling is used to generate precise dimensions and thus create workpieces with an accurate fit. By contrast, the objective of actual deep rolling is to introduce workhardening and compressive residual stresses into near-surface regions in order to increase fatigue strength. In this goal-oriented rather than method-oriented classification, the transition from finish rolling to deep rolling is to be viewed as a fluent one, since increased strength is an additional aspect beyond mere smoothing. Deep rolling generally entails rolling off the tool and the workpiece against each other repeatedly at a defined pressure. This forces a continually increasing plastic deformation in the near-surface region [2.5,2.57–2.59]. If a translational motion is applied in addition to the rotational movement, this is called feed method when used on smooth or slightly notched components, and plunge method when used on heavily notched parts.

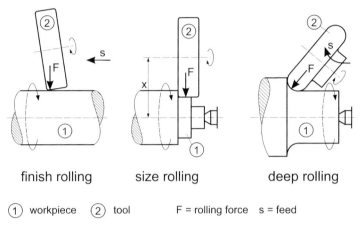

| finish rolling | size rolling | deep rolling |

① workpiece ② tool F = rolling force s = feed

Fig. 2.5: Methods of surface rolling: smooth rolling and deep rolling using the feed method (a), size rolling (b) and deep rolling using the plunge method (c) [2.56]

2.5.2
Application Examples

In the automotive industry, in general mechanical engineering and, to a certain extent, in aircraft construction, deep rolling is used on crankshafts, valve shafts, screws, bore-holes, axles, bolts and threaded parts. Due to the rotation of both the tool and the workpiece required in the method, its application is generally limited to components or treatment regions of rotational symmetry. Special tools permit deep rolling of the interior surfaces of through holes, stepped holes or blind holes with a sufficiently large diameter. Due to its simple technical realization in standard machine tools, deep rolling is easily implemented in the same clamping after a cutting treatment, especially if this is preceded by turning processes.

2.5.3
Devices, Tools and Important Parameters

Deep rolling may generally be done on standard machine tools. A distinction is made between tools of a roller-type design and tools which are spherical, the latter exhibiting greater flexibility [2.60]. Using profiled rolls permits specific work on threads as well as precise sizing of notches with a special shape. Another distinguishing feature is the way in which loads are applied during the deep rolling process. In mechanically prestressed tools, the deep rolling force is regulated by spring elements, whereas hydrostatic tools generate the deep rolling force by means of the operating fluid pressure. In both cases, the rolling force is increased and decreased gradually during the beginning and ending stages of the work process in order to prevent the occurrence of stress peaks in the component.

In contrast to the descriptions mentioned above, deep rolling of crankshaft bearings involves special machinery. For work on the seats of rolling bearings the part is clasped by a pincer-shaped tool that consists of two large supporting rollers and two milling rollers, the latter working on the radii which lead to the crank cheeks [2.61].

Fig. 2.6 summarizes the essential actuating variables of the deep rolling process, which may be classified according to workpiece, tool, method or device used.

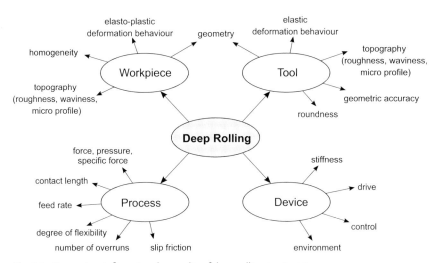

Fig. 2.6: Parameters influencing the results of deep rolling treatments

2.6
Laser Peening

2.6.1
Definition and Delimitation of Procedure

Laser shock treatment exposes the surface of a workpiece to laser pulses with a pulse duration in the nanosecond range, and is used for the mechanical treatment of the near-surface regions of the workpiece reaching surface depths on the order of millimeters. Building on numerous 1970s papers on laser-induced shock waves – which, however, dealt first with determining the pressure created by pulsed laser beams [2.63–2.65] and only later focused on the effects of the shock waves on the material and on applications for treating materials [2.66,2.67] – work on the modification of materials based on measurements of hardness and residual stress analysis has been intensified only since the late 1980s and 90s. The first laser systems used for this research were Nd:glass [2.68–2.71] and Nd:YAG [2.72,2.73] types, the majority of which were designed for research exclusively and exhibited only low pulse repetition rates, while they could produce pulse energy levels beyond the capability of industrial laser systems. The three system types shown in Table 2.2 are the present competitors in the emerging field of industrial application. Apart from phase-conjugated Nd:glass lasers [2.74–2.76] and q-switched and simultaneously frequency-doubled Nd:YAG lasers [2.77–2.81], XeCl-excimer lasers [2.82] are favored.

Tab. 2.2: Laser types used for laser shock treatments

Laser Type	Nd:Glass	Q-switched Nd:YAG	XeCl-Excimer
Country	USA, F	J, D	J
Reference	[2.72–2.74]	[2.75–2.79]	[2.80]
Wave Length [nm]	1053 or 527	532	308
Pulse Energy [J]	20–50	0.2	2
Pulse Duration [ns]	10–30	8	50
Pulse Rise Time [ns]	0.2	10	20
Pulse Frequency [Hz]	6	10	20
Beam Dimensions	□ 8 x 8 mm², ⌀ 0.9–6 mm	⌀ 1 mm	□ 40 x 50 mm²
Sepcial Feature	SBS-Phase- Conjugation	Fiber Coupling	

Laser-induced compressive impacts are mostly created by ablation of materials applied to the workpiece or by ablation of near-surface regions, which is caused by absorption of intense laser radiation and is additionally increased by the vaporized material which forms on the workpiece surface [2.83]. For laser shock treatment use, a laser system should have a short rise time of the pulse on the order of a few nanoseconds, pulse durations on the order of tens of nanoseconds, a lowest possible beam divergence below one millirad, pulse energies of approx. one joule minimum, and wavelengths with a maximum of one micrometer [2.84].

2.6.2
Application Examples

To date, the laser types and the associated problems mentioned above have limited the number of technical applications that go beyond the experimental stage. At present, the most important application is found in aviation technology, where laser shock treatment is used to increase the fatigue strength and the foreign-object-damage tolerance of jet turbine vanes and blades [2.85]. In medical technology, laser shock treatment is used for treating the surfaces of knee and hip prostheses [2.76]. Applications for automotive technology have been tested [2.76]. In addition, laser shock treatment is used to reshape airplane wings and for labeling parts susceptible to tearing [2.76]. Finally, [2.77] reports its use in nuclear facilities as a way to increase the resistance to corrosion fatigue in weld joints.

2.6.3
Devices, Tools and Important Parameters

At present, there is only a small number of systems suitable for industrial use. Their fundamentally different types and some essential specifications are com-

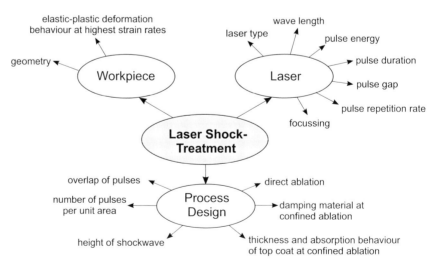

Fig. 2.7: Parameters influencing the results of laser shock treatments

pared in Table 2.2. In most cases, black adhesive strips/paint coatings with a water cover or water with a glass cover are used as an absorption and cover layer. The essential actuating factors are summarized in Fig. 2.7. They may be roughly divided into factors attributable to the workpiece, to the laser itself and to the process setup.

References

2.1 DIN 8200, Strahlverfahrenstechnik, Begriffe, Beuth-Verlag, Berlin, 1982.

2.2 DIN 8201, Strahlmittel, Einteilung – Bezeichnung, Beuth-Verlag, Berlin, 1975.

2.3 R. Clausen: Fertigungsplan und Fabrikeinrichtungen, ZwF 74(1979) 7, pp. 154–156.

2.4 J. M. Duchazeaubeneix: Stressonic shot peening (Ultrasonic Process), In: A. Nakonieczny (ed.), Proc. Int. Conf. Shot Peening 7, Warschau, 1999, pp. 444–452.

2.5 I. Horowitz: Oberflächenbehandlung mittels Strahlmitteln – Handbuch über Strahltechnik und Strahlanlagen – Band 1: Die Grundlagen der Strahltechnik, Vulkan-Verlag, Essen, 1982.

2.6 H. Wohlfahrt: The influence of peening conditions on the resulting distribution of residual stress, In: H. O. Fuchs (ed.), Proc. Int. Conf. Shot Peening 2, American Shot Peening Society, Paramus, 1984, pp. 316–331.

2.7 T. Ito, T. Mizuno, T. Takahashi, J. Kurosaki, T. Togashi: Peening characteristics of cemented carbide peening shot and life improvement of cold forging die, In: A. Nakonieczny (ed.), Proc. Int. Conf. Shot Peening 7, Warschau, 1999, pp. 371–377.

2.8 A. Niku-Lari: Oberflächen-Bearbeitung durch Kugelstrahlen, Vergleichende Versuche mit Strahlmitteln, Untersuchungsbericht (aus dem Französischen), CETIM, Senlis, 1980.

2.9 Metal Improvement Company: Militärspezifikation: Kugelstrahlen von Metallteilen, MIL-S-13165B, 1979.

2.10 Military Specification "Shot Peening of Metal Parts", MIL-S-13165B, Amendment-2, 1979.

2.11 Standards SAE J 444, SAE J 441, 1980.

2.12 U.S. Specification AMS 2430.

2.13 R. G. Bosshard, K.-P. Hornauer: Controlled shot peening, complete plants and examples of application, In: H. Wohlfahrt, R. Kopp, O. Vöhringer (eds.), Shot Peening, Proc. Int. Conf. Shot Peening 3, DGM Informationsgesellschaft, Oberursel, 1987, pp. 63–70.

2.14 H. Barzoukas, J. Jauffret: Peening with ceramic shot, In: K. Iida (ed.), Proc. Int. Conf. Shot Peening 4, Japan Society of Precision Engineering, Tokyo, 1990, pp. 47–56.

2.15 T. Frey, W. Pfeiffer: Tragfähigkeitssteigerung keramischer Werksoffe durch Kugelstrahlen, Tribologie und Schmierungstechnik(2001).

2.16 W. Pfeiffer, M. Rombach: Residual Stresses and Damage in Ceramics Due to Contact Loading, In: T. Ericson, M. Oden, A. Anderson (eds.), Proc. Int. Conf. on Residual Stresses 5, Linköping University, Linköping, 1997, pp. 302–307.

2.17 W. Pfeiffer, T. Frey: Shot peening of ceramics: Damage or benefit?, In: Jahrestagung der DKG, Bayreuth, 2001.

2.18 T. Hirsch, H. Wohlfahrt: Zahnfussdauerfestigkeit kugelgestrahlter Zahnräder, Antriebstechnik 25(1986) 9, pp. 73–80.

2.19 T. Hirsch, H. Wohlfahrt, E. Macherauch: Fatigue Strength of Case Hardened and Shot Peened Gears, In: H. Wohlfahrt, R. Kopp, O. Vöhringer (eds.), Shot Peening, Proc. Int. Conf. Shot Peening 3, DGM-Informationsgesellschaft, Oberursel, 1987, pp. 547–560.

2.20 H. Wohlfahrt: Practical aspects of the application of shot peening to improve the fatigue behaviour of metals and structural components, In: H. Wohlfahrt, R. Kopp, O. Vöhringer (eds.), Shot Peening, Proc. Int. Conf. Shot Peening 3, DGM-Informationsgesellschaft, Oberursel, 1987, pp. 563–584.

2.21 M. M. Woelfel: Maximization of flexural fatigue strength of heat treated steel specimens using multiple peening techniques, In: H. Wohlfahrt, R. Kopp, O. Vöhringer (eds.), Shot Peening, Proc. Int. Conf. Shot Peening 3, DGM-Informationsgesellschaft, Oberursel, 1987, pp. 125–132.

2.22 H. Wohlfahrt: Kugelstrahlen und Dauerschwingverhalten, In: A. Niku-Lari (ed.), Proc. Int. Conf. Shot Peening 1, Pergamon, Paris, 1981, pp. 675–702.

2.23 H. Wohlfahrt, J. Heeschen: Possibilities for the improvement of the fatigue strength of butt welded joints of high strength structural steels, In: Institution of Mechanical Engineers "Fatigue of Engineering Materials and Structures", Vol. 2, 1986, pp. 451–458.

2.24 A. Bignonnet, L. Picouet, H. P. Lieurade, L. Castex: The application of shot peening to improve the fatigue life of welded steel structures, In: C. Noordhoek, J. deBack (eds.), Proc. Int. ECSC Offshore Conf. Steels in Marine Structures, Elsevier, Amsterdam, 1987.

2.25 H. P. Lieurade, A. Bignonnet: Fundamental aspects of the effect of shot peening on the fatigue strength of metallic parts and structures, In: H. Wohlfahrt, R. Kopp, O. Vöhringer (eds.), Shot Peening, Proc. Int. Conf. Shot Peening 3, DGM-Informationsgesellschaft, Oberursel, 1987, pp. 343–359.

2.26 J. C. Straub, D. May: Stress Peening, The Iron Age(1949), pp. 66–70.

2.27 R. L. Mattson, J. G. Roberts: The effect of residual stresses induced by strain peening upon fatigue strength, In: G. M. Rassweiler, W. L. Grube (eds.), Symposium internal stresses and fatigue in metals, New York, 1959, pp. 338–357.

2.28 J.L. Xu, D. Q. Zhan, B. J. Shen: The fatigue strength and fracture morphology of leaf spring steel after prestressed shot peening, In: A. Niku-Lari (ed.), Proc. Int. Conf. Shot Peening 1, Pergamon, Paris, 1981, pp. 367–373.

2.29 G. Kühnelt: Der Einfluss des Kugelstrahlens auf die Dauerfestigkeit von Blatt- und Parabelfedern, In: A. Niku-

Lari (ed.), Proc. Int. Conf. Shot Peening 1, Pergamon, Paris, 1981, pp. 603–611.

2.30 R. Zeller: Verbesserung der Ermüdungseigenschaften von Bauteilen durch optimiertes Kugelstrahlen unter Zugvorspannung, In: 1991, pp. 93–103.

2.31 R. Zeller: Kugelstrahlen unter Zugvorspannung, Materialprüfung 35 (1993), pp. 218–221.

2.32 F. Engelmohr, B. Fiedler: Erhöhung der Dauerfestigkeit geschmiedeter Pleuel durch Kugelstrahlen unter Vorspannung, Materialwissenschaft und Werkstofftechnik 22 (1991), pp. 211–216.

2.33 F. Engelmohr, B. Fiedler: Festigkeitsstrahlen unter Vorspannung, Auswirkungen auf Eigenspannungszustand und Schwingfestigkeit von Bauteilen, In: 1991, pp. 77–91.

2.34 E. Müller: Der Einfluss des Plastizierens und des Kugelstrahlens auf die Ausbildung von Eigenspannungen in Blattfedern, Hoesch Berichte aus Forschung und Entwicklung (1992), pp. 23–29.

2.35 B. Kaiser: Randschichtverfestigung und Schwingfestigkeit hochfester Parabelfedern, VDI Bericht 852, VDI-Verlag, Düsseldorf, 1991, pp. 587–600.

2.36 C. G. Robinson, E. Smart: The use of specialised shot peening techniques an tapered leaf suspension springs for road vehicles, In: H. O. Fuchs (ed.), Proc. Int. Conf. Shot Peening 2, American Shot Peening Society, Paramus, 1984, pp. 79–83.

2.37 J. M. Potter, R. A. Millard: The effect of temperature and load cycling on the relaxation of residual stresses, Advances in X-Ray Analysis 20(1976), pp. 309–320.

2.38 L. Bonus, E. Müller: Spannungsstrahlen von Fahrzeugtragfedern – Untersuchungen der Relaxationsverhaltens von spannungsgestrahlten Schraubenfedern, Draht 47(1996) 7/8, pp. 408–410.

2.39 E. Müller, L. Bonus: Lebensdauer spannungsgestrahlter Schraubenfedern unter Korrosion, Draht 48(1997) 6, pp. 30–33.

2.40 A. Wick, V. Schulze, O. Vöhringer: Kugelstrahlen bei erhöhter Temperatur mit einer Druckluftstrahlanlage, Materialwissenschaft und Werkstofftechnik 30(1999), pp. 269–273.

2.41 A. Tange, K. Ando: Study on the shot peening processes of coil spring, In:

Proc. Int. Conf. Residual Stresses 6, IOM Communications, Oxford, 2000, pp. 897–904.

2.42 M.-C. Berger, J. K. Gregory: Selective hardening and residual stress relaxation in shot peened Timetal 21s, In: C. A. Brebbia, J. M. Kenny (eds.), Surface Treatment IV – Computer Methods and Experimental Measurements, Southampton, 1999, pp. 341–348.

2.43 M. Schilling-Praetzel, F. Hegemann, P. Gome, G. Gottstein: Influence of Temperature of Shot Peening on Fatigue Life, In: D. Kirk (ed.), Proc. Int. Conf. Shot Peening 5, Oxford, 1993, pp. 227–238.

2.44 M. Schilling-Praetzel: Einfluss der Werkstücktemperatur beim Kugelstrahlen auf die Schwingfestigkeit von Drehstabfedern, Dissertation, RWTH Aachen, 1995.

2.45 A. Tange, H. Koyama, H. Tsuji: Study on warm shot peening for suspension coil spring, SAE Technical Paper Series 1999–01–0415, International Congress and Exposition, Detroit, 1999, pp. 1–5.

2.46 A. Tange, K. Ando: Study on the shot peening processes of coil spring, In: Proc. Int. Conf. Residual Stresses 6, IOM Communications, Oxford, 2000, pp. 897–904.

2.47 L. Bonus: Versuchsbericht Warmverdichten am Funker 2000, Abteilung EF12 Hoesch-Hohenheimburg, Federnwerke, 1989.

2.48 J. Ulbricht, H. Vondracek: Möglichkeiten zur Federwegvergrößerung bei Drehstäben, Estel-Berichte (1976) Heft 3/76, pp. 125–132.

2.49 MAN: Die Trucknology Generation – TG-A der MAN Nutzfahrzeuge, Automobiltechnische Zeitschrift 102 (2000) 9, pp. 666–674.

2.50 A. Rössler: Schwingfestigkeitsverhalten von SiCr-Federstählen nach einer mechanischen Oberflächenbehandlung und Warmauslagerung, Dissertation, Technische Universität München, 2001.

2.51 A. Rössler, J. K. Gregory: Einfluss einer Kugelstrahlbehandlung mit anschließender Wärmebehandlung auf das Schwingfestigkeitsverhalten hochfester SiCr Federstähle unter Umlaufbiegung, In: P. Mayr, S. Hock (eds.),

Ermüdung hochharter Stähle, AWT, Wiesbaden, 2001, pp. 89–103.

2.52 A. Rössler, J. K. Gregory: Einfluss des Bake Hardening Effekts auf das thermische und zyklische Abbauverhalten kugelstrahlinduzierter Eigenspannungen, Materialwissenschaft und Werkstofftechnik 32(2001), pp. 725–736.

2.53 A. Wick, V. Schulze, O. Vöhringer: Effects of warm peening on fatigue life and relaxation behaviour of residual stresses of AISI 4140, Materials Science and Engineering A293(2000), pp. 191–197.

2.54 A. Wick, V. Schulze, O. Vöhringer: Effects of stress- and/or warm peening of AISI 4140 on fatigue life, Steel Research 71(2000) 8, pp. 316–321.

2.55 VDI-Richtlinie 3177, Oberflächen-Feinwalzen, VDI-Verlag, Düsseldorf, 1983.

2.56 K.-H. Kloos, J. Adelmann: Schwingfestigkeitssteigerung durch Festwalzen, Materialwissenschaft und Werkstofftechnik 19(1988), pp. 15–23.

2.57 E. Broszeit: Grundlagen der Schwingfestigkeitssteigerung durch Fest- und Glattwalzen, Zeitschrift für Werkstofftechnik 15(1984), pp. 416–420.

2.58 A. Ostertag: Festwalzen und Glattwalzen zur Festigkeitssteigerung von Bauteilen, ECOROLL AG Wergzeugtechnik, Celle, 1983.

2.59 H.-H. Gerlach: Dissertation, TH Hannover, 1961.

2.60 A. Ostertag: Festwalzen im Einstich- und Vorschubverfahren, Nr. 6090, ECOROLL AG Wergzeugtechnik, Celle, 1999.

2.61 C. Achmus: Messung und Berechnung des Randschichtzustands komplexer Bauteile nach dem Festwalzen, Dissertation, TU Braunschweig, 1998.

2.62 B. Scholtes: Eigenspannungen in mechanisch randschichtverformten Werkstoffzuständen, Ursachen-Ermittlung-Bewertung, DGM-Informationsgesellschaft, Oberursel, 1990.

2.63 N. C. Anderholm: Journal of Applied Physics 16(1970) 3, pp. 113.

2.64 J. D. O'Keefe, C. H. Skeen: Journal of Applied Physics 44(1973) 10, pp. 4622.

2.65 B. P. Fairand, A. H. Clauer: Journal of Applied Physics 50(1979) 3, pp. 1497.

2.66 A. H. Clauer, B. P. Fairand, B. A. Wilcox: Pulsed laser induced deformation in

Fe-3 wt-% Si alloy, Metallurgical Transactions 8A(1977), pp. 119–125.

2.67 P. Bournot, et al.: Spie 801, Den Haag, 1987, pp. 308.

2.68 C. Dubouchet: Traitements thermo-méchaniques de surfaces métalliques à l'aide de lasers CO_2 continues et de laser impusionnel, Thèse PhD, Université de Paris-Sud Centre d'Orsay, 1993.

2.69 P. Peyre: Traitement mécanique superficiel d'alliages d'Aluminium par ondes de choc-laser, Charactérisation des effets induites et application à l'amélioration de la tenue fatigue, Thèse PhD, Université de Technologie de Compiégne, 1993.

2.70 P. Peyre, R. Fabbro: Laser scock processing: a review of the physics and applications, Optical and Quantum Electronics (1995) 27, pp. 1213.

2.71 J.-E. Masse: Charactérisation mécanique de surfaces d'échantillons d'aciers traités par choc-laser, Thèse PhD, Nationale École Supérieure d'Arts et Métiers, Centre d'Aix-en-Provence, 1994.

2.72 J. Fournier: Génération d'ondes de choc par laser pulsé de fortes énergies, Applications mécaniques et métallurgiques, L'École Polytechnique, 1989.

2.73 P. Ballard: Contraintes résiduelles induites par impact rapide, Application au choc-laser, Thèse PhD, Ecole Polytéchnique, 1991.

2.74 J. J. Daly, J. R. Harrison, L. A. Hackel: New laser technology makes lasershot peening commercially affordable, In: A. Nakonieczny (ed.), Proc. Int. Conf. Shot Peening 7, Warschau, 1999, pp. 379–386.

2.75 A. H. Clauer: Laser Shock Peening for fatigue resistance, In: J. K. Gregory, H. J. Rack, D. Eylon (eds.), Proc. Symp. Surface Performance of Titanium, Cincinnati, 1997, pp. 217–230.

2.76 A. H. Clauer, D. F. Lahrman: Laser Shock Processing as a Surface Enhancement Process, In: Durable Surface Symposium, International Mechanical Engineering Congress & Exposition, Orlando, 2000.

2.77 M. Obata, Y. Sano, N. Mukai, M. Yoda, S. Shima, M. Kanno: Effect of laser peening on residual stress and stress corrosion cracking for typ. 304 stainless

steel, In: A. Nakonieczny (ed.), Proc. Int. Conf. Shot Peening 7, Warschau, 1999, pp. 387–394.

2.78 Y. Sano, N. Mukai, K. Okazaki, M. Obata: Residual stress improvement in metal surface by underwater laser irradiation, Nuclear Instruments and Methods in Physics Research B 121 (1997), pp. 432–436.

2.79 T. Schmidt-Uhlig, P. Karlitschek, G. Marowsky, Y. Sano: New simplified coupling scheme for the delivery of 20 MW Nd:YAG laser pulses by large core optical fibers, Applied Physics B 72 (2001), pp. 183–186.

2.80 P. Karlitschek, G. Hillrichs: Active and passive Q-switching of a diode pumped Nd:K-GW-laser, Applied Physics B 46 (1997), pp. 21–24.

2.81 T. Schmidt-Uhlig, P. Karlitschek, M. Yoda, Y. Sano, G. Marowsky: Laser shock processing with 20 MW laser pulses delivered by optical fibers, The European Physical Journal, Applied Physics (2000) 9, pp. 235–238.

2.82 K. Eisner: Prozesstechnologische Grundlagen zur Schockverfestigung von metallischen Werkstoffen mit einem kommerziellen Excimerlaser, Dissertation, Universität Erlangen-Nürnberg, 1998.

2.83 L. Berthe, R. Fabbro, P. Peyre, E. Bartnicki: Laser shock processing of materials: experimental study of breakdown plasma effects at the surface of confining water, Laboratoire pour l'Application des Laser de Puissance, 2002.

2.84 M. Freytag, H. W. Bergmann: Randschichtverfestigung durch Laser-Shock-Processing, In: H. Wohlfahrt, P. Krull (eds.), Mechanische Oberflächenbehandlungen: Grundlagen-Bauteileigenschaften-Anwendungen, Wiley-VCH Verlag, Weinheim, 2000, pp. 167–177.

2.85 S. D. Thompson, D. W. See, C. D.Lykins, P. G. Sampson: Laser shock peening vs shot peening a damage tolerance investigation, In: J. K. Gregory, H. J. Rack, D. Eylon (eds.), Proc. Symp. Surface Performance of Titanium, Cincinnati, 1997, pp. 239–251.

3
Surface Layer States after Mechanical Surface Treatments

3.1
Shot Peening

3.1.1
Process Models

a)   Elementary Processes

By means of the elastic and plastic impacts between the peening medium and the workpiece, shot peening transforms part of the kinetic energy of the shot, which is contingent on particle mass and velocity, into
- elastic and plastic deformation work, preferably in the workpiece, but possibly also in the shot,
- work effecting a change of lattice defects, preferably in the workpiece, but possibly also in the shot,
- energy of newly created surfaces,
- deformation- and friction heat,
- kinetic energy of abrased particles,
- and, in special cases, into work which induces phase transformations.

The surface layer states created by shot peening are a consequence of local, inhomogeneous plastic deformations of the workpiece surface which occur as a result of the impact of the individual shot particles on the workpiece. Fig. 3.1 shows the fundamental correlations using the example of the depths of penetration, as determined by [3.1] by measuring the depth of impact craters on AISI 4140 after peening with bearing balls. Penetration depth grows practically linearly with shot velocity, the slopes increasing with shot diameter. When workpiece hardness increases, however, penetration depth decreases hyperbolically with impact velocity. Conversely, penetration depth increases degressively as shot hardness is increased. This effect, too, is more pronounced at great shot velocities.

Understanding and assessing the way shot peening treatments affect the surface layer states requires a closer examination and separation of the deformation processes taking place. As shot peening creates strain rates on the order of 10^{+4} to

Modern Mechanical Surface Treatment. Volker Schulze
Copyright © 2006 WILEY-VCH Verlag GmbH & Co. KGaA, Weinheim
ISBN: 3-527-31371-0

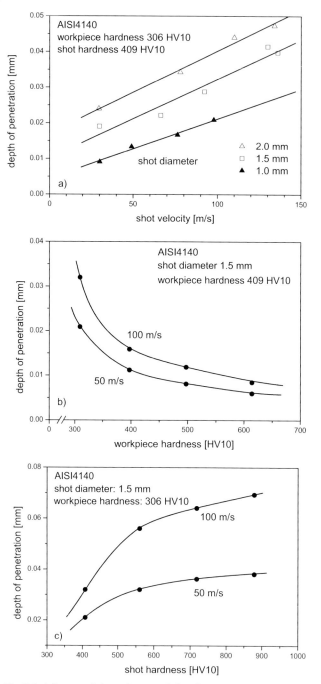

Fig. 3.1: Influence of shot velocity and shot diameter (a), workpiece hardness (b) and shot hardness (c) on remaining penetration depth after shot impact on AISI 4140, acc. [3.1]

10^{+5} 1/s due to the high impact velocities, the quantitative dependence of deformation behavior on temperature and strain rate is an important factor which shall be described in the following, based e.g. on [3.2–3.4]. The elasto-plastic deformation of steels is determined by different mechanisms, which are dependent on temperature T and strain rate $\dot{\varepsilon}$. At $T \leq 0.3\ T_M$ (T_M = melting or solidus temperature), the dependency of the flow stress on temperature and strain rate may be attributed to short range obstacles being overcome by thermally activated slip of dislocations [3.5–3.8]. In accordance with

$$\sigma = \sigma_G(structure) + \sigma^*(T, \dot{\varepsilon}, structure) \tag{3.1}$$

below a critical, strain rate dependent temperature T_o, flow stress σ is additively composed of an athermal component σ_G (temperature-dependent only to the extent of shear modulus G or Young's modulus E) and a thermal component σ^* (comp. Fig. 3.2) [3.2]. The former is based on the effect of long-range slip obstacles, among which are the stress fields of other dislocations, the grain boundaries, the stress fields of dissolved foreign atoms, or particles which need to be cut or looped. The second component σ^*, in contrast, results from the effects of short-range slip obstacles, such as the periodic lattice potential (Peierls' potential), which can be overcome by the aid of thermal lattice energy fluctuations [3.9,3.10]. The free activation enthalpy ΔG available at a thermal flow stress σ^* is linked to strain rate, temperature and the Boltzmann constant k by the equation

$$\dot{\varepsilon}(T, \sigma) = \dot{\varepsilon}_o \exp\left(\frac{-\Delta G(\sigma^*)}{kT}\right). \tag{3.2}$$

$\dot{\varepsilon}_o$ is determined by the density of slip dislocations, the Burgers-vector, the Debye frequency and the lattice structure. The stress dependence of the activation enthalpy directly follows the shape of the force-distance curve of the respective

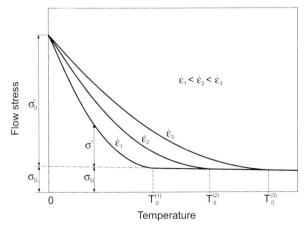

Fig. 3.2: Influence of temperature and strain rate on flow stress (schematic), acc. [3.4]

obstacle. Numerous theoretical and experimental findings present an approach to this correlation by means of power laws of the form [3.6–3.8]

$$\Delta G(\sigma^*) = \Delta G_0 \left[1 - \left(\frac{\sigma^*}{\sigma_0^*} \right)^{\frac{1}{m}} \right]^{\frac{1}{n}}. \tag{3.3}$$

According to [3.2], this leads to

$$\sigma^* = \sigma_0^* \left[1 - \left(\frac{T}{T_0} \right)^n \right]^m \tag{3.4}$$

as the correlation between the thermal flow stress component, temperature and strain rate $\dot{\varepsilon}$, which defines the critical temperature

$$T_0 = \frac{\Delta G_0}{k \ln(\dot{\varepsilon}_0 / \dot{\varepsilon})}. \tag{3.5}$$

At particularly high strain rates above $\dot{\varepsilon} \approx 10^{+4}\, 1/s$, the dislocation movement is additionally obstructed by interactions of the slip dislocations with electrons and phonons due to electron or phonon viscosity and phonon scattering [3.11–3.14]. This leads to stress-proportional dislocation velocities, which affect the flow stress according to

$$\sigma = \sigma_{th} + a B \dot{\varepsilon}. \tag{3.6}$$

σ_{th} is a temperature dependent critical stress, a is a proportionality constant and B is the so-called damping constant. Due to this so-called viscous damping, the sensitivity of the flow stress to the strain rate shows particularly high increases at great strain rates.

According to Fig. 3.3, there is a distinction to be made in shot peening between the stretching of the direct surface caused by forces perpendicular and parallel to the surface, Hertzian pressure with maximum plastic deformations beneath the surface, and heat development, which, on the one hand, reduces strength in regions close to surface and, on the other hand, can effect states of compressive stress due to the obstruction of thermal expansion [3.15]. However, as [3.16] esti-mates temperature to increase by a maximum of 20 °C – assuming that the entire kinetic energy of the peening medium is transformed into heat in shot peening treatments on steel using high impact velocities – thermally induced effects are of secondary importance in conventional shot peening. In particular, high coverages can lead to low cycle fatigue processes in regions close to surface of the workpiece, and cyclic deformation behavior can thus be involved as well. An assessment of the significance of the individual effects is aided by the quantitative process descriptions provided in the following section. Depending on whether it is Hert-zian pressure or surface stretching that predominates, the greatest compressive residual stresses occur either beneath the surface or directly at the surface. According to [3.15], the hardness of the workpiece compared to that of the shot is the crucial factor in this context. The effects of Hertzian pressure are pre-

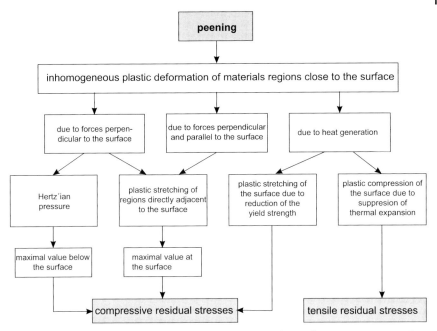

Fig. 3.3: Scheme for the evaluation of deformation processes during shot peening, acc. [3.260]

dominant particularly in hard material states (in steels: hardness > 600 HV) during peening processes utilizing shot of similar hardness, whereas surface stretching effected by the shot impacts is particularly significant in soft workpiece states (in steels: hardness < 300 HV) in peening processes using shot which is considerably harder. However, in medium hard material states (in steels: 300 HV < hardness < 600 HV) which are peened using slightly harder shot, the effects of surface stretching and Hertzian pressure are combined.

The Bauschinger effect [3.17] can be a highly significant factor during unloading from the maximum load of shot impact. In elasto-plastically deformed material states which are deformed in the opposite direction after original loading, the Bauschinger effect frequently results in a reduction of the flow stress necessary for initiating plastic deformation. There may even be backward plastic deformation occurring already during unloading, and an altered deformation behavior is subsequently observed. Various properties, originating in stress values, strain values, deformation work or secant stiffness, are utilized for a quantitative description of the Bauschinger effect [3.18]. Describing its micro-structural causes has been the objective of numerous models, each reflecting the level of knowledge of their time about the microstructural processes of plastic deformation. They can be divided into grain anisotropy models, dislocation models and back stress models. Grain anisotropy models are based on the ideas of Masing [3.19,3.20] and Heyn, [3.21,3.22]. In their view, the reduction of flow stress at load reversal is caused by residual stresses, which occur during deformation and which result from varying flow stresses or varying workhardening behavior of the grains and grain regions present in the polycrystal [3.23,3.24]. By contrast, the dislocation models which go back to Sachs and Shoji [3.25] are based on the assumption that the cause of the Bauschinger effect is to be found in the slip planes themselves, as single crystals of brass, too, exhibit a noticeable Bauschinger effect. Orowan [3.26] postulates that back stresses lead to dislocation movement in the opposite direction, possibly already during unloading, due to the equilibrium with external loading which is required at load reversal. A possible cause being discussed are the interactions in heterogeneous materials between dislocations and precipitations, dispersions and coarse secondary phases, which lead to the formation of back stresses – e.g. due to the newly created boundary area after cutting processes [see e.g. 3.27] – or the dissolution of previously formed dislocation segments after Orowan looping processes [see e.g. 3.27]. Alternatively, the residual stress fields of accumulated dislocations in the vicinity of phase boundary areas are viewed as the cause of the Bauschinger effect [3.28–3.33].

The Bauschinger effect generally increases with the concentration of dissolved, resting foreign atoms. In this process, the development of planar dislocation structures which show elevated back stresses in comparison to arrangements within dislocation cells, is supported in part by a decrease in stacking fault energy [3.34,3.35]. By contrast, the Bauschinger effect is reduced when diffusing foreign atoms cause effects of static and dynamic strain aging which block slip dislocations and thus impede backward deformation [3.36,3.37]. When dislocation density rises, the Bauschinger effect usually increases as well, due to growing dislocation-induced back stresses [3.36,3.38]. However, the extent of the Bauschinger effect depends to a considerable degree on the stability of the dislocation network in its pre-deformation state [3.39], which is difficult to approach on an experimental level. Apart from the dislocation density itself, which is proportional to the built-up back stresses in single-phase materials, the free path lengths of dislocations are essential. However, there is only little and moreover, contradictory, information on the influence of grain size. While [3.25,3.40,3.41] report the Bauschinger effect to be essentially equal in single- and poly-crystalline material states, [3.35,3.42,3.43] observe an increasing Bauschinger effect as grain size decreases. As described above in the context of using back stress models to explain the Bauschinger effect, increasing contents of coherent or incoherent particles and coarse secondary phases cause increases of back stresses – induced by dislocations or by the material state – and thus of Bauschinger effect [see e.g. 3.44]. Particularly in quenched and tempered steels, the Bauschinger effect is very pronounced, due to a small grain size, fine carbide precipitations and the high content of dissolved carbon in high strength states [3.45].

b)   **Quantitative Description of Processes**

The quantitative process descriptions presented in the literature can be grouped into analytical and numerical models, which focus on the impact of a single peening medium particle and/or multiple particle impact. The analytical process descriptions are based, in essence, on the fundamental approach of Hertz [3.46,3.47]. For elastic contact of a sphere with a radius R impacting vertically on a flat plate, he states a distribution of surface pressure as described by

$$p(r) = \frac{2}{3}\bar{p}\sqrt{1 - \left(\frac{r}{a_s}\right)^2} \qquad (3.7)$$

with a mean of

$$\bar{p} = \frac{F}{\pi a_s^2} = \sqrt[3]{\frac{16}{9\pi} E^{*2} \frac{F}{R^2}}. \qquad (3.8)$$

Here, F is the impact force,

$$a_s = \sqrt[3]{\frac{3FR}{4E^*}} \qquad (3.9)$$

is the contact radius, and

$$\frac{1}{E^*} = \frac{(1 - v_1^2)}{E_1} + \frac{(1 - v_2^2)}{E_2} \qquad (3.10)$$

is the reciprocal value of the effective Young's modulus, which results from Young's moduli E_i and Poisson's ratios v_i of the plate and the sphere. The depth distribution of the individual stress components in the impact center is a result of

$$\sigma_z(z) = -\frac{3}{2}\bar{p}\frac{1}{1 + \left(\frac{z}{a_s}\right)^2}$$

$$\sigma_r(z) = \sigma_\theta(z) = -\frac{3}{2}(1 + v)\bar{p}\left(1 - \frac{z}{a_s}\tan^{-1}\frac{a_s}{z}\right) + \frac{3}{4}\bar{p}\frac{1}{1 + \left(\frac{z}{a_s}\right)^2}, \qquad (3.11)$$

from which the depth distribution of equivalent stress

$$\sigma_{eq}(z) = \left|\frac{3}{2}(1 + v)\bar{p}\left(1 - \frac{z}{a_s}\tan^{-1}\frac{a_s}{z}\right) - \frac{9}{4}\bar{p}\frac{1}{1 + \left(\frac{z}{a_s}\right)^2}\right| \qquad (3.12)$$

follows according to v. Mises. Accordingly, the equivalent stress reaches its maximum value under contact at a distance to surface of $z = 0{,}47\,a_s$, plastic deformation setting in at a mean surface pressure of

$$\bar{p} = 1{,}075\,R_\gamma. \qquad (3.13)$$

The depth distribution of the stress components and the distribution of equivalent stress according to v. Mises are shown in Fig. 3.4.

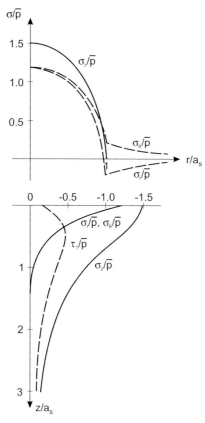

Fig. 3.4: Surface- and depth distribution of stress components and equivalent stresses according to v. Mises during elastic Hertzian contact of a sphere and a plane plate acc. [3.47]

The impact of the sphere is taken into account by Tabor [3.48,3.49], assuming an ideal elasto-plastic behavior, a complete transfer of kinetic energy from the sphere to the plate and a plastification up to the surface, by solving the equation of movement

$$\frac{4}{3}\pi R^3 \rho v dv + dW_{el} + dW_{pl} = 0. \tag{3.14}$$

This leads to the mean plastic deformation

$$\bar{\varepsilon} = 0,2\frac{a_s}{R} \tag{3.15}$$

beneath the contact center point when the contact radius a_s is reached, and the mean surface pressure

$$\bar{p} = 2,8R_y \tag{3.16}$$

that occurs. The stored elastic work immediately preceding the sphere's rebound upon reaching the final value of the contact radius is

$$W_{el} = 2,35\pi^2 a_s^3 \frac{R_y^2(1-\nu^2)}{E}, \tag{3.17}$$

and the plastic work achieved at that point is

$$W_{pl} = 0,7\pi \frac{a_s^4}{R} R_y. \tag{3.18}$$

These equations, however, do not yet provide an assessment of the distribution of plastic strains. From the Hertzian formulae, in combination with Boussinesque's solution for a single force within a semi-infinite space [3.51], [3.50] derive the thickness of the plastified surface layer

$$z_{pl} = 1,816a_s \tag{3.19}$$

at the mean surface pressure required for achieving plastification up to the surface, which again is

$$\bar{p} = 2,8R_y. \tag{3.20}$$

Based on [3.50] and assuming ideal plastic deformation behavior, [3.52–3.54] state the sphere's penetration depth related to sphere radius as

$$\frac{\bar{z}}{R} = \sqrt[4]{\frac{2}{3}} \sqrt{\frac{\rho v_0^2}{\bar{p}}} \tag{3.21}$$

and the thickness of the plastified zone related to sphere radius as

$$\frac{z_{pl}}{R} = \sqrt{\frac{\bar{z}}{R}} \tag{3.22}$$

as a solution of the equation of movement. This is the first time that a correlation is made between sphere velocity and the thickness of the plastified surface layer, ρ being sphere density and v_o being its initial velocity. Analogous approaches describing penetration depth and the thickness of the plastified zone are pursued in [3.55–3.58], supplemented by the inclusion of corrective factors in Equations 3.21 and 3.22 which reflect the dependency of deformation behavior on the strain rate.

[3.59] is the first examination to state distributions of residual stress on the basis of Hertzian pressure. It also shows that when yield strength has been exceeded, increasing plastic deformation causes the distribution of contact stress to become less inclined compared to the half-sphere-shaped distribution according to Hertz. [3.60] additionally estimates the thickness of the plastically deformed zone under varying sphere impact velocities. While both studies show the calculated maximum residual stresses to be well in accordance with experimental results, the surface residual stresses deviate very strongly from the measured values. Again based on Hertz's theory, [3.61] develop an analytical model for calculating the residual stresses in shot peening. However, it neglects to fulfill the stress equilibrium across the diameter. Assuming that elasto-plastic material behavior in the workpiece shows multilinear workhardening, their calculations of residual stresses close to the surface correspond well with experimental results.

Other analytical models describing the residual stresses in shot peened surface layers do not examine the impact of individual particles, but presuppose a temporally and locally constant distribution of strains or stresses across the surface of the workpiece, for which radial symmetry and the disappearance of stress components perpendicular to the surface may be assumed. The sum of the plastic strains caused by shot peening treatments is thus viewed as non-location-dependent, and the effects of different sequences of plastic deformation development are neglected. The first such model is described in [3.62], and it starts from the assumption that the plastic deformations in the shot-peened surface layer run from the surface following a cosine shape according to

$$\varepsilon_{p.o}(z) = \varepsilon_{p.o,\,max} \cos\left[\frac{\pi}{2}\frac{(z/z_{pl} - a)}{(1 - a)}\right]. \tag{3.23}$$

Here, $\varepsilon_{p.o,max}$ is the plastic strain at the surface contingent on the geometry and solidification behavior of the workpiece, z is the distance to surface coordinate, z_{pl} is the depth of the plastic zone, h is workpiece thickness, and a is a coefficient which is dependent on the material of the workpiece. It has a value between 0 and 1 and describes the depth of the plastic deformation maximum in relation to the plastic zone's thickness. When multiplied with Young's modulus E, this plastic strain results in a so-called source stress, which allows for an equilibrium of forces and momenta across the cross section when it is superposed with a normal stress which is constant across the cross section and linearly distributed across the cross section. For a flat specimen of the length l, exhibiting a deflection f after shot peening,

$$\varepsilon_{p.o,\,max} = \frac{\pi f h}{l^2 \beta(1 - -2\beta + 4\beta/\pi)} \tag{3.24}$$

is stated as the maximum plastic strain. β is the ratio of plastically deformed zone z_{pl} to workpiece thickness h. Combining this with the information on specimen dimensions, plastic zone depth and specimen deflection, the distribution of residual stresses can be stated as

$$\sigma^{rs}(z) = E\varepsilon_{p.o,\,max}\left[\frac{12\beta(1 - a)}{\pi h} C_1\left(z - \frac{h}{2}\right) + \frac{2}{\pi}\beta(1 - a)C_2 - \frac{\varepsilon_{p.o}(z)}{\varepsilon_{p.o,\,max}}\right]. \tag{3.25}$$

Based on this, [3.63] provides diagrams for estimating maximum strains and maximum residual stresses for different deflections as well as for different ratios of plastified zone to workpiece thickness. Fig. 3.5 is an exemplary representation of the influence of specimen thickness on the resulting depth distribution of

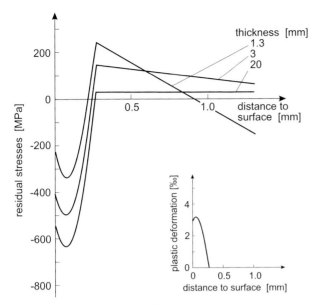

Fig. 3.5: Influence of specimen thickness on the depth distributions of residual stresses estimated by [3.63]

residual stress. The region affected by plastic deformation shows a maximum of compressive residual stresses beneath the surface, below which they change into courses of bending stress. The steepness of the latter increases with decreasing sample thickness, resulting in lower residual stress levels close to the surface. [3.52–3.54] proceeds in a manner similar to [3.62] in calculating the distribution of residual stress, stating comparable correlations.

It was first mentioned by [3.52] that the cyclic deformation behavior of the workpiece, too, is important for the shot peening process. This is based on the assumption that the increments of plastic strain will decrease with each impact and saturation or stabilization effects will occur. Additionally, it is pointed out that due to the loading and unloading processes during repetitive impacts the tendency of the workpiece to exhibit Bauschinger effect is a factor. Under the assumption that all surface points are concurrently and similarly loaded and thus, a flat strain state exists, [3.64,3.65] calculate the areas of plastic strains and residual stresses from the deformation field determined for elastic material behavior according to Hertz. The surface layer of the workpiece is divided into a region which, following initial plastic deformation, is deformed elastically only, and a region close to the surface which experiences cyclic elasto-plastic deformation in kinematic hardening and softening and which shows a stabilized deformation behavior. Thus, apart from quasi-static workhardening behavior, cyclically stabilized workhardening behavior, too, is relevant to describing the state of the surface layer. Approaches utilized for the simplified examination of mechanical compatibility are found in [3.66]. Corresponding to the division into two zones, residual stress distribution shows constant values within the zone of cyclic elasto-plastic deformation and decreasing values within the zone of cyclic elastic deformation. In order to take into account the drop in residual stress amounts close to the surface as observed in high strength material states, the existence of an additional zone of cyclic and entirely elastic deformation must be assumed. [3.67,3.68] took up this model, inserted improved material models and arrived at similar results. [3.69] experimentally determined a mean of 15 shot impacts for every surface point. Therefore, in contrast to the cyclic stress-strain curve determined by standard methods, they inserted the stress-strain correlation at a cycle number of 15 into the model introduced by [3.64,3.65]. In [3.70] the approaches described were developed into an easily usable software package, which allows residual stress distributions to be inferred from the Almen intensity and the shot diameter. A primary program is initially used to determine the deflection of an Almen strip under similar model assumptions as in the description of the shot peening process itself. This allows inferring the shot velocity from the shot diameter and the Almen intensity. Shot velocity is then used as an input parameter in the actual calculation, in which the axial-, bending- and source stresses caused by plastic deformation close to the surface are superimposed in a similar manner as in the models of [3.52,3.58] and the resulting system of equations is solved iteratively. To take account of multiple peening runs, there may be multiple runs of the calculation. Finally, simple specimen geometries also allow for any shot peening-induced deviations of shape to be estimated. [3.71–3.75] introduced a model building on the approaches of [3.68,3.76] which takes into account friction effects, different impact angles and the hardness ratio of the shot to the workpiece. In this model, friction leads to surface residual stresses decreasing close to the surface, i.e. to residual stress depth distributions which show a pronounced peak of compressive residual stresses beneath the surface. By contrast, impact angles which deviate from the surface normal reduce the depth effect of plastic deformation and thus of residual stresses. The model assumptions were first introduced in [3.73] as a software named "Shotpeen", whose successor, "Peenstress", has seen commercial use in recent years [3.75]. [3.75] list applications on steels, nickel-, titanium- and aluminum-base alloys, which are well in accordance with the residual stress values determined experimentally. Fig. 3.6 is an exemplary depiction, showing the depth distribution of shot peening-induced residual stresses in Udimet 720 [3.72]. The surface value and the maximum value of the residual stresses are described well, as are the depths of the maximum and of the position of zero residual stresses.

Supplementing the model approaches introduced so far, the literature also provides a large number of numerical approaches for describing surface layer development during shot peening. The first known study to focus on this was [3.77] in 1971, describing the flattening of contact stress distribution in comparison to Hertz's elastic model for ideal elasto-plastic material behavior. [3.78–3.80] carried out similar calculations on elasto-plastic material behavior. However, the verification of their results is limited to Hertz's solution for the purely elastic case. [3.78,3.80] measured the residual stresses by superposing the numerical solution with an elastic, analytically derived stress distribution. [3.81–3.87] also assumed

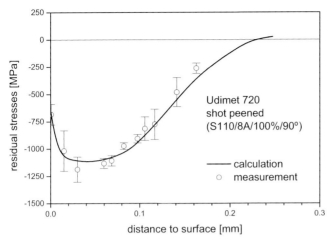

Fig. 3.6: Comparison of depth distributions of residual stresses calculated and experimentally determined by [3.72]

areal loads as external loading, derived from primary calculations or from Hertz's contact formulae. [3.81] even included a description of multiple impacts. [3.88,3.89], too, calculated single and multiple contacts, using Finite-Element- and Discrete-Element calculations coupled by interaction laws. However, the Finite-Element method has become the method of choice for most research teams working in this field.

The kinematics of the shot impact are considered in Finite-Elements calculations. This was done in [3.90] regarding frictionless single impact, and in [3.90–3.93] regarding frictional single impact. The interaction of the spheres is included in the simulation in [3.94–3.97] as double or multiple impact. Examinations of the influence of the strain rate on the peening result, however, figured prominently for the first time in the calculations of [3.98], and the kinematic workhardening that permits the Bauschinger effect to occur was introduced by [3.99].

Numeric descriptions of the shot peening process were carried out in [3.100], which examined the influence of material behavior and peening parameters on the residual stress states of the two steel varieties, AISI4140 with $R_{po,2} = 800$ MPa – described as medium-hard in the following, thus deviating from [3.100]'s classification as soft – and 54SiCr6 with $R_{po,2} = 1680$ MPa – described as hard in the following. Elasto-plastic material behavior is approximated multilinearly and work-hardening is preferably viewed as isotropic. In 2-dimensional ABAQUS-based Finite-Element calculations describing single impact, [3.100] finds that for purely elastic material behavior, Hertz's equations are confirmed for contact pressure distribution, while regarding elasto-plastic material behavior, contact pressure distribution drops off, and mean contact pressure is roughly $\bar{p} = 3\ R_y$, thus corresponding to the analytic model of [3.50]. The surface value and the maximum value of residual stresses are barely influenced by sphere radius, while the maximum- and permanent penetration depths, the contact radius and the depths of the residual stress maximum and of the plastically deformed zone increase proportionally to sphere diameter. According to Fig. 3.7 the maximum- and permanent penetration depths $\bar{z}$ and z_{end}, as well as the thickness of the plastically deformed zone z_{pl}, increase linearly with rising sphere impact velocity. By comparison, contact radius a_c increases marginally. The resulting values are always higher for the medium-hard material than for the hard material state. As sphere impact velocity grows, residual stress levels at the surface increase significantly, and maximum residual stress values increase only slightly, as shown in Fig. 3.8. In the medium-hard

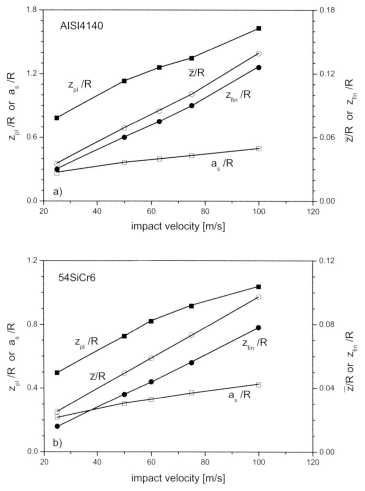

Fig. 3.7: Influence of shot impact velocity on maximum penetration depth, contact radius, remaining penetration depth and thickness of plastically deformed zone for medium hardness AISI4140 (a) and high hardness 54SiCr6 (b) determined by [3.100] in simulations of single shot impacts

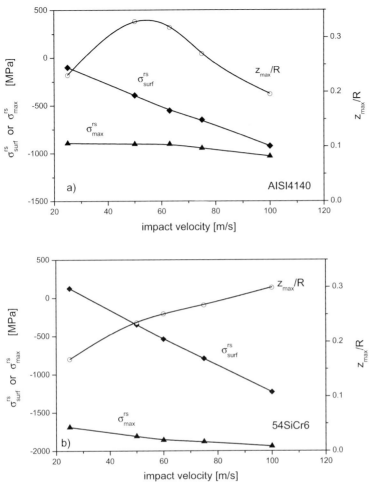

Fig. 3.8: Influence of shot impact velocity on surface- and maximum values of residual stresses as well as on depth of maximum residual stresses for medium hardness AISI4140 (a) and high hardness 54SiCr6 (b) determined by [3.100] in simulations of single shot impacts

material state this leads to near-congruence of both values for the highest sphere impact velocity examined. The hard material state, on the other hand, exhibits greater residual stresses, which are clearly higher beneath the surface than they are at the surface, even at great sphere impact velocities. The depth of the residual stress maximum in the hard material state increases with rising sphere impact velocity, and in the medium-hard material state it shows a maximum at medium speeds. As the surface values and maximum values of the residual stresses approximate each other in the medium-hard work material state, the plateau which is typical for these material states forms. When friction effects are taken

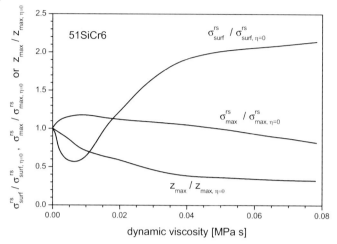

Fig. 3.9: Parameter study by [3.100] on the influence of viscous damping on the residual stress state after single shot impact on hard 54SiCr6

into account, the residual stresses close to the surface are shifted toward tensile values. The influence of plastic deformations of the sphere was additionally examined by varying the relation of the yield strengths of the sphere and the workpiece. With decreasing sphere hardness, this results in decreasing penetration depths and contact radii, and residual stresses can be tensile at the surface. A relation of strengths between the sphere and the workpiece larger than 2 results in constant values of the stated factors – in agreement with the results of [3.91,3.92] discussed above – which are congruent with the values for purely elastic sphere behavior. Accordingly, no plastic deformation occurs for spheres whose hardness is at least double the hardness of the workpiece. In parameter studies to take account of the strain-rate-dependence of material behavior using a viscous damping term, [3.100] finds a pronounced increase of close-to-surface residual stresses at rising strain-rate sensitivity, following an initial decrease – as shown in Fig. 3.9. By contrast, the maximum residual stresses increase slightly initially – while their depth clearly decreases – before they gradually fall.

In ABAQUS simulation calculations – again, 2-dimensional – for describing multiple impacts on the same position, [3.100] finds that with a growing number of sphere impacts, the maximum and permanent penetration depth, contact radius and thickness of the plastically deformed zone show decreasing amounts of change, due to an increasing workhardening of the workpiece. While the residual stresses at the surface become greater with every impact – again, the amount of change decreases –, there is hardly any change in maximum residual stress values and their depth. For increasing sphere impact velocities, the discussed findings regarding single impact are also obtained after multiple impacts. In accordance with experimental findings for peening treatments with complete coverage, the aforementioned effect of sphere impact velocity found in the single impact simulation is thus reduced, because at increasing sphere impact velocities, the same coverage is achieved by a decreasing number of sphere impacts.

Finally, [3.100] also estimates the surface layer state after multiple impacts at different positions. He starts from a 3-dimensional model in ABAQUS which describes single sphere impact. The results are then applied – consecutively at different positions, as displacements – to a larger model within Adina. By selecting suitable neighboring positions and utilizing the symmetry possible herein, it is possible to define small areas within which an arithmetic averaging of stress and deformation values at a constant distance to surface permits statements on the surface layer state after shot peening treatments. It becomes evident that the residual stresses are smaller than under single impact and that they are in greater accordance with the experiment. At the same time, residual stress fluctuations close to the surface can be assessed.

[3.101], also using the Finite-Element method, simulated the surface layer formation during shot peening. Here, the ABAQUS/Explicit software package was utilized with a user-defined material routine which describes deformation behavior at high strain rates based on the theory of thermally activated dislocation movement. Calculations were carried out for a single central impact and for 6 or 12 impacts occurring within two concentric circles around the central point whose impact areas touch. In order to keep effects of sequence as small as possible, the impact sequence was carried out in such a way that the number of preceding impacts in the vicinity was statistically equally distributed for the individual impacts. For determining representative residual stress depth distributions, the residual stress values were averaged for the given depth over an area corresponding to the circle circumscribing the 7 inner impacts. The residual stress distributions in quenched and tempered AISI4140 were described well regarding their distribution at increasing distance to surface and the depth of the position of zero residual stresses. However, residual stress values were clearly too high overall. The Bauschinger effect was viewed as one possible cause, as it was not yet implemented in the material model used, while bearing great significance regarding quenched and tempered steel. Selected characteristics of the depth of the residual stresses obtained with this method were compiled and comparatively assessed in [3.101].

As Fig. 3.10 shows, surface- and maximum residual stress values hardly change with increasing sphere diameter, while the depth of the zero crossing and the maximum value of the residual stresses are clearly increased, since the kinetic energy of the impacting spheres increases. Accordingly, the influence of impact velocity – shown in Fig. 3.11 – is similar, albeit that the residual stress values decrease slightly at the smallest and greatest impact velocities. Fig. 3.12 shows the influence of the impact angle, which is indicated as the angle to the surface normal. When the angle is increased, the residual stresses in the direction of the impact component parallel to the surface, as well as the depths of the zero crossing and of the maximum value of the residual stresses, are reduced because the impact component perpendicular to the surface decreases. By contrast, the residual stresses in the transversal direction remain approximately constant, since the

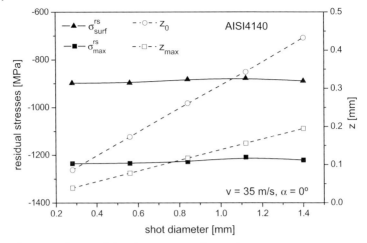

Fig. 3.10: Influence of shot diameter on surface- and maximum values of residual stresses as well as on depth of maximum and zero residual stresses for quenched and tempered AISI4140, determined by [3.101] in simulations of multiple shot impacts

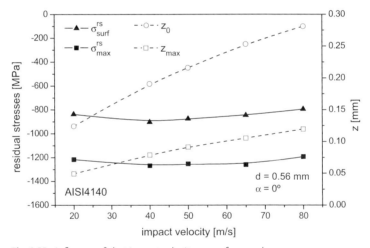

Fig. 3.11: Influence of shot impact velocity on surface- and maximum values of residual stresses as well as on depth of maximum and zero residual stresses for quenched and tempered AISI4140, determined by [3.101] in simulations of multiple shot impacts

supportive effect of the surrounding material during impact is barely changed. In parameter studies using the methods discussed above, [3.101] also studied the isolated influence of individual material parameters, as Fig. 3.13 shows exemplarily for yield strength and workhardening behavior. While an increasing yield strength results in rising residual stress surface values and, particularly, in rising maximum values of residual stress, the penetration depths of residual stresses and the depths of the residual stress maxima decrease. By contrast, as workhardening rates rise – as expressed by the workhardening coefficient used in Fig. 3.13b – the surface values of residual stresses decrease and their maximum values increase. Thus, the effect of Hertzian pressure becomes more important than the effect of plastic stretching at the surface. Residual stress penetration depth and the depth of the residual stress minimum, by contrast, are only marginally decreased. Regarding the results of a variation of the workhardening coefficient, it must be noted that raising the workhardening coefficient also significantly increases the flow stress achieved for the strong plastic strains in the simulation, and thus the effects of workhardening behavior and of the strength level appear jointly.

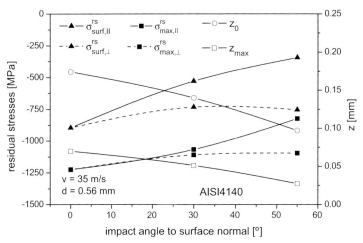

Fig. 3.12: Influence of impact angle to surface normal on surface- and maximum values of residual stresses as well as on depth of maximum and zero residual stresses for quenched and tempered AISI4140, determined by [3.101] in simulations of multiple shot impacts

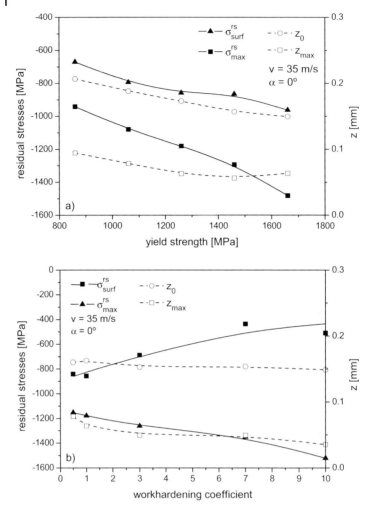

Fig. 3.13: Influence of yield strength (a) and work hardening coefficient (b) on surface- and maximum values of residual stresses as well as on depth of maximum and zero residual stresses for quenched and tempered AISI4140, determined by [3.261] in simulations of multiple shot impacts

3.1.2
Changes in the Surface State

Due to the successive, statistically irregular impact of individual spheres on the workpiece surface, the surface layer properties do not remain constant along the surface in any case. This must be considered when assessing shot peened material states. Fig. 3.14 shows this by means of high-lateral-resolution measurements

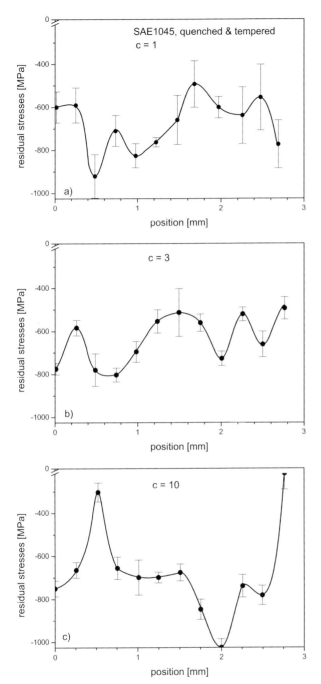

Fig. 3.14: Local residual stress distribution at the surface of a quenched and tempered specimen of SAE1045 shot peened using different coverages (S330, $p = 3$ bar) [3.102]

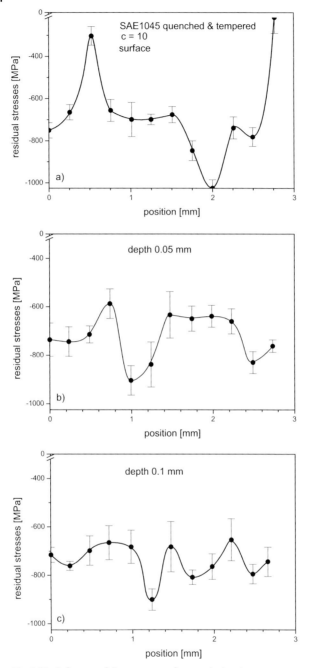

Fig. 3.15: Influence of distance to surface on the local residual stress distribution at the surface of a quenched and tempered specimen of SAE1045 shot peened using a coverage of 10 (S330, $p = 3$ bar) [3.102]

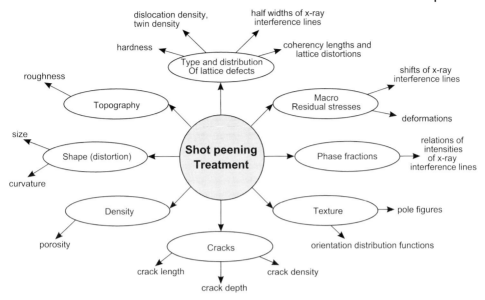

Fig. 3.16: Workpiece properties affected by shot peening treatments and methods for their experimental determination [3.140]

of residual stress values at the surface of quenched and tempered SAE1045 subjected to different shot peening treatments [3.102]. The area covered by the X-ray beam was approximately circular, with a diameter of 0.23 mm. The measurements were carried out along a line drawn arbitrarily on the surface, at distances of 0.25 mm. At a degree of coverage of 1.25, residual stresses between approximately -930 and -500 MPa are found. Although their fluctuation decreases during 3-fold coverage, during 10-fold coverage it increases to a degree at which marginal local residual stresses are facing compressive residual stresses above 1000 MPa. [3.102] introduced similar findings also for a normalized and for a quenched state of SAE1045. Fig. 3.15 shows that this effect is not restricted to the surface values of residual stresses, but can be observed beneath the surface as well. However, the residual stress state appears to homogenize with increasing distance to surface. The average values of the measured residual stresses always correspond to the residual stresses determined by measurements which integrate over greater areas of the material [3.102]. Effects of these inhomogeneities in the surface layer state on material behavior are to be expected particularly either when the highest loadings are concentrated on small material regions, as is the case e.g. in sharply notched components, or when very large shot diameters are used, e.g. in order to achieve a high penetration depth of the residual stresses. Due to the elementary processes stated above, alterations of the properties summarized in Fig. 3.16 occur during shot peening which can be understood by measuring the properties indicated.

The following sections will focus individually on the influence of the peening parameters on the integral values of these properties.

a) Influences on Shape

Shot peening is usually not aimed at altering dimensions. In thin-walled components, however, this may occur, due to deep plastic deformations during shot impact. By contrast, alterations of shape are the specific objective of peen forming, which today constitutes an important process for producing thin-walled structures of complex shape [e.g. 3.90,3.103–3.105].

b) Influences on Topography

In essentially all cases, shot peening is associated with changing surface topography. In Fig. 3.17, images obtained using a confocal white light microscope show the surface of shot peened AISI4140 in different heat treatment states, as an exaggerated three-dimensional representation and as a profile graph. The impact points of the shot are clearly visible, their size and depth decreasing from the normalized to the quenched and tempered and to the quenched state.

The fundamental connections between selected peening parameters and the resulting topography have already been pointed out in Fig. 3.1 by looking at deformation depth after single sphere impact. Concerning technically relevant shot peening treatments, [3.106] presents an overview of the different roughnesses caused by different shot peening treatments of AISI4140 steel. Fig. 3.18 shows that shot peening treatments increase the roughness which results from prior sanding treatments. This increase is the more pronounced the lower workpiece hardness is. While roughness increases by about a factor of 10 in the normalized state, hardly any changes in roughness are observed in the quenched state or in the state quenched and tempered at 180 °C. Fig. 3.18 also indicates that the created roughness intensifies with increasing shot size and hardness. In this respect, and in accordance with [3.1], marked effects on topography are still observable even when shot hardness is lower than workpiece hardness. The causes of this correlation are higher workpiece hardness – which increases strength while at the same time it decreases the amount of deformation under similar shot conditions –, the kinetic energy of the peening particles – which increases with rising shot hardness – and the tendency of the shot to show decreased plastic deformation as its hardness rises. Fig. 3.19 confirms this by showing the influence of peening pressure on the roughness of AISI4140. The roughness of the softer work material states, in particular, increases with the growth of the peening particles' kinetic energy when peening pressure rises. By contrast, Fig. 3.20 shows that when the mass flow rate is varied, a decrease of roughness can be observed, which in turn is more pronounced in softer work material states. The probable cause is the rising probability of mutual impacts of the spheres in the peening nozzle and in the supply hoses, resulting in decreasing kinetic energy of the peening particles. By comparison, [3.107] used AZ31 magnesium alloy to show that when the Almen inten-

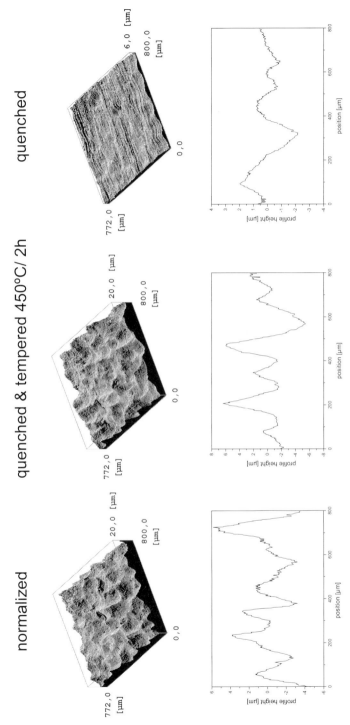

Fig. 3.17: Influence of heat treatment state on surface topography for shot peened AISI410 determined using confocal white light microscopy [3.262]

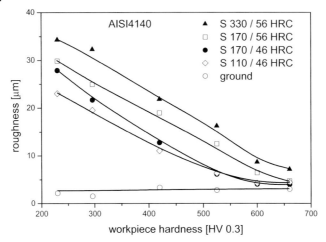

Fig. 3.18: Influence of shot on roughness vs. bulk material hardness for shot peened AISI4140 ($p = 1.6$ bar, $\dot{m} = 1.5$ kg/min) [3.106]

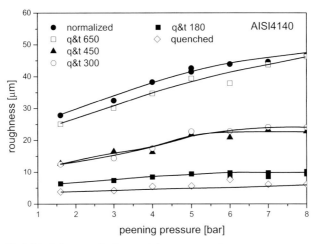

Fig. 3.19: Influence of bulk material state on roughness vs. peening pressure for shot peened AISI4140 (S170 46 or 56 HRC, $\dot{m} = 1.5$ kg/min) [3.106]

sity of the peening treatment remains the same, medium-sized peening medium diameters allow for minimal roughness (comp. Fig. 3.21).

Ultrasonic shot peening is due to achieve lower levels of surface roughness than comparable conventional shot peening treatments, as the peening medium deviates slightly from exact spherical shape [3.108].

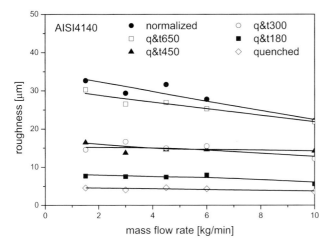

Fig. 3.20: Influence of bulk material state on roughness vs. mass flow rate for shot peened AISI4140 (S170 46 or 56 HRC, $p = 3$ bar) [3.106]

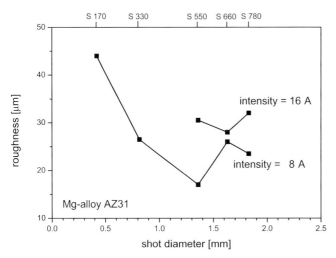

Fig. 3.21: Influence of Almen intensity on roughness vs. shot diameter for shot peened AZ31 magnesium alloy [3.107]

c) **Influences on Residual Stress State**

[3.106] presented residual stress distributions found in states of AISI4140 after various kinds of shot peening treatment. According to Fig. 3.22, the magnitude of surface residual stresses increases with rising workpiece hardness from the normalized state to the 450 °C-quenched and tempered state. With an additional increase of hardness, surface residual stresses drop by up to 100 MPa. With

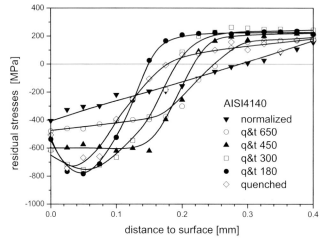

Fig. 3.22: Influence of bulk material state on residual stress distribution for shot peened AISI4140 (S170 46 HRC, $p = 1.6$ bar, $\dot{m} = 1.5$ kg/min) [3.106]

increasing hardness of the peened specimens, the penetration depth of residual stresses grows smaller, and only the state quenched and tempered at 180 °C exhibits an even lower penetration depth than the quenched state. There is a characteristic compressive residual stress maximum which forms under the surface with rising workpiece hardness and which shows values ranging from -800 to -750 MPa, its depth of 0.05 mm remaining roughly constant. Fig. 3.23 represents the same facts by showing the magnitudes of surface residual stresses $\left| \sigma_{surf}^{rs} \right|$, the

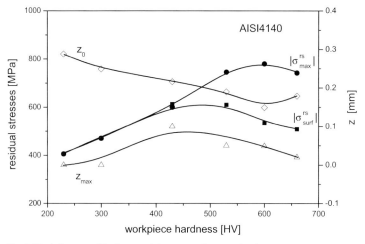

Fig. 3.23: Influence of bulk material state on the magnitudes of surface- and maximum residual stresses as well as depth of maximum- and zero residual stresses for shot peened AISI4140 (S170 46 HRC, $p = 1.6$ bar, $\dot{m} = 1.5$ kg/min) [3.106]

magnitudes of maximum residual stresses $|\sigma^{rs}_{max}|$, depths of residual stress maximums z_{max} and the depth locations of the residual stress zero crossings z_o contingent on workpiece hardness. In accordance with [3.109], greater compressive residual stresses are found for the state annealed at the lowest temperature than for the quenched state, which according to [3.102] is typical for low shot hardnesses and is due to the increases of strength in this annealing stage, which in turn can be traced back to the arrangement of dislocations into energetically more favorable structures and to the formation of carbon clusters which pin these dislocations [3.110].

According to [3.15] the observed findings are an expression of the two competing processes of plastic deformation during shot peening. At low workpiece hardness, the plastic stretching of close-to-surface regions dominates, compressive residual stresses thus assuming their maximum at the immediate surface. Only under extreme peening conditions a compressive stress maximum can occur beneath the surface. At medium workpiece hardness, plastic stretching of regions at the immediate surface and Hertzian pressure are roughly equivalent, and thus, great compressive residual stresses at the surface appear alongside a marked compressive residual stress maximum beneath the surface. At high workpiece hardness, however, the effect of Hertzian pressure dominates, and thus, a strongly pronounced compressive stress maximum beneath the surface is registered together with relatively small compressive residual stress values directly at the surface.

For shot peened AISI4140 in a quenched state, Fig. 3.24 depicts the distributions (Fig. 3.24a) and their characteristics (Fig. 3.24b) to show that in this heat treatment state, the maximum value and the surface value of compressive residual stresses are determined more by shot hardness rather than by shot diameter. By contrast, the position of the sub-surface compressive residual stress maximum and the position of the residual stress zero crossing increase, with the exception of the smallest shot diameter S110 which probably must be viewed as an outlier value. These findings, too, can be interpreted using the model of [3.15]. Accordingly, for soft workpiece states the thickness of the layer containing compressive residual stresses increases with peening intensity. By contrast, for medium workpiece hardnesses, increases of shot velocity and of shot diameter result in significantly greater depths of the compressive residual stress maximums, in conjunction with only slight increases or even decreases of the compressive residual stress maximums. An increase of shot hardness, however, leaves the compressive residual stress maximums unchanged. For hard workpiece states, increases of shot velocities result in barely lowered depths of the compressive residual stress maximum – while compressive residual stress maximums clearly increase –, as this case hardly permits any control over the size of the contact zone. An increase of shot hardness creates raises the surface value and the maximum value of compressive residual stresses.

These effects are confirmed by [3.111]'s examinations on the AlZn4.5Mg2 aluminum alloy – with a hardness of ca. 125–150 HV –, the results of which are represented in Fig. 3.25. Shot peening treatments employing aluminum granals with a hardness of approx. 20 HV lower than that of the workpiece cause compressive

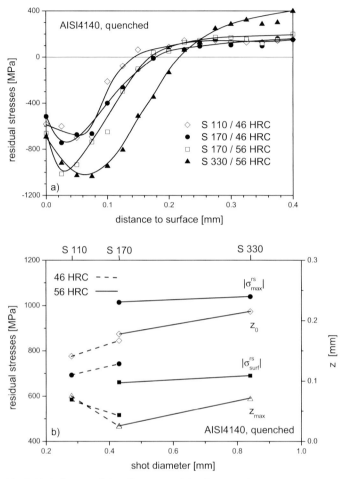

Fig. 3.24: Influence of shot diameter and -hardness on the residual stress distribution (a) and its characteristics (b) for shot peened AISI4140 ($p = 1.6$ bar, $\dot{m} = 1.5$ kg/min) [3.106]

residual stress maxima beneath the surface, in conjunction with comparatively extensive penetration depths of the residual stresses. Peening with glass beads possessing significantly smaller dimensions, but a significantly higher hardness between 800 and 1000 HV, effects distinctly greater surface- and maximum compressive residual stresses, yet lower penetration depths. Similar depth distributions of residual stress are observed after peening treatments using steel shot. By contrast, ceramic beads with 800–1000 HV of hardness, while arriving at surface residual stresses and penetration depths similar to those of the granals, result in markedly reduced maximum values of compressive residual stress, in comparison with steel and glass, yet at similar depths.

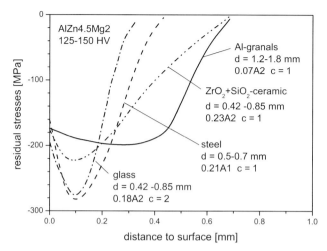

Fig. 3.25: Influence of shot type on residual stress distribution for AlZn4.5Mg2 [3.111]

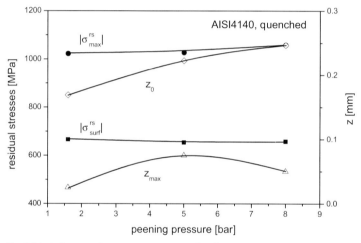

Fig. 3.26: Influence of peening pressure on the characteristics of residual stress distribution for shot peened AISI4140 in a quenched state (S170 56 HRC, $\dot{m} = 1.5$ kg/min) acc. [3.106]

Fig. 3.26 shows that with quenched AISI4140, raising the shot velocity in compressed-air peening systems does not result in any changes of the surface value of the residual stresses, but – as long as shot velocity is not reduced by collisions – results in increasing depths of the residual stress zero crossing and of the residual stress maximum, the value of the latter remaining unchanged [3.106]. The same study contains similar reports regarding the normalized state of AISI4140, in which, however, peening pressure in excess of about 5 bar appears to lead to a saturation of the achievable depth effect. On an industrially normalized and a quenched and tempered state of AISI4140, [3.112] examined the residual stress distributions after shot peening treatment using S110 shot with 2-fold coverage and variation of velocity, mass flow rate, nozzle diameter and nozzle distance. In accordance with aforementioned examinations, it became evident that there is no significant influence on surface- or maximum residual stress values in either heat treatment state. By contrast, and as depicted in Fig. 3.27, the depths of the residual stress zero crossing – which, geometry preventing correction of layer removal, are substituted here by the depths at residual stress values of -200 and -400 MPa – increase strongly with growing impact velocity and less markedly with growing nozzle diameter or distance. This is in agreement with parallel measurements of Almen intensity, which increases in the same manner.

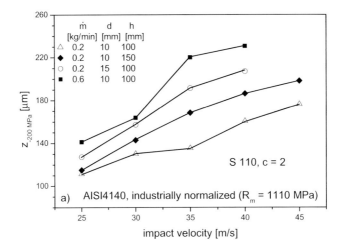

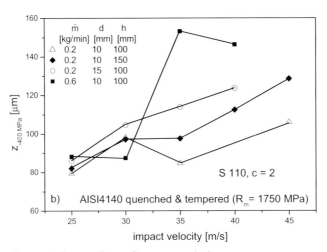

Fig. 3.27: Influence of mass flow rate, nozzle diameter and nozzle distance on depths of −200 and −400 MPa for industrially normalized (a) and quenched and tempered (b) AISI4140 [3.112]

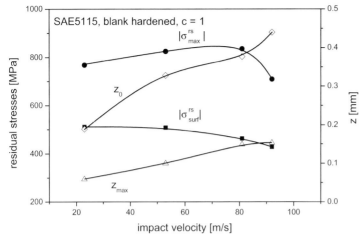

Fig. 3.28: Influence of drop velocity on the characteristics of residual stress distribution for shot peened SAE5115 in a blank hardened state acc. [3.16, 3.113–3.117]

In rotating wheel peening devices, peening pressure is replaced by the number of revolutions of the wheel, and therefore, release velocity becomes the essential factor determining shot velocity. By example of blankhardened and consecutively shot peened SAE5115, Fig. 3.28 shows that with increasing shot velocity and at virtually constant surface values, the depth as well as the magnitude of the compressive residual stress maximum and -penetration depth increase, given that no over-peening effects occur, as is the case here at the highest shot velocity of $v_{rel} = 92$ m/s, giving rise to surface cracks and other forms of surface layer damage [3.113,3.114]. Comparable effects are observed in the casehardened state [3.114,3.115] and in the casehardened and annealed state [3.116], as well. By contrast, according to [3.117], for shot peening treatment of normalized SAE5115, increases of shot velocity only marginally affect the maximum values of the compressive residual stresses close to the surface, but achieve a marked gain in compressive residual stress penetration depth. Thus, these effects are comparable to those of peening pressure increases in compressed-air peening systems, and the fit seamlessly into the model assumptions of [3.15].

Fig. 3.29 is an exemplary portrayal of the influence of mass flow rate $\dot{m}$ on the residual stress distributions in AISI4140 in a state quenched and tempered at 450 °C [3.106]. With increasing mass flow rate, decreasing residual stress penetration depths are observed, whereas the surface values of the residual stresses rise initially and the compressive residual stress maximum values are approximately constant in the plateau region that is characteristic for quenched and tempered material states. This may be attributed – as is the case for the corresponding effects on surface topography – to the increase of collision processes among the shot particles and the resulting drop in the peening particles' impact energy. Fig. 3.30 shows particularly impressive evidence for this, summarizing the residual stress

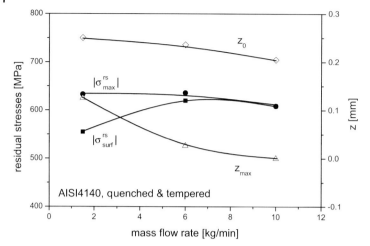

Fig. 3.29: Influence of mass flow rate on the characteristics of residual stress distribution for shot peened AISI4140 in a state quenched and tempered at 450 °C (S170 46 HRC, $p = 3$ bar) [3.106]

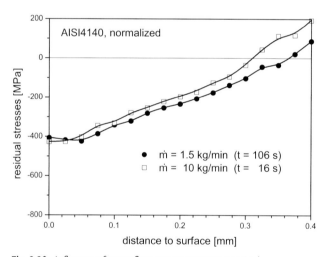

Fig. 3.30: Influence of mass flow rate at constant nominal coverage on residual stress distribution for normalized AISI4140 (S170 46 HRC, $p = 3$ bar) [3.106]

distributions for a normalized AISI4140 after two peening treatments. Different mass flow rates were used in the treatments, yet the achieved coverages are nominally equal due to a suitable selection of peening durations. The shot peening treatment with the higher mass flow rate resulted in a smaller penetration depth of the residual stresses, due to the decreased impact energy [3.106].

The impact angle of the shot on the peened surface, too, is crucial for the resulting surface layer states. Using normalized SAE1045 shot peened at different angles, [3.118] was able to demonstrate that the penetration depth of the compressive residual stresses decreases together with the decrease of the impact velocity's normal component, i.e. with the increase of angle a to the surface normal (comp. Fig. 3.31). In addition, increasing tangential components of the impact forces result in rising deviations of the residual stress state from the rotation symmetry present at perpendicular incidence. In a longitudinal direction, in which the shot exhibits a tangential velocity component, the residual stress value decreases as the angle to the surface normal increases. By contrast, in the transversal direction,

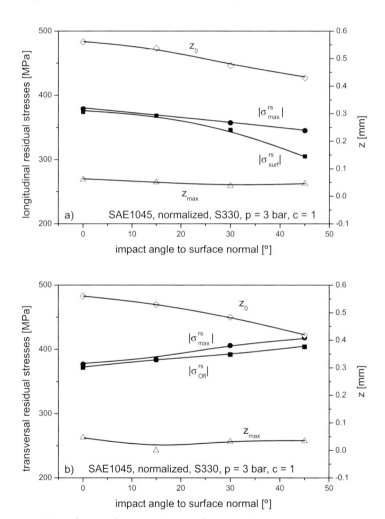

Fig. 3.31: Influence of impact angle to surface normal on the characteristics of residual stress distribution for shot peened SAE1045 in a normalized state acc. [3.118]

which is perpendicular to the direction discussed above, the surface values and the maximum values of the residual stress magnitudes increase. The explanation offered is that due to the shear stresses on impact, the equivalent stress and thus plastic deformation are increased in regions close to the surface. With the exception of the transversal residual stress amounts, the results are in accordance with the findings obtained by means of the simulation calculations of [3.101] shown in Fig. 3.12. [3.1,3.87,3.119] show that deformation depth at varying impact angles is also dependent on the workpiece hardness and on shot velocity. The fact that deformation depth exhibits a maximum at angles between 5 and 20° to the surface normal indicates the apparent complexity of these processes, and it necessitates further theoretical and experimental studies, particularly regarding the issue of dividing the local plastic strains into their components.

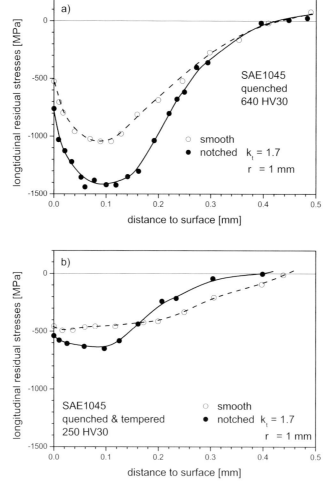

Fig. 3.32: Residual stress distributions after identical shot peening treatments of smooth and notched specimens of quenched (a) and quenched and tempered (b) SAE1045 [3.120]

The aforementioned effects of the impact angle on the residual stress state also figure in the shot peening of notched geometries. According to [3.120], maximum magnitudes of residual stresses are present in the notch ground, which may be caused, on the one hand, by the smaller impact angles on the notch flanks, and on the other hand, by the obstruction of plastic deformation which occurs at the notch ground surface or in regions directly beneath it. The significant role of the obstruction of plastic deformation is evidenced by the fact that there are greater notch ground residual stresses in the direction of loading than in the transversal direction [3.120,3.123]. Accordingly, no rotationally symmetric residual stresses are found in curved workpiece surfaces. Fig. 3.32 compares the shot peening-induced residual stress states in smooth and in notched specimens of quenched and of quenched and tempered SAE1045. In both cases the longitudinal residual stresses are stated for the notched specimens. Accordingly, notched specimens exhibit higher magnitudes of residual stress in regions close to the surface than smooth specimens do. In quenched and tempered specimens, this relation is reversed at higher distances to surface. However, it must be noted that [3.124], looking at similar degrees of material hardness, found residual stresses unchanged in comparison with smooth specimens, and [3.125] found residual stresses in notched flat specimens of TiA16V4 which were reduced in comparison with smooth specimens. Systematic examinations of the correlations between residual stress development and workpiece geometry in shot peening are required to determine whether the obstruction of plastic deformation, as applied to the argument in a rather general form, is sufficient.

Fig. 3.33 uses phase-specific measurements on iron-copper-titanium carbide alloys in a shot-peened state – with varied iron and copper contents – to show that macroscopic residual stresses are always within the compressive range and decrease as the copper content increases. Yet phase-specific residual stresses are within the compressive range only in the two metallic phases, while in the TiC phase they are in equilibrium with great tensile residual stresses [3.126]. At ca. 10% copper content, the latter have the effect that the residual stresses present overall in the TiC phase are already tensile when there is no loading taking place. Therefore, additional tensile loading promotes delamination and the initiation of cracks.

The fundamental influence of workpiece hardness on residual stress surface- and maximum values, as well as the location of the residual stress maximum and zero crossing, is represented once more in Fig. 3.34 for different groups of materials. Fig. 3.35 is a schematic summary of the dependence of residual stress distribution on the peening parameters [3.127].

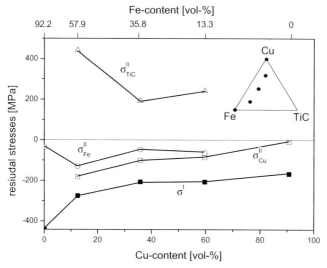

Fig. 3.33: Macro- and phase specific residual stresses due to shot peening in sintered materials with different contents of copper, iron and titanium carbide [3.126]

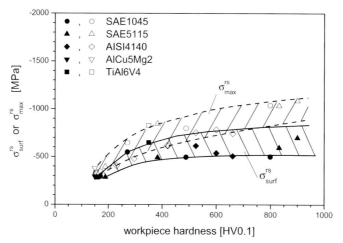

Fig. 3.34: Influence of bulk material hardness on the magnitude of surface- and maximum residual stresses after shot peening acc. [3.127]

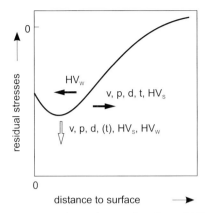

Fig. 3.35: Scheme showing the effects of different parameters on the residual stress state after shot peening [3.127]

d) Influences on Workhardening State

The changes in the workhardening state close to the surface, too, are quite essential for the mechanical properties of shot peened material states. Fig. 3.36 shows the influence of the heat treatment state on the distribution of the half widths of x-ray interference lines, using AISI4140. As the distance to surface decreases, the half widths increase in the normalized state and in the state quenched and tempered at 650 °C, which exhibit the smallest half widths due to the low number of defects in their core region. On the other hand, the half widths of the state quenched ad tempered at 450 °C show virtually no dependence on distance to surface. By contrast, the states with greater core values of the half widths – caused, in

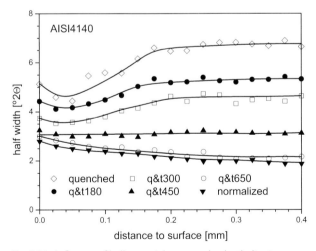

Fig. 3.36: Influence of bulk material state on the depth distribution of half widths for shot peened AISI4140 (S170 46 HRC, $p = 1.6$ bar, $\dot{m} = 1.5$ kg/min) [3.106]

particular, by high dislocation densities – initially show half widths decreasing towards the surface. After passing through a minimum close to the surface, they show surface values that are slightly higher than the minimum, yet still markedly lower in comparison to the core value. Fig. 3.37 depicts the results of detailed x-ray line profile analyses of the same material to show that the mean strains, which correspond to micro-residual stresses, are elevated due to shot peening when hardness is low and are reduced when hardness is high. At a bulk material hardness of 450 HV0.3 there are no shot peening-induced changes of mean strains. On the other hand, the domain sizes, also described as the sizes of coherently scattering regions, show lower values with increasing hardness, which show

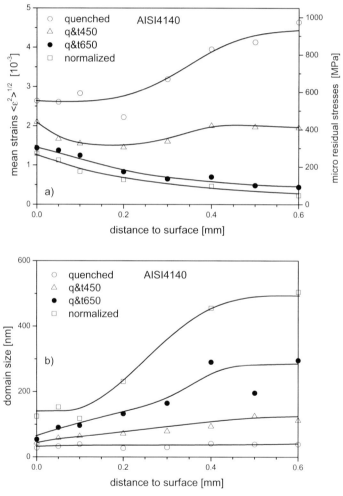

Fig. 3.37: Depth distributions of mean strains (a) and domain sizes (b) in different states of AISI4140 after shot peening (S330 56 HRC, 3-fold coverage) [3.130]

minima after shot peening which are nearly independent of bulk material hardness. The ratios of the mean strains to the domain sizes, which according to [3.128] are proportional to dislocation density, exhibit depth distributions in shot peened material states as shown in Fig. 3.38. These depth distributions show that the increase of the half widths in the shot peening-affected surface region of soft work material states are caused by an increase of dislocation density which occurs during deformation, and that the reduction of half widths found in harder material states occurs alongside a decrease of dislocation density. The great dislocation densities found in these material states after quenching or quenching and tempering cannot be increased any further during shot peening. Instead, dislocation rearrangements into energetically more favorable arrangements and dislocation annihilations occur in these cases. Furthermore, the dissolved carbon – supported by the mechanical energy provided – is able to change position within octahedral voids with lower lattice distortions. This process, known as the Snoek effect [3.129] can contribute – at least in the states of highest hardness – to a decrease of the half widths, in addition to the dislocation rearrangements and -annihilations mentioned above [3.130,3.131]. [3.132,3.133] found similar effects in shot peened surface layers. Regarding the compressive deformation of quenched and of quenched and tempered steels with low annealing temperatures, [3.134] reported initial deformation-induced decreases of the half widths and mean strains in spite of strong macroscopic workhardening. By way of TEM examinations he confirmed that deformation-induced rearrangements of initially statistically distributed, high-density dislocations into energetically more favorable arrangements – such as tangles, walls or cells – may occur, resulting in the formation of hard and soft regions and, thus, in an increase of flow stress.

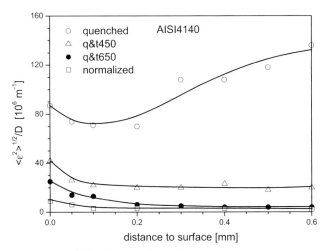

Fig. 3.38: Depth distributions of the relations of mean strains and domain sizes, which are proportional to dislocation density, in different states of AISI4140 after shot peening (S330 56 HRC, 3-fold coverage) [3.130]

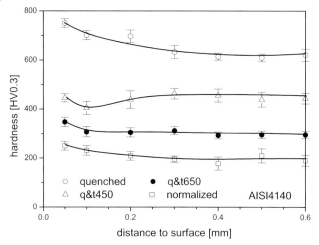

Fig. 3.39: Depth distribution of microhardness in different states of AISI4140 after shot peening (S330 56 HRC, 3-fold coverage) [3.130]

Depicting the influence of the heat treatment state of AISI4140 on the distributions of micro-hardness after shot peening, Fig. 3.39 shows that the distributions of micro-hardness – which, due to their simple determination, are often used to characterize shot peening-induced workhardening states – only inadequately describe the microstructural changes of the surface zone. Except for the state quenched and tempered at 450 °C, shot peened surface layers show increases of micro-hardness. While the properties determined using x-rays describe the microstructural state after shot peening, determining micro-hardness presupposes that plastic deformation occurs and is influenced by the residual stress state present. Experimental findings and theoretical considerations of [3.135] show that in hardness measurements, compressive loading stresses reduce the resulting shear stresses. Therefore, smaller regions beneath the penetrating body are plastically deformed. This causes greater hardness values when compressive loading stresses occur and when, in a qualitatively similar way, compressive residual stresses occur. Thus, increased values of hardness are observable in hard material states that incorporate great compressive residual stresses, despite the fact that the aforementioned dislocation rearrangements cause microstructural softening.

The influence of the peening parameters – peening pressure, mass flow rate and velocity – on the workhardening state of the surface layer are evidenced mostly by changes of the depth at which the workhardening state of the bulk material is reached, starting from the surface. As this transition is difficult to determine – changes in the respective measurement parameters in relation to the bulk material's value being rather small – there are hardly any statements known regarding the effects of said parameters on the workhardening state which go beyond what has been described by way of the residual stress depth distributions.

Fig. 3.40 schematically summarizes the dependencies of the depth distribution of the half widths on the peening parameters [3.127].

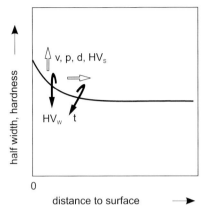

Fig. 3.40: Scheme showing the effects of different parameters on the half widths after shot peening [3.127]

e) Influences on Microstructure

The micro-structural changes within shot peened surface layers are very difficult to access by way of experimental examination, due to their low thickness and the incorporated property gradient. For these reasons, there are only very few known examinations of this subject, a fraction of which utilize the technology of cross-sectional preparation of TEM foils, developed specifically for this purpose [3.136,3.137].

The depiction in Fig. 3.41 of transmission electron microscopic examinations, carried out by [3.138] on normalized AISI4140 using conventional methods, shows that the initial dislocation structure with a low dislocation density, which occurs close to the surface in the original state, is transformed by shot peening to a statistically distributed dislocation structure with a highly elevated density. This dislocation structure is typical of deformation with high strain rates, which can reach up to $5 \cdot 10^3 l/s$ in shot peening [3.138]. This is caused by a strain rate-induced obstruction of the cross slip of screw dislocations, suppressing the formation of spatial dislocation arrangements like those found in dislocation cells. Corresponding TEM examinations on quenched and tempered AISI4140, depicted in Fig. 3.42, show that the dislocation structures found close to the surface after shot peening are not very different from those found in the initial state, because the dislocation arrangement is largely determined by the heat treatment-induced distribution of finest carbides [3.138]. On normalized SAE1045, Fig. 3.43 shows that deformation is not restricted to the ferritic regions, but that statistically distributed dislocations of increased density are also present in the pearlitic regions between the Fe_3C lamellae [3.139]. The fact that the dislocation arrangement does not always inevitably have to be statistical is shown in Fig. 3.44 on the microstructure close to the surface of shot peened AlMg1 and AlMg5 aluminum alloys [3.140].

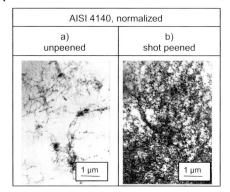

Fig. 3.41: Transmission electron microscopic view of the microstructure at a distance to surface of about 0.08 mm in unpeened (a) and shot peened (b) specimens of normalized AISI4140 (S170 54–58 HRC, $p = 1.0$ bar, single coverage) [3.138]

AISI 4140, quenched and tempered	
a) unpeened	b) shot peened

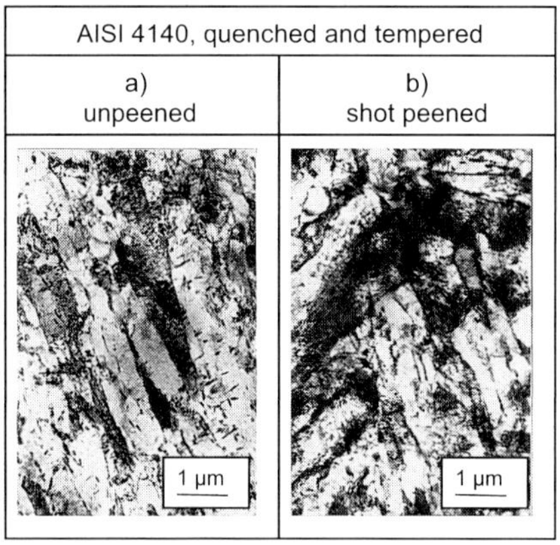

Fig. 3.42: Transmission electron microscopic view of the microstructure at a distance to surface of about 0.15 mm in unpeened (a) and shot peened (b) specimens of quenched and tempered AISI4140 with an initial hardness of 320 HV30 (S170 54–58 HRC, $p = 1.6$ bar, single coverage) [3.138]

While homogeneously distributed dislocations of increased density, along with a few dislocation tangles, are found in AlMg5, AlMg1 exhibits sharply pronounced dislocation cell structures. This is caused by the stacking fault energy which, being triple the amount of that present in AlMg5, strongly restricts the splitting of dislocations and thus eases the cross slip of screw dislocations. [3.141,3.142] recorded TEM micro-structure depth profiles of normalized SAE1045 in a shot peened state, resulting in tangled dislocations with densities of up to $8 \cdot 10^{11} \, 1/\mathrm{cm}^2$ and significant local differences of the dislocation densities. Pronounced bending lines point toward high inhomogeneous micro residual stresses, as well. Twin formation on AZ31 magnesium alloy is reported in [3.143].

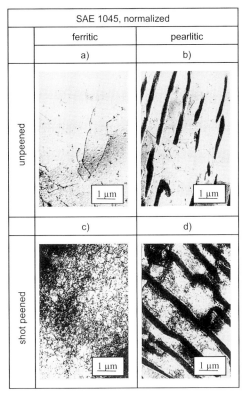

SAE 1045, normalized	
ferritic	pearlitic
a)	b)
c)	d)

Fig. 3.43: Transmission electron microscopic view of the microstructure at a distance to surface of about 0.08 mm in ferritic (a, c) and pearlitic (b, d) regions of unpeened (a, b) and shot peened (c, d) specimens of SAE1045 in a normalized state (S170 54–58 HRC, $p = 1.6$ bar, single coverage) [3.139]

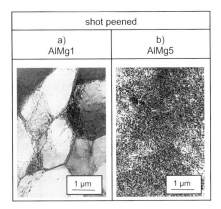

shot peened	
a) AlMg1	b) AlMg5

Fig. 3.44: Transmission electron microscopic view of the microstructure at a distance to surface of about 0.12 mm in shot peened specimens of AlMg1 (a) and AlMg5 (b) (S170 54–58 HRC, $p = 0.24$ bar, single coverage) [3.140]

Shot peening can lead to an increase of density in regions close to the surface in metallic sintered materials. [3.140] reports on densities close to the surface reaching 7.72 g/cm^3 in shot peened sintered iron with a core density of 7.37 g/cm^3. [3.144] observes decreases of porosity close to the surface in unalloyed sintered steels, caused by ultrasonic shot peening. The higher the initial porosity is, the deeper-reaching the decreases become.

f) Influences on Phase Fractions

Phase transformations due to shot peening treatments are of particular significance regarding steels with observable transformations of austenite into martensite. Among these are casehardened steels [3.145–3.149], quenched high carbon tool steels and metastable austenitic steels [3.150–3.153]. [3.154] reports that contents of retained austenite – which are transformed into martensite during shot peening (comp. Fig. 3.45) – increase with rising austenitizing temperature in AISID3 tool steel. In relation to the initial value, the transformations of retained austenite are the smaller the higher the austenitizing temperature is. The transformed fraction increases towards the surface. In the immediate vicinity of the surface, however, high contents of retained austenite are preserved in states with high initial content of retained austenite. Thus, a maximum of phase transformation occurs beneath the surface, and is shifted to higher distances to surface when the initial contents of retained austenite increase. While the reduced tendency of martensite formation at increasing austenitizing temperature may be explained by the increasing chemical stabilization of the austenite, [3.154] attributes the reduction of transformations found at the immediate surface to temperature increases during shot peening beyond the critical temperature M_D, above which

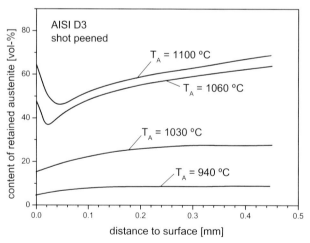

Fig. 3.45: Influence of austenitizing temperature of AISID3 tool steel on the depth distributions of retained austenite contents after shot peening [3.127]

no deformation-induced martensite formation can occur. These temperature increases are not expected to occur to the same extent beneath the surface, where there are smaller deformations.

In TEM examinations on AISI 304, [3.142,3.143] found strain-induced martensite contents of up to 40 % and a two-phase nano-crystalline layer consisting of austenite and martensite, with a thickness of 1–2 μm, grain sizes smaller than 20 nm within immediate proximity to the surface, and grain sizes of ca. 50–200 nm in the underlying region. This layer is an indication of static re-crystallization occurring during the shot peening process. Recrystallization effects were also found after ultrasonic shot peening of AISI 316L steel [3.155], in α-iron [3.155], and in nickel-base alloys [3.156–3.158].

g) Influences on Texture

The strong plastic deformations occurring close to the surface during shot peening treatments cause a preferential orientation of the crystallites, which creates characteristic peening textures in originally texture-less material states, or permits an alteration of existing textures. This is represented in Fig. 3.46 by means of {110}- pole figures of normalized SAE1045 before and after shot peening [3.159]. The essentially texture-less normalized state receives a faint {110}-fibre texture, i.e. there is a preferential arrangement of the {110}-slip planes parallel to the peened surface.

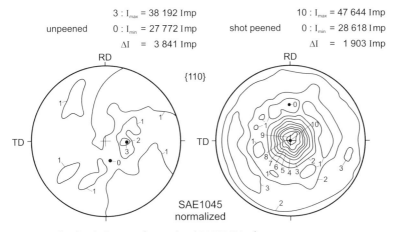

Fig. 3.46: {110}-pole figures of normalized SAE1045 in the unpeened (a) and shot peened (b) state [3.159]

3.2
Stress Peening

3.2.1
Process Models

a) Elementary Processes

In order to gain an understanding of the processes taking place in stress peening, it is advisable to divide the stress peening treatment into three phases, which are
- elastic prestressing,
- shot peening during loading, and
- unloading after the peening treatment is concluded.

In the first phase, elastic prestressing takes place by means of loading stresses, showing a uniaxial homogeneous distribution in the case of tensile prestress, a uniaxial inhomogeneous distribution in the case of bending prestress, and a biaxial inhomogeneous distribution in the case of torsional prestress, given that notch effects need not be considered. In the actual peening process in the second phase, the elementary processes are fundamentally the same as in conventional peening. However, the plastic deformations required for inducing residual stresses benefit from the introduced loading stresses which approximate the yield strength. Simple models [3.62,3.160,3.161] assume that the same rotationally symmetric stress state as the one found in conventional peening processes is present after the second phase of stress peening, independent of prestress. During unloading in phase three, stresses are thereby shifted in such a way that the remaining residual stresses in the direction of the tensile prestress component are greater than in conventional shot peening and are reduced in the transversal direction. The essence of these statements may be applied to all of the varieties of prestress described above. When equivalent stress reaches the level of flow stress, which is very likely anytime when components of compressive residual stress increase due to unloading, backward plastic deformations can occur. Thus, the formation of residual stresses due to supposedly elastic unloading – as would be expected for the stress state present after the second phase – is counteracted. This effect becomes even more pronounced as load reversal additionally gives rise to a significant Bauschinger effect.

b) Quantitative Descriptions of Processes
All models for a description of the residual stress states after stress peening treatments assume the residual stress state which results from conventional peening treatments to be the stress state which is present after the second phase. In a uniaxial surface-core model with constant stress values in the surface layer, with the diameter A_{surf} and in the core, with the diameter A_{core}, and assuming ideal elasto-plastic material behavior, [3.160,3.161] postulated that there are always equal total

strains present in the surface layer and in the core. The residual stresses in the surface region then increase during stress peening treatment

$$\sigma_{surf}^{rs} = \sigma_{surf,o}^{rs} - \frac{1}{1 + \dfrac{A_{surf}}{A_{core}}} \sigma_{pre} \qquad (3.26)$$

linearly to tensile prestress σ_{pre}, in which $\sigma_{surf,o}^{rs}$ is surface layer residual stress after conventional peening. By example of quenched and tempered SAE1045 of 525HV hardness, Fig. 3.47 shows that experimental results in fact indicate such behavior, given that the loading stress applied as prestress is not excessive and therefore minor backward plastic deformations occur during unloading, which would reduce residual stresses [3.160]. Again using the example of a surface-core model incorporating constant stress values within the surface layer and the core, appropriate partitioning of the cross-sectional areas and ideal elasto-plastic material behavior for peening treatments of spring wires under torsional prestress, [3.162] assumed that the unloading occurring after the actual peening process is purely elastic and that it leads to an equilibrium of momenta. With this model, he arrived at correlations similar to the one indicated in Equ. 3.26, which are valid also for torsional prestresses. Finally, [3.62] also developed a model for residual stress formation in peening treatments under torsional prestress. It is assumed that in the model assumption of [3.52–3.54] – described in Equations 3.20–3.22 –, torsional prestress reduces the surface pressure required from the outside to achieve plastification of the entire cross section, and thus the thickness of the plastically deformed surface layer is increased while the surface pressure remains the same. Given that the stress state after the second phase of the peening process corresponds to the residual stress state present after conventional peening, and that the momentum effective from the outside is reduced to a point at which it is completely

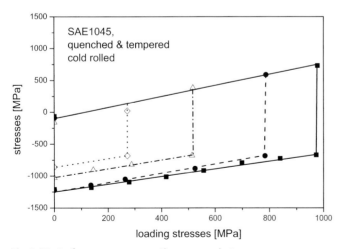

Fig. 3.47: Surface stresses vs. tensile prestress during stress peening for quenched and tempered SAE1045 [3.160]

taken over by the purely elastically deformed core region, a shear-compatible and purely elastic calculation of unloading leads to an equilibrium of momenta between the surface layer and the core and to clearly increased residual stress values within an enlarged surface region when the multiaxiality of stresses is taken into account. This is in contrast to peening treatments lacking prestress.

3.2.2
Changes in the Surface State

a)   Influences on Shape

There are no known studies on the effects of stress peening regarding changes of shape. In addition to the effects mentioned in the context of conventional shot peening, one must consider that during the application of loading, and also when the loading stresses are active during the actual shot peening treatment, unintended plastic deformations may potentially occur and result in changes of shape. Furthermore, when yield strength is exceeded during unloading from high prestresses following the actual peening process, backward plastic deformations will occur. These, however, should rather serve to reduce possible shot peening-induced changes of shape. Thus, there are two competing processes that need to be taken into account when assessing the risk of distortion in stress peening.

b)   Influences on Topography
The effects of prestresses on the topography of stress-peened surfaces are fundamentally the same as those found in conventional shot peening. In homogeneously, uniaxially prestressed quenched and tempered AISI4140 specimens [3.163] finds no significant influence of prestress on mean roughness. According to [3.164], elliptical impact craters – their half axis enlarged in the direction of prestress versus the transversal direction – are observable after the stress peening of uniaxially tensile prestressed specimens. However, this effect of mechanical stretching of regions close to the surface, which is greater in the direction of pre-stress than in the transversal direction, stands in opposition to the model assumptions stated above, according to which there should be an occurrence of circular impact craters before unloading. They change shape only by elastic, and sometimes also plastic, backward deformation during unloading, and therefore the ellipses should be facing in the opposite direction.

c)   Influences on Residual Stress State
In homogeneously uniaxially prestressed specimens of 450 °C-quenched and tempered AISI4140, [3.163] found growing tensile prestress to increase residual stresses at the surface in the direction parallel to prestress, while the positions of the zero crossings of residual stress are increased. This is shown in Fig. 3.48. By contrast, the residual stresses in the direction transversal to that of prestress decrease slightly with growing prestress. In normalized and 650 °C-quenched and

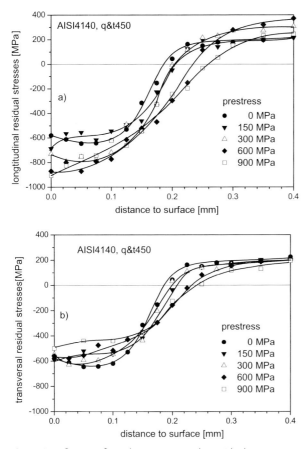

Fig. 3.48: Influence of tensile prestress on the residual stress distribution parallel (a) and perpendicular (b) to the prestressing direction for quenched and tempered AISI4140 [3.163]

tempered specimens of the same steel, however, the application of tensile prestress effects smaller residual stress changes.

The most pronounced increases of residual stress values in the direction of prestress are found in 300 °C-quenched and tempered and in quenched material states. According to [3.163], the residual stress maxima exhibit changes which are linear to prestress roughly up to half the yield strength. A degressive course is found at greater prestress levels. Fig. 3.49 summarizes how the residual stress maximum values in the longitudinal and transversal directions and the positions of the residual stress zero crossings change when prestress is increased in relation to the respective yield strength of the different heat treatment states. With rising workpiece hardness, the longitudinal and transversal residual stresses show an increasing sensitivity to prestress in relation to yield strength. The longitudinal residual stresses increase by about three times the amount by which the transversal residual stresses decrease. When the ratio of prestress to yield strength grows, all of the heat treatment states see equal increases of the depths of the residual stress zero crossings.

[3.165] introduced measurements of the residual stresses in the direction of prestress in quenched and tempered spring steel for different amounts of bending prestress. As is the case for homogeneous prestresses, increases in surface values, maximum values and penetration depths of the compressive residual stresses are registered when positive bending prestress is increased, as shown in Fig. 3.50. For comparison, it also shows the residual stress distribution that was measured on the compressively prestressed side of the specimen which had received the highest prestress, and it shows low tensile residual stresses close to the surface.

As opposed to the tensile and compressive residual stresses discussed thus far, the application of torsional prestress involves applying a biaxial loading state. The principal prestresses are located at 45° and 135° to the torsional axis and have the same value, but reversed mathematical signs. Accordingly, in comparison to tensile or bending prestresses, stress peening creates differences between the main components of residual stress – which correspond to the principal stress orientations of the prestress state – to be more pronounced. Fig. 3.51 shows results obtained on highest strength spring steel with an ultimate tensile strength of 2150 MPa. Excessive torsional prestresses can cause significant compressive residual stresses of roughly 1200 MPa in the 45°-direction which received tensile prestress during the peening process. However, even tensile residual stresses may develop in the 135°-direction which received compressive prestress during the peening process [3.166]. Fig. 3.52 shows that when appropriate torsional prestresses are applied, the distributions of residual stress measured in the 45°-direction after stress peening follow the correlations found after peening treatments using tensile prestresses [3.166]. Torsional prestresses shift the surface values and maximum values of residual stress and the positions of zero residual stresses toward higher values. As is the case in stress peening using tensile prestress, great prestresses during torsional stress peening cause degressive dependencies of the maximum compressive residual stresses on prestress related to yield strength (comp. Fig. 3.53, [3.162,3.167]).

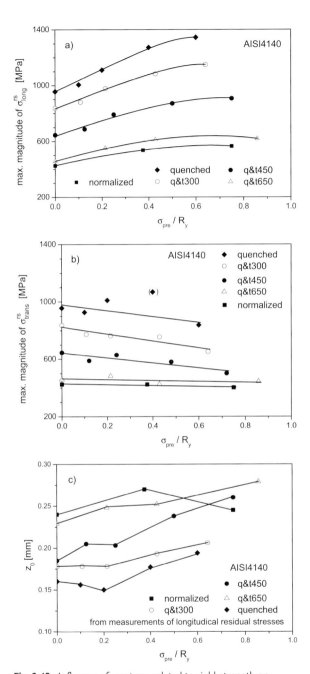

Fig. 3.49: Influence of prestress related to yield strength on the maximum longitudinal (a) and transversal (b) residual stresses and the depths of zero residual stresses (c) for AISI4140 after different heat treatments [3.163]

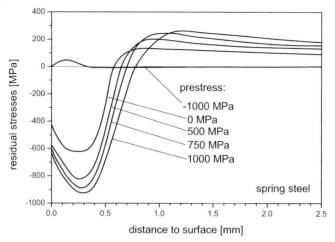

Fig. 3.50: Influence of bending prestresses on the resulting depth distributions of the residual stresses in the prestressing direction after stress peening of a quenched and tempered spring steel [3.165]

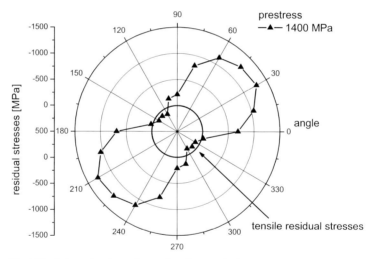

Fig. 3.51: Orientation-dependency of surface residual stresses in a highest strength spring steel ($R_m = 2150$ MPa) after stress peening with an inappropriately high torsional prestress [3.166]

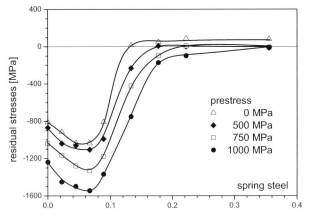

Fig. 3.52: Influence of torsional prestress on the residual stress distribution in a direction rotated 45° to the torsional axis for a high strength spring steel (R_m = 2000 MPa) [3.166]

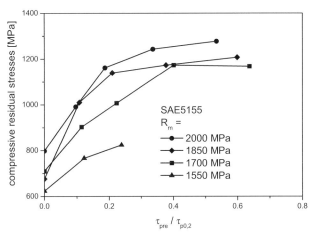

Fig. 3.53: Influence of ultimate tensile strength on maximum compressive residual stresses vs. torsional prestress related to torsional yield strength for stress peening of SAE5155 acc. [3.162, 3.167]

Fig. 3.54 shows the surface values of residual stress and the depths of the positions of zero residual stresses in quenched and tempered AISI4140, which was stress peened at different ratios of torsional prestress to torsional yield strength [3.168]. It becomes evident that the residual stress amounts in the 45° direction of prestress during the peening process increase up to τ_{pre} / τ_y = 0.8, and thereafter they show clear decreases due to backward plastification during unloading. In the 135° direction of peening prestress, the residual stress amounts decrease strongly in comparison to specimens subjected to tensile or bending prestress. This is due to the fact that only torsional prestress has a loading stress component in this

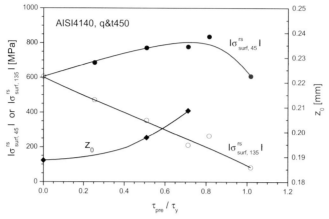

Fig. 3.54: Influence of torsional prestress related to torsional yield strength on surface residual stresses in a direction rotated 45° and 135° to the torsional axis and on depths of zero residual stresses for quenched and tempered AISI4140 [3.168]

direction and thus, the stress level shifts toward smaller amounts during unloading. The depths of the positions of zero residual stress increase in agreement with the examinations mentioned above.

d) Influences on Workhardening State

Fig. 3.55 shows that the distribution of half widths in AISI4140 quenched and tempered at 450 °C , peened using different amounts of prestress, exhibit no significant dependence on prestress [3.161]. This is equally valid for the softer and the harder material states of AISI4140, the residual stress states of which were mentioned above. Evidently the changes of the amount of plastic deformation close to the surface are too small to be registered by means of half width measurements.

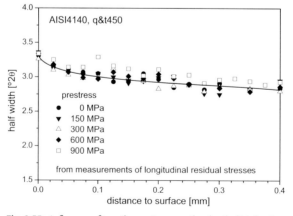

Fig. 3.55: Influence of tensile prestress on the depth distribution of half widths of quenched and tempered AISI4140 [3.163]

3.3
Warm Peening

3.3.1
Process Models

a) Elementary Processes

In order to understand the processes occurring during hot peening, it is helpful to differentiate three phases of the process:
- heating to peening temperature
- shot peening at an elevated temperature
- cooling to room temperature

Aside from material states with low creep resistances and high initial residual stresses, which might experience thermal relaxation, no noteworthy changes of state occur in the first phase. The same elementary processes as in conventional peening occur during the actual shot peening process. However, it must be taken into account that the strength of the workpiece material may be reduced and the workhardening rate may be more pronounced at higher temperatures. The former may lead to higher values and penetration depths of the compressive residual stresses. Additionally, regarding strain aging materials such as steels and peening temperatures that lie within the range of dynamic strain aging, altered deformation processes and therefore, changed dislocation arrangements, are expected to occur in the plastically deformed surface layer. In the third phase, the cooling phase, a still sufficient mobility of dissolved atoms may cause static strain aging processes, again, in strain aging materials.

The following will focus in detail on the effects of static and dynamic strain aging, their microstructural causes and the models proposed to describe them. Static strain aging is the elastic interaction of diffusing dissolved atoms with resting dislocations. [3.169–3.174] distinguish four stages with different repercussions on the uniaxial deformation behavior observed after pre-deformation, unloading and the concluding annealing treatments which use increasing annealing times and -temperatures during additional deformation. While stage 1 merely experiences the appearance of a yield point phenomenon , i.e. an initial increase of flow stress and a Lüders strain, additional effects found in stage 2 range from permanent flow stress increases to great extents of additional deformation, which can also cause increases of tensile strength. In stage 3 there are also increased workhardening rates and decreased elongations to fracture. All of the aforementioned effects decrease in stage 4, for which the term over aging is used. As concentrations of the strain aging dissolved elements increase, the effects of strain aging become more pronounced and the rates at which they develop increases. Accordingly, at low levels of dissolved content the only observable effects are those of stage 1, and these may occur spontaneously even in higher concentrations [3.170]. As established by [3.170] for carbon- and nitrogen-induced strain aging in steels, growing contents of other elements have an effect – which is always one of reduction – on the intensity of strain aging, namely if they are elements which interact strongly with the strain aging element. While increasing dislocation densities also result in an increase of strain aging effects, they are found to diminish with increasing grain size [3.170]. Precipitations which contain significant amounts of strain aging elements reduce the level of dissolved content, thereby reducing the effects of static strain aging. The four stages of strain aging are caused by the pinning of the dislocation structure, which increases with longer annealing times [3.170]. In stage 1 the dislocations are pinned by accumulating carbon atoms as so-called Cottrell-atmospheres or by finest carbides forming close to dislocations – where they can break away after an initial excess of stresses. In stage 2, dislocations are permanently immobilized, resulting in the formation of new dislocations rather than tearing processes. By contrast, in stage 3 the precipitations surrounding the dislocations become so large that they themselves increase the workhardening rate, which is possible even without prior saturation. In stage 4 the precipitations present at the dislocations may either coarsen within the grains (Ostwaldt-ripening) or they may dissolve and precipitate again at the grain boundaries. Here, increased stresses may become necessary to activate the dislocation sources. At the same time, however, the stresses required for moving previously pinned dislocations are reduced. Thus, the three essential mechanisms of static strain aging are the stress-induced arrangement of dissolved interstitial atoms in suitable voids – termed Snoek Effect [3.129,3.175,3.176] –, the formation of Cottrell atmospheres in the dilatation region of dislocations [3.177], and the precipitation of intermetallic phases in dislocation proximity. A description of the respective effects utilizing simple mathematical approaches was first presented by Cottrell and Bilby [3.178] for carbon in steels. They assumed the growth of flow stress to be proportional to the increase of concentration in the dislocation region, and showed that an Arrhenius term arises with the enthalpy of migration of the interstitial element within the matrix and that, in respect to the dependence on time, a proportionality to $t^{2/3}$ results. However, the changes in availability of the strain aging element within the matrix and the saturation of the dislocations are neglected. Harper [3.179] assumed the number of precipitating atoms to be proportional to the available content, and he arrived at formulations that are hardly applicable due to their complexity [3.180]. In order to incorporate the effects of over-aging, [3.181–3.184] viewed the re-solution of interstitial atoms available at dislocations to the grain boundaries and to the occurring carbides as mechanisms to be considered. Using a model building on this, [3.185] was able to describe aging and its maximum value in bake-hardening steels.

In contrast to the static strain aging discussed thus far, dynamic strain aging is the elastic interaction of diffusing foreign atoms with moving dislocations. It is manifested in the jagged curves of quasi-static deformation behavior which are usually observed between initial and final plastic strains [3.170]. Flow stresses occur here which show smaller decreases at rising temperatures than would be expected in the absence of strain aging effects, or even show constant levels or increases. Steels, in particular, show increases of the workhardening rate and of tensile strength along with strongly reduced ductility. This is termed blue brittleness [3.170]. Observing the microstructural level, there are greater increases of dislocation density than there are after deformation at room temperature, given comparable plastic strains [3.186–3.188]. The dislocation structures are more diffuse [3.189], which is caused by an increasingly planar slip due to a prevention of cross slip processes. The reasons for this can be a complete immobilization of dislocations caused by Cottrell atmospheres or by precipitations with a subsequent re-formation of slip dislocations [3.190], a reduction of mutual annihilations of dislocations [3.190], blocked dislocation sources [3.191], or internal friction losses on moving dislocations due to Cottrell atmospheres or the Snoek Effect [3.192]. The effects of dynamic strain aging are determined by similar mean velocities of the diffusing strain aging atoms and mean velocities of the shifted dislocations. The velocity of the strain aging atoms is determined, in particular, by temperature and its strong influence on the diffusion coefficient. On the other hand, the mean dislocation velocities increase as the strain rate increases, while they decrease as dislocation density increases. Accordingly, there is a temperature range in which the strain aging effects are particularly pronounced. This helps to explain aforementioned initial and final strains, which are observed as plastic deformation increases, i.e. as dislocation densities increase and dislocation velocities therefore decrease. The intensity of dynamic strain aging increases with rising levels of dissolved content of the strain aging element [3.193]. By contrast, increasing contents of other elements have an effect only if interaction occurs with the strain aging element [3.194, 3.195]. In this case, the intensity of strain aging effects decreases, or they are shifted toward higher temperatures. Finally, a smaller grain size intensifies these strain aging effects, as the regions of inhomogeneity become more localized.

b)   Quantitative Descriptions of Processes

At present, there is a lack of satisfactory research regarding the quantitative modeling of the stress state after warm peening treatments. Only [3.196] stated a simple model which assumes that the stress state resulting from the impact process may be described by the existence of a compressive yield strength at a high, yet temperature-independent level of compression. However, he assumes the sum of this loading stress state and the induced residual stresses to be determined by the initial flow stress. Given the assumptions that flow stress decrease is linear to rising temperature and that the drop of the compressive yield strength is also linear – albeit significantly less pronounced –, i.e. that greater workhardening occurs when the temperature increases, [3.196] predicts growing surface residual stresses and increasing penetration depths of the residual stresses, which is in accordance with experimental results.

3.3.2
Changes in the Surface State

a) **Influences on Shape**

There is no information on the effects of warm peening on changes of shape. As the temperatures are fairly low for creep- or relaxation processes, it may be assumed that there are no effects to be expected in addition to those which occur in conventional shot peening.

b) **Influences on Topography**

The effects of peening temperature on the topography of warm peened surfaces are principally the same as in conventional peening. However, due to the elevated temperatures and thus slightly reduced warm strength, [3.163] finds roughness increasing with peening temperature in quenched and tempered AISI4140 (comp. Fig. 3.56).

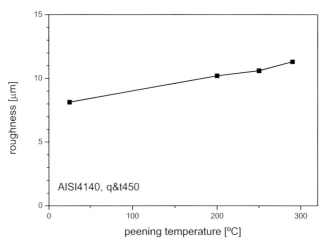

Fig. 3.56: Influence of peening temperature on roughness of warm peened AISI4140 in a state quenched and tempered at 450 °C acc. [3.163]

c) **Influences on Residual Stress State**

Peening temperature only marginally influences the residual stress state of warm peened AISI4140 in a 450 °C-quenched and tempered heat treatment state. Fig. 3.57a [based on 3.167] shows that the residual stress values of ca. -600 MPa close to the surface increase to slightly over -700 MPa and that the increases of the residual stress zero crossings are barely noticeable – aside from the state peened at 310 °C, where the peening parameters were altered slightly. By contrast, [3.196,3.197] finds that the compressive residual stress values close to the surface

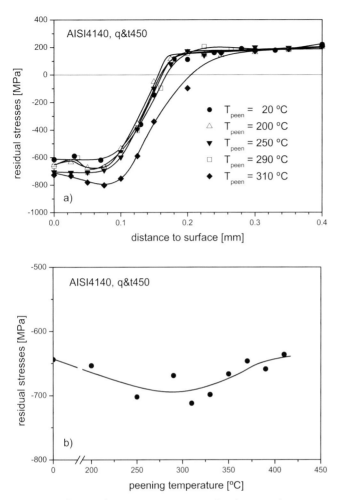

Fig. 3.57: Influence of peening temperature on distribution and surface values of residual stresses in warm peened AISI4140 in a state quenched and tempered at 450 °C acc. [3.163]

increased noticeably during warm peening of 60SiCr7 spring steel in a rotating wheel device. The longitudinal residual stresses – represented in Fig. 3.57b as a function of peening temperature – show slight initial increases as peening temperature rises and drop again above 330 °C, as a thermally-induced reduction of residual stresses occurs due to relaxation processes.

d) Influences on Workhardening State

Fig. 3.58a shows the depth distributions of the half widths in the states of AISI4140 discussed above [based on 3.163]. At peening temperatures up to 310 °C, warm peening results in an increase of the half widths within the entire plastically deformed surface layer, while peening temperature does not show any significant effect. As seen in Fig. 3.58b, the half widths close to the surface begin to decrease again already at peening temperatures above 250 °C. [3.163] supposes that the increased half widths are created by dynamic strain aging processes which occur during peening and lead to more diffuse dislocation structures, the mobility of which is additionally obstructed by static strain aging processes. As the peening process proceeds, and if plastic deformation is induced again in the regions treated in this way, new dislocations are created. The result is an altered dislocation structure with increased dislocation-induced lattice distortions, noticeable by the increases of the half widths. The decrease of the half widths which is registered at higher temperatures can be attributed – as in the case of the residual

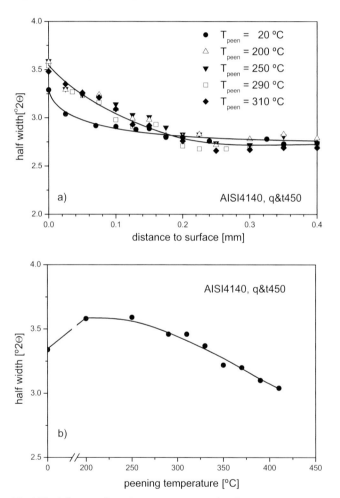

Fig. 3.58: Influence of peening temperature on distribution and surface values of half widths in warm peened AISI4140 in a state quenched and tempered at 450 °C acc. [3.163]

stresses – to thermal effects caused by relaxation processes and the rearrangement of dislocations into energetically more favorable structures.

[3.198–3.200] reported on measurements of micro-hardness within the surface regions of a high strength spring steel – roughly equivalent to a SAE9260 (comp. Fig. 3.59) – peened at various temperatures. At peening temperatures of 300 °C and above, the levels of micro-hardness found in the entire surface layer were increased in comparison to conventional peening. This may be regarded as an effect of the dynamic and static strain aging effects which obstruct the plastic deformation that is required for measuring micro-hardness. When the spring steel experiences increases of bulk hardness, or decreases of the annealing temperature, the increase of micro-hardness induced by warm peening becomes more pronounced due to the increasing carbon content which determines the strain aging tendency.

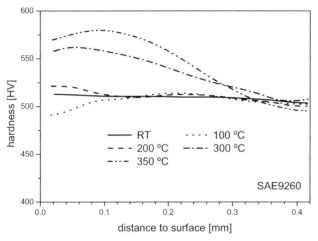

Fig. 3.59: Influence of peening temperature on distribution of microhardness in warm peened SAE9260 spring steel [3.198]

3.4
Stress Peening at elevated Temperature

The surface layer changes caused by stress peening treatments at elevated temperatures, according to [3.201,3.202], are explained by the combination of the effects of prestressing and of peening temperature. No particular models appear to exist in respect to this issue. While the roughness of quenched and tempered AISI4140 is approximately equivalent to that of warm peened states treated at the same temperature – the influence of prestresses being insignificant –, the residual stress values and the positions of zero residual stresses are increased still further in comparison to stress peened states. Close to the surface, the half widths show the same values found in warm peened states, exhibiting comparatively large depths until they drop down to the level of the bulk material state. Figs. 3.60 and 3.61 compare this with the previously discussed examples of quenched and tempered AISI4140 in modified shot peening treatments [3.163].

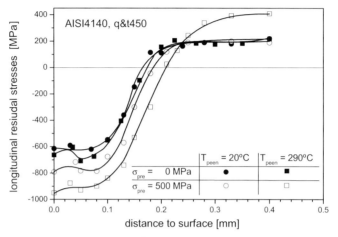

Fig. 3.60: Depth distribution of residual stresses after various modified shot peening treatments for AISI4140 in a state quenched and tempered at 450 °C [3.163]

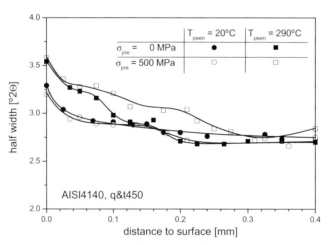

Fig. 3.61: Depth distribution of half widths after various modified shot peening treatments for AISI4140 in a state quenched and tempered at 450 °C [3.163]

3.5
Deep Rolling

3.5.1
Process Models

a)   Elementary Processes

The mechanical elementary process of deep rolling is the surface pressure created between the workpiece and the tool in the contact zone. It causes triaxial stress states, which change with the distance to surface. They are dependent on contact geometry, and are described in Sect. 3.1.1b for the elastic contact between a sphere and a flat plate (Hertzian pressure). They cause smoothing- and friction effects at the immediate surface. When the yield strength is exceeded by the resulting equivalent stress, local plastic deformations occur, creating residual stresses and the associated microstructural workhardening- or worksoftening effects. The plastic deformations cause heat to develop. However, its scope and possible effects on process results have not yet been examined. Another aspect remaining to be examined is the influence on process temperature exerted by the coolant lubricants which are often utilized to reduce friction.

b)   Quantitative Descriptions of Processes

The first model-like descriptions of deep rolling were developed by [3.203,3.204] to calculate the stress distribution induced by Hertzian pressure in curved contact areas – which may also be multiply-curved. These approaches view material behavior as elastic only and therefore they cannot predict residual stress distributions. Fig. 3.62 is an exemplary representation of the stress distributions under the conditions summarized in the caption. However, by varying essential factors it is possible to make statements in respect to the position of the maximum and the equivalent stress gradient present around this maximum. Thus, the change from point- to line- or area contact shifts the position of maximum equivalent stress away from the surface. However, this occurs together with an increase of the equivalent stress surface value. Likewise, when the axis ratio of the contact area increases as the shape of contact area changes from circular to elliptical or even to rectangular, the equivalent stress surface value increases, while its maximum value decreases. In addition, the ratio of the Young's moduli in the workpiece and the tool were varied. Accordingly, stiffer tools cause greater depths of the equivalent stress maximum. While the amount of equivalent stress decreases slightly, its surface value increases. Growing friction coefficients cause greater tangential forces, yielding notable increases of equivalent stress at the surface. Thus, the maximum value of equivalent stress may even be shifted to the surface. This model also serves to explain the influence of the number of rolling passes on the process result. Even if the model does not include plastic deformations, it is evident that due to previous rolling passes, compressive residual stresses decrease

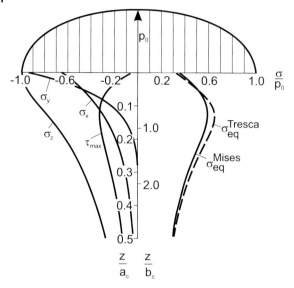

Fig. 3.62: Elastically calculated stress distributions acc.
[3.203] in the contact area between tool and workpiece in a
deep rolling treatment (tool: $R_1 = 45$ mm, $R_2 = \infty$; workpiece:
$R_1 = 35$ mm, $R_2 = 164.5$ mm; relation of axes of contact
ellipse $= 6$)

the equivalent stress locally and reduce further plastic deformation there. By contrast, the equivalent stress increases in the regions surrounding the plastified
zone, thus causing the latter to expand. Plastic deformation causes workhardening, which results in residual stress distribution reaching saturation as the number of rolling passes grows.

The only known descriptions of the deep rolling process which include plastic
deformation and thus, are able to represent the formation of residual stresses, are
Finite-Element simulations. This is due to the combination of the non-linearities
of material behavior, surface layer conditions and large displacements. The implicit codes of [3.205] and [3.206–3.208] constitute the first approaches. The model of
[3.205] was aimed at estimating the residual stress states in casehardened and subsequently deep rolled gears. According to its author's own judgment, it did not
succeed in calculating residual stress distributions to a degree corresponding sufficiently to measured values, since the model is 2-dimensional. By contrast, the
background to [3.206]'s model is a deep rolling treatment of the transitional radii
between a crankshaft's main hub and its cheeks consisting of AISI4140 ($R_m = 1066$
MPa). It was formulated as a 3-dimensional contact problem in the MARC Finite-
Element package. It calculates the deep rolling treatment on a mesh describing a
quarter of the circumference. Friction is neglected in order to reduce calculation
time. The roller is assumed to be purely elastic, the workpiece is assumed to be
elasto-plastic with isotropic hardening. The influences of strain rate and temperature, and the influence of cyclic softening due to multiple rolling passes, are

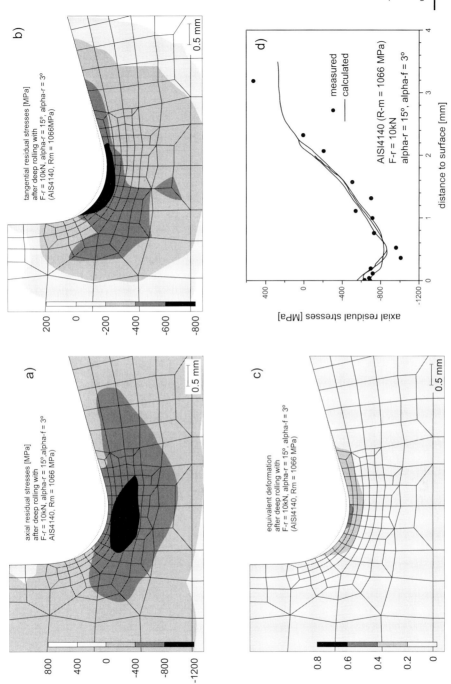

Fig. 3.63: Fields of axial (a) and tangential (b) residual stresses and equivalent plastic strains (c) in the notch region and distribution of axial residual stresses in the notch ground (d) calculated using the finite element method [3.206]

represented as mean effects. They are achieved by way of a 15 % reduction of the flow curve determined according to [3.209], based on compressive tests carried out on deep rolled and subsequently trepanned cylinders. [3.206] presented residual stress distributions in the axial and tangential directions and distributions of plastic deformations as examples for his calculations, as shown in Fig. 3.63. Maximum residual stresses are found in an axial direction beneath the surface of the hub in the region of the relief groove (Fig. 3.63a). By contrast, the tangential residual stresses, depicted in Fig. 3.63b, are greatest in the former contact area while showing clear effects on a region which extends into the cheek. The same is true for the equivalent strains shown in Fig. 3.63c. The distribution of residual stress in the notch ground exhibits a typical residual stress maximum beneath the surface and low gradients toward the surface and up to the position of zero residual stresses. According to Fig. 3.63d, the simulation results obtained for different positions are well in accordance with experimental results, particularly regarding axial residual stress depth distributions at the notch.

Achmus [3.210–3.212] proceeded in a similar manner. His use of ABAQUS/ Explicit and an explicit integration yields an advantage in terms of computation time, compared to the implicit algorithm. As shown in [3.213], there is agreement with neutronographic residual stress measurements for distances to surface greater than ca. 0.5 mm. It is demonstrated that the roller may be assumed to be rigid, without skewing the residual stress states in an inadmissible way.

Meanwhile the process simulation of deep rolling has been extended by [3.214] and it is used as a basis for assessing the effects on the fatigue limit of crankshafts. The reduction of residual stress during loading and the influence of residual stress on the effective stress intensity of an assumed surface crack is calculated to this end (comp. Sect. 6.2.1c and 6.3.5).

3.5.2
Changes in the Surface State

a) Influences on Shape

Using appropriate process parameters in deep rolling permits setting specific final dimensions, e.g to allow for defined reworking of a notch radius. Size rolling makes use of this. However, there is a danger of lamination, which presents likely starting points for cracks during future fatigue loading. For the treatment of crankshafts, deep rolling is even used as part of the alignment process in the form of "roll trueing" [3.205]. An aspect which must be considered in respect to long, slender elements is that the forces of deep rolling may result in changes of shape if they are not appropriately absorbed by way of local support of the workpiece. Here, the use of support rollers on the opposite side of the main rollers may be advisable.

b) Influences on Topography

Deep rolling, as well as the related smooth rolling, causes a specific smoothing of the surface. Fig. 3.64 shows that with increasing rolling pressure, or rolling force, roughness initially decreases, then passes through a minimum and ultimately increases again slightly [3.142]. While the initial reduction of roughness is a result of the smoothing of the surface profile created in the preceding cutting process, the increase of roughness at great rolling forces may be attributed to damage occurring in regions close to the surface of the workpiece, e.g. surface cracks. The minimum occurs also when the number of rolling passes is increased, while the optimal number of rolling passes decreases as the deep rolling force increases. Thus, for economic reasons an attempt is commonly made to reduce handling time by applying great deep rolling forces. However, this bears the danger of the number of rolling passes becoming so small that, particularly in ductile work materials, the plastic flow processes in the surface layer have not yet reached a saturated state and, therefore, the residual stress state and the workhardening state are not yet at optimum values in respect to fatigue resistance properties. In principle, it should be noted that the minimum values of roughness which may be achieved are essentially limited by the roughness of the tool.

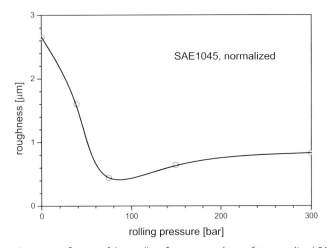

Fig. 3.64: Influence of deep rolling force on roughness for normalized SAE1045 [3.142]

c) Influences on Residual Stress State

Fig. 3.65 shows typical residual stress distributions following different deep rolling treatments on round specimens of normalized SAE1045 [3.142]. As rolling pressure is increased, the maximum values of the axial residual stresses which are achievable close to the surface, i.e. the residual stresses in the transversal direction to that of deep rolling, show a slight increase, and the depths of the zero crossing grow noticeably. By comparison, the distributions of tangential residual stresses, i.e. the residual stresses in the direction of deep rolling, depicted on the right-hand side of Fig. 3.65, show values which are ca. 40 % lower and exhibit signifi-

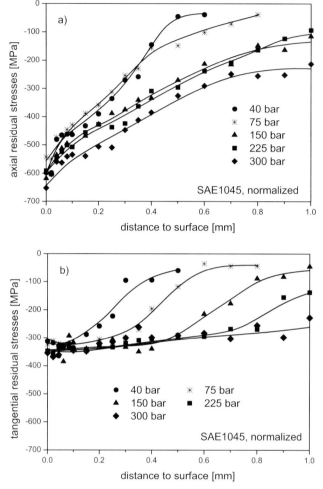

Fig. 3.65: Influence of deep rolling force on the distribution of residual stresses in the axial (a) and the tangential (b) direction for normalized SAE1045 [3.142]

cantly more pronounced increases of depth effect. Evidently, the plastic stretching in the transversal direction outweighs the plastic stretching which occurs in the direction of rolling.

Fig. 3.66 compares the distributions of residual stresses which occur in quenched and tempered AISI4140 after deep rolling with those found after shot peening. In comparison to the shot peened state, the deep rolled state shows slightly higher amounts of residual stress within proximity of the surface. At 0.9 to 1.0 mm, the residual stresses after deep rolling extend noticeably deeper into the interior of the material than the residual stresses do after shot peening, dropping to near-zero values at a distance to surface of only 0.25 mm. Fig. 3.67 shows

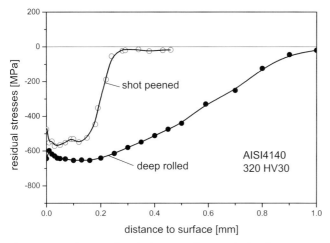

Fig. 3.66: Distribution of residual stresses for quenched and tempered AISI4140 after deep rolling or shot peening [3.138]

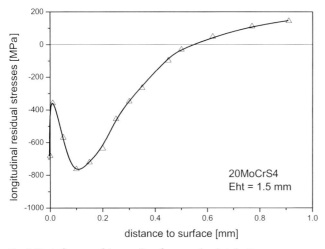

Fig. 3.67: Influence of deep rolling force on the distribution of axial residual stresses in case hardened 20MoCrS4 [3.215]

that deep rolling treatments on casehardened 20MoCrS4, can also result in the formation of the compressive residual stress maximums typical of shot peening treatments [3.215]. Depending on workpiece hardness, the effects of Hertzian pressure thus are another essential factor in deep rolling. It should further be noted that the residual stress amounts may exhibit another increase within immediate proximity of the surface. This increase of the residual stress amounts close to the surface is attributed to friction and slip processes [3.102].

The residual stress amounts achievable by deep rolling also pass through a maximum when the deep rolling force is increased. This is shown in Fig. 3.68 for

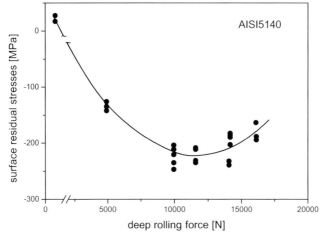

Fig. 3.68: Surface residual stresses vs. deep rolling force in quenched and tempered AISI5140 [3.216]

AISI5140 by example of the dependence of the surface residual stresses on the deep rolling force [3.216]. This is an expression of the surface damage mentioned above in the discussion of the effects on topography.

d) Influences on Workhardening State

Fig. 3.69 shows the influence of rolling pressure on the distribution of the half widths in normalized SAE1045 [3.142]. Starting from the value of the untreated material, the half widths increase continually towards the surface, suggesting a plateau region at medium distances to surface. As rolling pressure intensifies, the

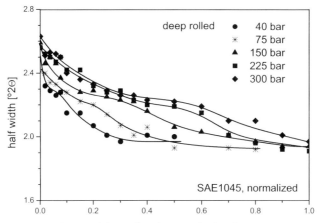

Fig. 3.69: Influence of deep rolling force on the distribution of half widths in normalized SAE1045 [3.142]

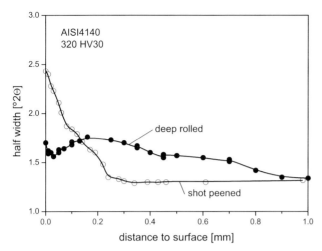

Fig. 3.70: Distribution of half widths for quenched and tempered AISI4140 after deep rolling or shot peening [3.138]

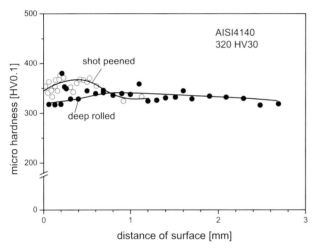

Fig. 3.71: Distribution of microhardness for quenched and tempered AISI4140 after deep rolling or shot peening [3.138]

thickness of surface regions increases and higher values are observed close to the surface. By example of quenched and tempered AISI4140, the distributions of the half widths are represented in Fig. 3.70 for a deep rolled state and for a shot peened state. Corresponding to the respective residual stress depth distributions, shown in Fig. 3.66, the increases of the half widths found close to the surface after deep rolling occur within surface regions that are clearly thicker. Below a distance to surface of 0.2 mm, the half widths in the deep rolled state fall marginally, before increasing again slightly at the immediate surface. The growth of the half widths close to the surface is noticeably smaller in the deep rolled state than it is in the shot peened state. Fig. 3.71 shows the same material state, indicating that micro-hardness values decrease slightly towards the surface after deep rolling, while they exhibit a small increase after shot peening.

e) Influences on Microstructure

The transmission electron microscopic examinations by [3.138], shown in Fig. 3.72, demonstrate that deep rolled specimens of normalized-state AISI4140 do not contain statistically distributed dislocations with an elevated density – as found in

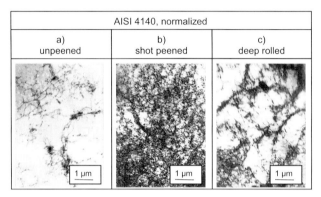

AISI 4140, normalized		
a) unpeened	b) shot peened	c) deep rolled

Fig. 3.72: Transmission electron microscopic view of the microstructure at a distance to surface of about 0.08 mm in unpeened (a), shot peened (b; S170 54–58 HRC, $p = 1.6$ bar, single coverage) and deep rolled (c; $p_o = 1450$ MPa, $n = 20$) specimens of normalized AISI4140 [3.138]

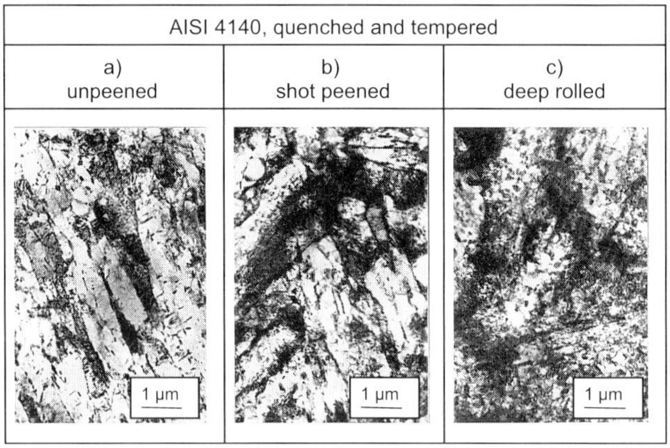

AISI 4140, quenched and tempered		
a) unpeened	b) shot peened	c) deep rolled

Fig. 3.73: Transmission electron microscopic view of the microstructure at a distance to surface of about 0.15 mm in unpeened (a), shot peened (b; S170 54–58 HRC, $p = 1.6$ bar, single coverage) and deep rolled (c; $p_o = 2750$ MPa, $n = 20$) specimens of quenched and tempered AISI4140 with an initial hardness of 320 HV30 [3.138]

shot peened specimens –, but show pronounced cell structures, again with elevated dislocation densities. These differences are attributed to the greater amounts of local plastic deformation and the significantly lower strain rates during deep rolling [3.138]. Therefore, a smaller degree of obstruction of cross slip occurs in deep rolling, in comparison to shot peening, and thus aids the formation of spatial dislocation arrangements, as constituted by the dislocation cells. Corresponding TEM examinations of quenched and tempered AISI4140, depicted in Fig. 3.73, show that the dislocation structures found close to the surface after deep rolling and after shot peening show only slight differences in the quenched and tempered state, as the dislocation arrangement is largely determined by the heat treatment-induced distribution of fine carbides [3.138].

According to Fig. 3.74, deep rolling can achieve increases of density in alloyed sintered iron. These are still detectable at more than 1 mm distance to surface, due to the deeper reach of deformation versus that caused by shot peening.

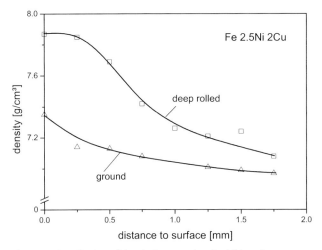

Fig. 3.74: Distribution of density in ground and additionally deep rolled specimen of sintered Fe2.5Ni2Cu [3.263]

f)   Influences on Phase Fractions

As is the case in shot peening, deformation-induced phase transformations can be observed during the deep rolling of steels containing retained austenite. This is verified in Fig. 3.75 by the changes in the distribution of retained austenite in casehardened 20MoCrS4 [3.215], which show a maximum at a distance to surface of 0.13 mm. Evidently, the Hertzian pressure maximum causes not only maximum amounts of residual stress within this region, but also a maximum of phase transformation. As in shot peening, it is possible to detect deformation-induced α'-martensite and nano-crystalline surface layers in deep rolled AISI304, as reported by [3.142].

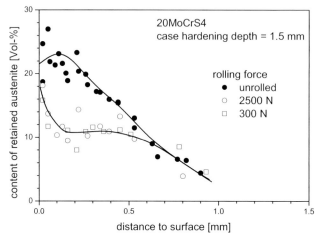

Fig. 3.75: Influence of deep rolling treatments on the distribution of retained austenite in case hardened 20MoCrS4 [3.215]

g)   Influences on Texture

There are only few known facts regarding deep rolling-induced alterations of texture. In sintered Fe2.5Ni2Cu [3.217] find that the initial grinding-induced texture is destroyed by just one rolling pass. A faint rolling texture forms after 40 rolling passes.

h)   Consecutive Shot Peening and Deep Rolling Treatments

[3.138] also examined consecutive shot peening and deep rolling treatments on AISI4140 in normalized and quenched and tempered states. It became evident that the depth effect of the residual stresses and of the workhardening – which is characterized by the half widths of the x-ray interference lines – is always determined by the deep rolling process. However, the residual stress- and workhardening states close to the surface are established by the last step of the treatment. For example, when quenched and tempered specimens are deep rolled and consecutively shot peened, lower surface values of residual stress are observed than in specimens which are shot peened first and then deep rolled. In an analogous manner, a final deep rolling process reduces the half widths which occur close to the surface after the peening process, while a final shot peening process significantly increases workhardening close to the surface. As shown in Fig. 3.76, supplementary transmission electron microscopic examinations carried out on normalized AISI4140 demonstrate that the microstructure close to the surface is determined by the final treatment step. Thus, a shot peening treatment which succeeds deep rolling can dissolve the previously created dislocation cell structure, and a final deep rolling treatment can transfer the randomly arranged dislocations into a cell structure.

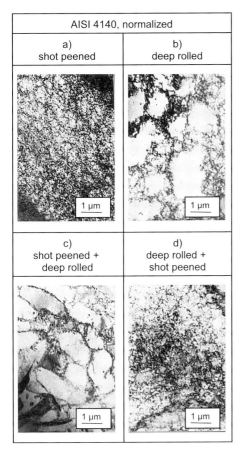

Fig. 3.76: Transmission electron microscopic view of the microstructure at a distance to surface of about 0.07 mm in shot peened (a) or deep rolled (b), as well as consecutively shot peened and deep rolled (c) or deep rolled and shot peened (d) specimens of normalized AISI4140 (S170 54–58 HRC, $p = 1.0$ or $p_o = 1450$ MPa, $n = 20$) [3.138]

3.6
Laser Peening

3.6.1
Process Models

a) Elementary Processes

Laser shock treatment uses laser pulses with pulse durations within the nanosecond range to modify the surface layers of workpieces by means of pressure bursts, affecting distances to surface within the millimeter range. These pressure

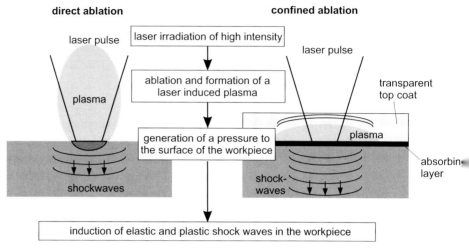

Fig. 3.77: Direct and confined ablation in laser shock treatments (schematic) acc. [3.219]

bursts are created by the ablation of material regions close to the surface due to the absorption of intense laser radiation. The pressure is additionally increased by material vapors forming on the workpiece surface [3.218]. A distinction is made between direct ablation and confined ablation (comp. Fig. 3.77).

In direct ablation, atom layers close to the surface are vaporized, creating plasma which can expand unimpededly into the surrounding atmosphere and which creates a shock wave within the workpiece. The regions around the vaporized material also absorb some laser radiation, causing a thermal loading of the workpiece's surface layer [3.219,3.220]. Due to the unhindered expansion of the plasma, the obstruction of thermal expansion by the material surrounding the laser-irradiated area becomes the dominant factor. This causes plastic compression of the surface region and, following temperature equilibration, results in the creation of tensile residual stress states in the surface layer, as shown in Fig. 3.78 [3.219].

By contrast, confined ablation is carried out by applying an absorbing layer and a transparent cover layer to the workpiece surface. Incoming laser radiation is transmitted through the cover layer and hits the absorbing layer, where it is absorbed and creates a plasma. The expansion of the plasma is impeded by the cover layer, and therefore it exerts pressure on the workpiece surface and on the cover layer [3.221]. Already during irradiation, shock waves are created in the workpiece and in the cover layer. Due to the plasma, the shock waves can still be transmitted into the workpiece for a certain duration after irradiation has ceased. After the plasma has recombined, the transmission of the impulse onto the workpiece surface occurs by expansion of the hot gas. This arrangement increases the achievable shock wave pressure by up to one order of magnitude, and it multiplies the duration in which the shock treatment is effective by several times, compared to direct ablation [3.219]. Due to the absorbing layer, laser irradiation does not

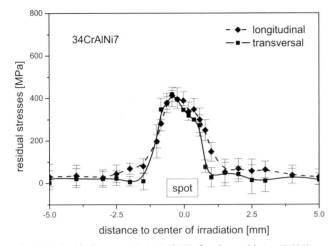

Fig. 3.78: Residual stresses in 34CrAlNi7 after direct ablation [3.219]

strike the surface of the workpiece directly. In confined ablation, the pressure of the shock wave is greater and the thermal loading of the surface zone of the workpiece is reduced, since irradiation reaches only the absorbing layer. Thus, the pressure wave is the dominant factor, causing plastic deformations when yield strength is exceeded. As this constitutes stretching of the surface regions in the direction parallel to the surface, compressive residual stresses develop – Fig. 3.79 shows an example. Particularly when high pressures and circular treatment areas are involved, so-called attenuation waves can form at the edges of the pressure-affected area, which combine in the center and cause decreases of compressive residual stresses or, in extreme cases, even create tensile residual stresses in the center of the treatment area.

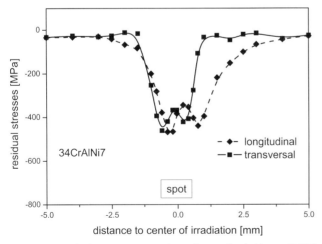

Fig. 3.79: Residual stresses in 34CrAlNi7 after confined ablation [3.219]

The spread of shock waves in materials is based on hydromechanical models, which are coupled with the mechanics of deformable solid state bodies [3.222,3.223]. One essential factor is the shock wave's pressure p, which spreads wavelike within the material. If the pressure is smaller than Hugoniot stress

$$p_H = \frac{1-v}{1-2v} R_y,$$ (3.27)

which – assuming uniaxial strain and validity of the shear stress hypothesis – describes the pressure required for plastic deformation to set in, a wave spreads within the material, at an elastic phase velocity of

$$c_{el} = \sqrt{\frac{E}{\rho}} \sqrt{\frac{(1-v)}{(1+v)(1-2v)}}$$ (3.28)

Greater pressures cause an additional, plastic wave which has a plastic phase velocity of

$$c_{pl} = \sqrt{\frac{E}{\rho}} \sqrt{\frac{1}{3(1-2v)}}$$ (3.29)

and therefore follows the elastic wave [3.223]. Pressures greater than twice the Hugoniot stress cause an elastic unloading wave to spread – together with the build-up of pressure – behind the plastic wave. The elastic unloading wave runs at the higher, elastic phase velocity and thus approaches the plastic wave running ahead. A plastic unloading wave may occur as well.

The plastic waves and the surface reflect the elastic waves repeatedly and therefore a complex superposition of waves is created. The result, given ideal elasto-plastic material behavior, is that the overall deformation retained in the plastified surface region becomes independent of depth. According to [3.222], the penetration depth of plastic deformation can be estimated as

$$z_{pl} = \frac{c_{pl} c_{el}}{c_{pl} - c_{el}} t_p Int \left(\frac{p_{max} + p_H}{2 p_H} \right),$$ (3.30)

wherein Int(x) stands for the integer part of x.

The mechanisms of deformation which occur in plastic shock wave deformation are highly dependent on the strain rate and are difficult to access experimentally. This is due, firstly, to the wave reflections causing microstructural changes at high degrees of strain stretching, and secondly, to the great influence exerted by the strain, which cannot be kept at a constant level in comparative examinations [3.225,3.226]. Apart from the athermal mechanisms listed above in the context of the description of the elementary processes of shot peening (see Sect. 3.1.1a), the mechanisms based on thermal activation of dislocation movement and the effects of electron- and phonon damping, it is also relativistic aspects and twin formation [see e.g. 3.227] that become particularly relevant as strain rate grows. The latter occurs as an alternative athermal deformation process whenever the stress required for geometrically unlimited dislocation movement is greater than twin initiation stress, which is equivalent to a critical twinning pressure [3.228]. Deformation is strongly determined by stacking fault energy, which constitutes a measure of the probability of cross slip of screw dislocations. At low levels of stacking fault energy, the splitting widths of the partial dislocations are high, cross slip is impeded and thus, there is a planar formation of dislocation structures. High stacking fault energy, on the other hand, sees the likelihood of cross slip increase, and the dislocation structure becomes cell-shaped [3.229–3.231]. Furthermore, [3.232] reported that stacking fault energy increases as the hydrostatic stress state increases. Thus, the pressure itself has an influence on the probability of cross slip. The formation of adiabatic shear bands is to be viewed as an additional deformation process. It occurs as an instability at high deformation velocities, when the removal of deformation heat is impeded in such a way that thermally-induced softening becomes dominant versus the workhardening created by deformation and strain rate [3.233–3.238]. As stated by [3.233], the microstructural changes which occur during high-strain rate deformation may be differentiated according to the systematics of lattice defects. Point defects appear in the form of growing amounts of vacancies and matrix interstitial atoms caused by increasing pressure. In addition, a number of them form area-shaped clusters that can be regarded as a preliminary stage of three-dimensional lattice defects [3.231]. Dislocation densities which increase roughly proportionally to the square root of pressure constitute one-dimensional lattice defects [3.231,3.239]. At high levels of stacking fault energy, they are arranged as cells – cell size being roughly proportional to the reciprocal value of the square root of pressure [3.231,3.240] – while at low levels of stacking fault energy, their arrangement is planar. Two-dimensional lattice defects are found as twins which appear above a critical twinning pressure. Their density increases proportionally to the square root of pressure, decreases proportionally to stacking fault energy and increases proportionally to grain size, according to [3.231]. In addition, the formation of adiabatic shear bands is observed. The formation of three-dimensional lattice defects due to phase transformations of retained austenite into martensite has been reported for austenitic steels [3.233].

b) Quantitative Descriptions of Processes

As laser shock treatment is only beginning to see technical application, there is only a small number of published studies which describe the process by means of analytical or numeric models. Regarding the technically relevant confined ablation, [3.221] examined laser shock treatments with pulse durations of t_p and stated three phases of impulse transfer into the workpiece.

- At $t < t_p$, plasma is created, exerting pressure on the workpiece and on the cover layer and initiating shock waves in the workpiece and the cover layer.
- At $t > t_p$, there is an adiabatic expansion of plasma, which results in a reduction of pressure, but maintains impulse transfer into the workpiece.
- At $t \gg t_p$, recombination of plasma occurs, and only the expanding hot gas effects an impulse transfer into the workpiece.

Given the simplified preconditions that laser intensity I, absorption and therefore, pressure, are constant during the pulse, that the portion of intensity a, which contributes to heating, and the portion $1 - a$, which causes ionization, are constant, that the plasma acts like an ideal gas and that the workpiece may be regarded as ideally elastic/plastic and the cover layer viewed as linearly elastic, [3.221] states pressure p in the first phase to be

$$p[GPa] = 0,01\sqrt{\frac{a}{2a+3}}\sqrt{Z}[g/cm^2s]\sqrt{I}[GW/cm^2]. \tag{3.31}$$

Here, $a \approx 0.1$ is valid, and Z is the weighted impedance of the workpiece (WP) and of the cover layer (CL), which is derived from the densities ρ and the shock wave velocities c as

$$\frac{1}{Z} = \frac{1}{2}\left(\frac{1}{\rho_{WP}c_{WP}} + \frac{1}{\rho_{CL}c_{CL}}\right). \tag{3.32}$$

By contrast, in the second phase, pressure results from

$$p(t) = p_0\left(1 + (\gamma + 1)\left(\frac{t - t_p}{t_p}\right)\right)^{-\frac{\gamma}{\gamma+1}}, \tag{3.33}$$

where p_o is pressure at the end of the pulse and γ is the adiabatic coefficient, for which $\gamma \approx 1.4$ is valid. From this follows the time $t_{1/2}$ required for pressure to be halved as

$$t_{1/2} = \frac{2^{\frac{\gamma}{\gamma+1}} + \gamma}{\gamma+1}t_p, \tag{3.34}$$

from which follows with $\gamma \approx 1.4$ $t_{1/2} \approx 1.2\ t_p$. Third-phase pressure development, by contrast, is negligible in the context of shock wave formation and is therefore not discussed in detail. Fig. 3.80 schematically depicts the distribution of pressure resulting from Eq. 3.32 and 3.33.

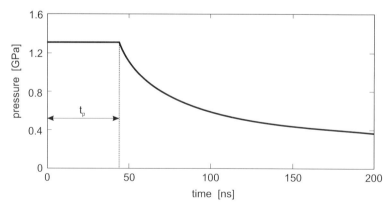

Fig. 3.80: Temporal development of pressure in laser shock treatments [3.219]

One additional consequence of [3.221]'s model reflection is a restriction of the model's applicability. Effective confined ablation is possible only when the plasma layer thickness

$$L(t) = \int_0^t (v_{CL} - v_{WP}) dt \qquad (3.35)$$

which is a result of plasma front velocities v_{CL} and v_{WP}, is greater than the roughness R_Z of the irradiated surface. Given

$$p = \rho_{CL} c_{CL} v_{CL} = \rho_{Wp} c_{WP} v_{WP}, \qquad (3.36)$$

therefore,

$$t_{p,\,min} = \frac{R_z}{p} \frac{\rho_{CL} \rho_{WP} c_{CL} c_{WP}}{\rho_{CL} c_{CL} + \rho_{WP} c_{WP}} \qquad (3.37)$$

must be valid for minimum pulse duration. Pressures above $2\,p_H$ see the occurrence of plastic saturation, up to the depth z_{pl} stated in Eq. 3.30. In this case,

$$\sigma_{surf}^{rs} = \left[1 - \frac{4\sqrt{2}}{\pi} (1+v) \frac{z_{pl}}{a} \right] R_Y \qquad (3.38)$$

is valid for surface residual stresses σ_{surf}^{rs} after irradiation of a square area with an edge length $a \gg z_{pl}$. The dependence of the surface residual stresses on the ratio z_{pl}/a for greater surface layer thicknesses is represented in Fig. 3.81 [3.219]. Testing the model described, [3.241] examined SAE5155 which was irradiated by a Nd:glass laser, using a glass cover layer. At 15 kbar of Hugoniot stress and 30 and 60 kbar of irradiation, his experimental results measured 0.65 and 2 mm for penetration depth, versus the calculated values of 0.615 and 1.9 mm. The same

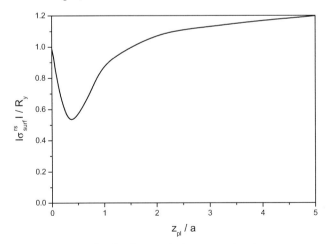

Fig. 3.81: Magnitude of surface residual stresses related to yield strength calculated acc. [3.241] vs. penetration depth of plastic deformation related to edge length of the irradiated area

irradiations yielded experimental compressive residual stress values at surface of 230 and 330 MPa, which correspond to the calculated values of 234 and 312 MPa. Accordingly, the model serves well in describing both the depth effect of plastic deformation and the surface values of residual stress.

Given repeated irradiation, n pulses result in

$$\sigma^{rs}_{surf,n} = \sigma^{rs}_{surf,n-1} - \left[1 - \frac{4\sqrt{2}}{\pi}(1+\nu)\frac{z_{pl}}{a}\right](R_\gamma + \sigma^{rs}_{surf,n-1}). \tag{3.39}$$

Accordingly, repeated irradiation achieves increases of compressive residual stresses at the surface, while flow stress dictates a saturation value. Due to the workhardening of material regions close to the surface, multiple irradiations result in plastic deformation being shifted to deeper regions of the material. Thereby, the thickness of the plastically deformed surface layer is increased.

3.6.2
Changes in the Surface State

a) Influences on Shape

Laser shock treatment creates plastic shock wave deformations close to the surface. They can affect work material regions within a surface distance of roughly 1 mm. Therefore, thin-walled components run the risk of experiencing macroscopic changes of shape. These can cause bending deflections toward the laser beam if the plastically deformed surface layers are sufficiently thin. When components are plastified completely, bending deflections away from the laser beam may occur. For flat specimens of quenched and tempered AISI4140 with a thickness of 5 mm, [3.242] reported deflections after laser shock treatment using a Nd:glass laser with a pulse energy of 25 J and pulse durations of 18 ns. The laser beam, with a diameter of 3.5 x 3.5 mm², was directed at the surface of the specimen through a water-damped absorption layer. Having received irradiation with 50 % overlap on one complete side, the 110 mm-long specimens exhibited deflections of 940 μm.

b) Influences on Topography

As the irradiation position is set discretely from pulse to pulse, laser shock treatments create periodic surface structures which are relatable to the edges of the areas hit by the individual pulses. This is shown in Fig. 3.82 for the surface of a flat specimen consisting of quenched and tempered AISI4140 which, like the specimen mentioned in the previous section, was treated with Nd:glass pulses with a beam diameter of 3.5 x 3.5 mm² at 50 % overlap. The surfaces, viewed using a confocal white light microscope, exhibited linear indents measuring roughly 3 μm. Their spacing of about 1.7 mm corresponds to the beam width halved by the 50 % overlap. Thus, after a treatment incorporating overlapping work positions, the positions of both the initial and the final pulse are detectable in the surface topography [3.242].

AISI4140, quenched & tempered

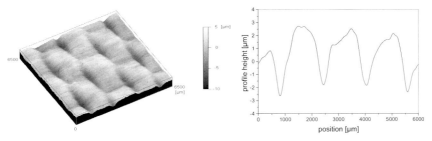

Fig. 3.82: Topography profile determined using confocal white light microscopy on a specimen made of quenched and tempered AISI4140 laser shock treated with 50% overlap [3.242]

However, it may be generally noted that laser shock treatment achieves high surface qualities. [3.243] demonstrated this by means of comparative measurements of roughness on polished and etched Ti(RT15) titanium specimens before and after laser shock treatments. While polished-state roughness increases significantly from $R_z = 0.4$ µm to $R_z = 1.2$ µm, it increases from $R_z = 3.7$ µm to $R_z = 3.9$ µm in the etched state and is therefore hardly altered. Obviously, surface roughnesses of particularly low degrees are raised, while medium roughnesses remain almost unchanged. There are similar reports by [3.244] regarding a titanium alloy on which the surface quality specified for the vanes of aircraft gas turbines was achieved by laser shock treatment, without reverting to the finishing treatment required after the usual shot peening processes. [3.245] demonstrated that the surface roughness measured on T62-state 2024 aluminum alloy shows laser shock treatment-induced decreases from 1.6 µm to about 0.8 µm at low power densities, and to just over 0.1 µm at a power density above ca. 1.1 GW/cm^2 (comp. Fig. 3.83).

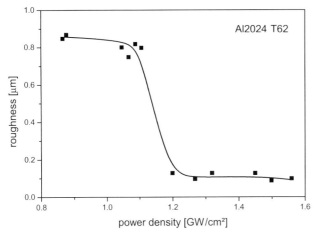

Fig. 3.83: Influence of power density on the roughness of 2024 – T62 aluminum alloy after laser shock treatment [3.245]

c) **Influence on Residual Stress State**

As discussed in the description of the elementary processes of laser shock treatment, the shock waves in laser shock treatment create compressive residual stresses, while the thermal effects on the workpiece surface layer induced by direct ablation cause tensile residual stresses. The primary focus of this description is mechanical surface treatment with the objective of increasing mechanical properties, and therefore, a further description of direct ablation is omitted. However, it should be noted that in confined ablation, too, any kind of thermal manipulation must be avoided. Consequently, the absorption layer must be thick enough to prevent thermal effects on the workpiece itself, and, particularly in cases of overlapping irradiation, the absorbing layer should not have sustained damage that may result in thermal effects on the workpiece during the final irradiation.

The surface residual stresses in laser shock-treated specimens fluctuate from pulse to pulse, due to the aforementioned discrete shifts of the irradiation position, as shown exemplarily in Fig. 3.84 for the quenched and tempered state of AISI4140 which was already shown in Fig. 3.82. All residual stresses are compressive, fluctuating between 400 and 750 MPa. Their peak levels above roughly 700 MPa show a spacing of roughly 3.5 mm, which is the full width of the beam. As opposed to the topographical effects, the resulting residual stresses are thought to be dominated by the last pulse, while the stress states created by the first pulse are of secondary importance. Unlike shot peening, which is also based on discretely distributed hits, laser shock treatment does not create observable oscillations of surface residual stresses that reach the tensile range (comp. Figs. 3.14, 3.15).

The influence of the overlap rate on surface residual stresses in 34CAlNi7, in laser shock treatment using a XeCl-excimer laser, is depicted in Fig. 3.85 [3.219]. While the compressive residual stresses still exhibit a clearly undulating profile at 33 % overlap, their level is constant at 50 % overlap. [3.219] deduced from this

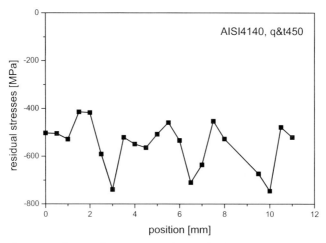

Fig. 3.84: Influence of position on surface residual stresses after laser shock treatment of quenched and tempered AISI4140 [3.242]

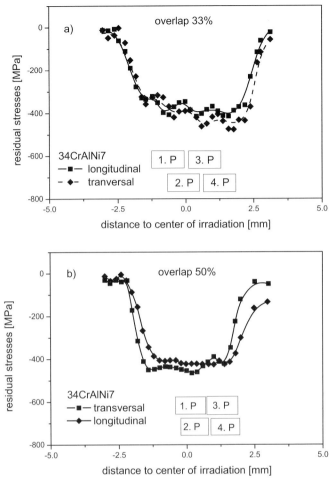

Fig. 3.85: Influence of degree of overlap (a: 33%, b: 50%) on the distribution of surface residual stresses in 34CrAlNi7 (XeCl-excimer-laser, adhesive tape, water) [3.219]

that an overlap rate of about 50 % is favorable. Yet this rate should not be raised any further, as this increases the aforementioned risk of damage to the absorbing layer and thus, the risk of thermal effects on the workpiece. The overlap required for minimizing residual stress oscillations presumably depends also on the laser type and on additional treatment parameters, or the results for surfaces created using a Nd:glass laser, depicted in Fig. 3.84, should not have exhibited any oscillations.

The influence of the number of pulses on residual stress depth distribution is shown in Fig. 3.86 for the same material and a similar treatment [3.219]. At low numbers of pulses, the residual stresses decrease continually from the maximum value of compressive residual stresses in the surface layer down to the position of

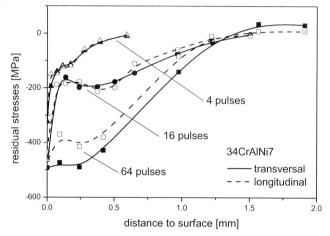

Fig. 3.86: Influence of number of pulses on the distribution of residual stresses for 34CrAlNi7 after laser shock treatment with 50% overlap in both feed directions (XeCl-excimer-laser, adhesive tape, water) [3.219]

zero residual stresses. As the number of pulses increases, so do the surface values and penetration depths. In the process, at first a great residual stress gradient forms close to the surface, followed by a region of small and undulating residual stresses, until they gradually disappear towards the zero crossing. Only after 64 pulses does a characteristic depth distribution of residual stresses occur, which decreases continually after showing a small plateau region close to the surface. Penetration depth increases as the number of pulses grows, as the shock waves always bring about additional plastic deformation. As soon as surface layer regions are sufficiently plastically deformed, have become workhardened and therefore have reached a saturation level of residual stress, further irradiation will induce more plastic deformation only at greater distances to surface. In this case, an increasing number of pulses causes an approximately linear increase of penetration depth.

In accordance with this, [3.246] found surface values and penetration depths of residual stresses in 40NiCrMo6 to increase with the number of laser pulses. Fig. 3.87 shows that in this case, the undulation of the residual stress distribution described above does not occur. In contrast to both previous examples, however, [3.247] found residual stress distributions in a soft 55Cr3 which attained their maximum surface value already after a single laser irradiation (comp. Fig. 3.88). Thus, they are in accordance with the model introduced in Sect. 3.6.1b.

Figs. 3.89–91 compare the residual stress distributions in laser shock-treated and shot peened states of quenched and tempered AISI4140, TiAl6V4 and Inconel 718 [3.242,3.248,3.249]. Except in the case of quenched and tempered AISI4140 the residual stress values close to the surface are slightly smaller after laser shock treatment than after shot peening, and they decrease continuously. In contrast to laser shock treatment, the effects of Hertzian pressure are present in the shot peened

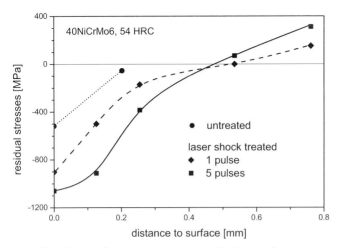

Fig. 3.87: Influence of number of pulses on the distribution of residual stresses in 40NiCrMo6 (Nd:glass-laser, water bath) [3.246]

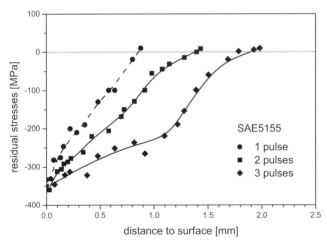

Fig. 3.88: Influence of number of pulses on the distribution of residual stresses in SAE5155 (Nd:glass-laser, water bath) [3.247]

states. Thus, they show a pronounced residual stress plateau in steel, and marked residual stress maxima beneath the surface in the other two materials. The resulting maximum values of residual stress in the titanium- and nickel-base alloys are approximately 50 % higher after shot peening than after laser shock treatment. Impressive increases in the penetration depth of the residual stresses are found after laser shock treatment.

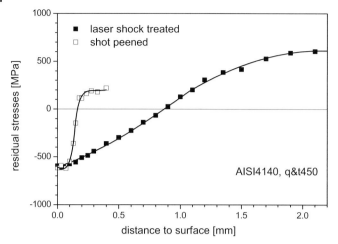

Fig. 3.89: Distribution of residual stresses after laser shock treatment (Nd:glass-laser, water bath) and shot peening (S170, 0.24mmA) of quenched and tempered AISI4140 [3.242]

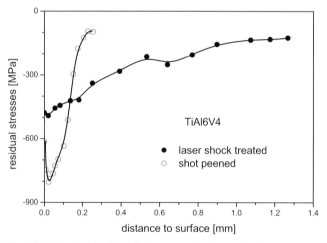

Fig. 3.90: Distribution of residual stresses after laser shock treatment (Nd:glass-laser, water bath) and shot peening (S110, 0.10 A) of TiAl6V4 [3.248]

In his examinations of austenitic AISI304 steel, which was laser shock treated using a q-switched Nd:YAG laser, [3.250] demonstrated the influence of the laser shock treatment parameters on the residual stresses close to the surface after irradiation, which was carried out in a meandrous pattern. Compressive residual stresses are measurable up to a distance to surface of about 1.2 mm. The residual stresses in the feed direction are slightly smaller than those in the transversal direction, presumably due to differences in the degree of overlap. When the diameter of the area irradiated by each laser pulse and the pulse energy are increased,

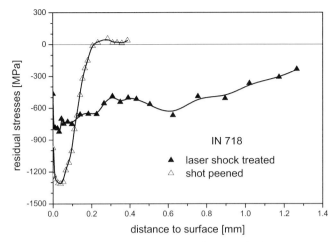

Fig. 3.91: Distribution of residual stresses after laser shock treatment (Nd:glass-laser, water bath) and shot peening (S110, 0.10 A) of IN718 [3.248]

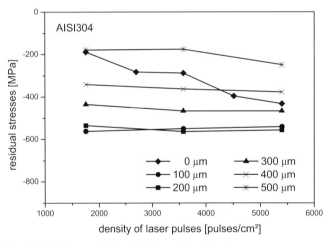

Fig. 3.92: Influence of pulse density on the residual stresses transversal to feed direction at different distances to surface for AISI304 (q-switched Nd:YAG-laser, water bath) [3.250]

the surface residual stresses increase until they reach optimum values. The residual stresses beneath the surface always show increases within the examined range of the parameters. By contrast, when pulse density is increased – as shown in Fig. 3.92 –, the surface residual stresses show continuous increases while the residual stresses beneath the surface remain approximately constant. In this laser configuration, water is used both as the absorbing layer and as the top coat, and therefore, there is no risk of thermal influence at higher pulse- and overlap rates.

Fig. 3.93 shows a compilation of the distributions of residual stresses achievable in different metallic materials in a water bath using Nd:glass lasers [3.251]. While aluminum alloys show the lowest values of surface residual stresses and the smallest penetration depths of compressive residual stresses, low alloyed SAE5155 steel and the nickel-base alloy Astroloy exhibit increased values of the surface residual stresses. Of all the materials examined here, AISI4135 steel with a hardness of 50 HRC shows the highest levels of both values. A comparison of the maximum compressive residual stress values and the respective material

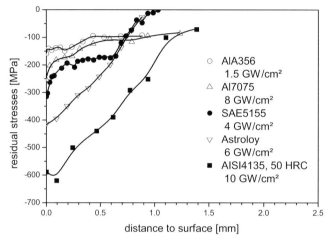

Fig. 3.93: Distribution of residual stresses in selected materials after laser shock treatment (Nd:glass-laser, water bath) [3.251]

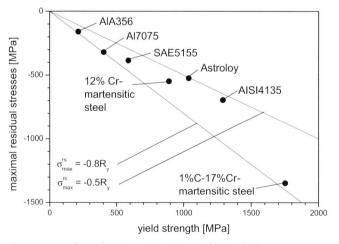

Fig. 3.94: Correlation between maximum possible residual stresses after laser shock treatment and yield stress for selected materials (Nd:glass-laser, water bath) [3.251]

hardness shows that the compressive residual stresses are within a range of 0.5 to 0.8 times the yield strength (see Fig. 3.94). The findings of [3.252] and [3.253] concur and show that aluminum alloys and austenitic AISI316L steel exhibit penetration depths that are significantly greater after laser shock treatment than they are after shot peening. However, the surface residual stresses are generally smaller.

d) Influences on Workhardening State

Fluctuations of treatment intensity are mentioned in previous sections. They appear as effects on topography and as an undulating distribution of residual stresses at the surface, as in the quenched and tempered AISI4140 steel shown in Figs. 3.82 and 3.89. These fluctuations are not found in the states of workhardening close to the surface. It should be noted that laser shock treatment causes only small changes of the half widths close to the surface in this material, anyway.

The workhardening states close to the surface found after laser shock treatment are extremely material-dependent. As Fig. 3.95 shows, exposure is described by the ratio of the surface coordinate x and the laser beam radius r_{Laser} between -1 and +1. The increases of hardness which are found close to the surface range between 0 and approximately 100 % of bulk material hardness [3.219,3.254,3.255]. As shown in the following section by a description of the microstructures close to the surface, the workhardening effect evidently is strongly dependent on the strength level of a given material and on the number of slip systems that can be activated. This number is determined by the crystal structure and by the stacking fault energy. Workhardening occurring close to the surface increases with the available number of activatable slip systems, from body centered cubic X2Cr11 and 34CrAlNi7 steels, to hexagonal titanium, to austenitic X6CrNiTi18 10 steel. By contrast, both steels with a body centered cubic matrix show decreases of work-

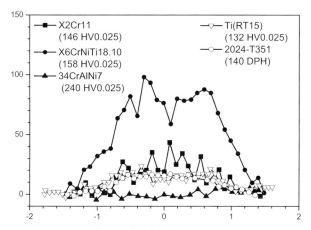

Fig. 3.95: Relative change of surface hardness vs, position related to radius of irradiated area for selected materials (XeCl-excimer-laser, adhesive tape, water) [3.219, 3.254, 3.255]

hardening in regions close to the surface when strength rises. In X2CrAlNi7 with $R_{p,0.2} = 380$ MPa, there is still an increase in hardness of about 15 %, while hardness is not affected in 34CrAlNi7 with $R_{p,0.2} = 650$ MPa. In the case of aluminum alloys, [3.252] found that strength has similar effects on the increase of hardness close to the surface found after laser shock treatments. While the low-strength T6-state A356 alloy shows increases in hardness of around 10 %, hardness growth is only about 5 % in the higher-strength 7351-state 7075 alloy.

As [3.253] demonstrated for austenitic AISI316L, surface hardness and the integral widths of the interference lines measured at the surface show increases when the number of pulses and power density are increased. However, they remain well below the values observed for the shot-peened state. This corresponds to e.g. [3.255], who found hardness levels in 2024 aluminum alloy increasing with rising peak pressure, while increasing the pulse rate caused increases of hardness in austenitic X2CrNi18–10.

Figs. 3.96–3.98 show results of laser shock treatment on AISI4140, TiAl6V4 and IN 718. The half widths close to the surface – and the plastic deformations derived from them – show continuous increases, starting from the bulk material and proceeding toward the surface. In comparison to the results for shot peened states, which are also depicted, the increase in the line width is noticeably smaller. In contrast to the high-strain rate deformation which occurs during shot peening, the induced by shock deformation evidently induces smaller degrees of deformation and workhardening effects.

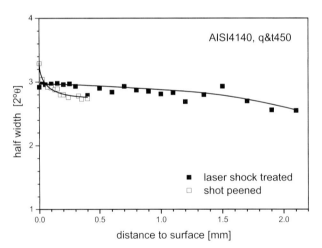

Fig. 3.96: Distribution of workhardening after laser shock treatment (Nd:glass-laser, water bath) and shot peening (S170, 0.24mmA) of quenched and tempered AISI4140 [3.242]

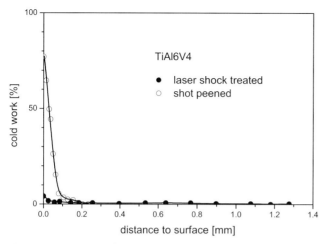

Fig. 3.97: Distribution of workhardening after laser shock treatment (Nd:glass-laser, water bath) and shot peening (S110, 0.10 A) of TiAl6V4 [3.248]

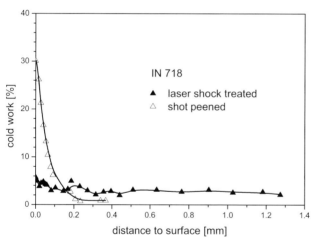

Fig. 3.98: Distribution of workhardening after laser shock treatment (Nd:glass-laser, water bath) and shot peening (S110, 0.10 A) of IN718 [3.248]

e) Influences on Microstructure

Due to the presence of shock loading in the region close to the surface, the typical formation of twins is pointed out in numerous studies. For example, [3.256] used differential interference contrast in an optical microscope, crystal orientation contrast in a scanning electron microscope and transmission electron microscopic examinations on specimens of poly-crystalline, highly purified iron and molybdenum after laser shock treatment (comp. exemplary representation in Fig. 3.99). He demonstrated that laser shock treatment causes twin formation close to the surface, especially in the case of iron. However, the effect is far less pronounced in molybdenum. Here, slip possibilities are still available and are visible as traces of wave-like slip.

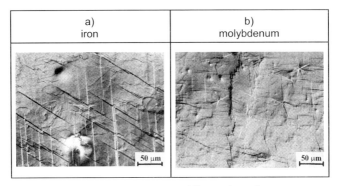

Fig. 3.99: Light microscopic view, using differential interference contrast, of microstructure in pure iron (a) and pure molybdenum (b) after laser shock treatment (XeCl-excimer-laser, glue band, water) [3.256]

In Ti(RT15), circular irradiation causes twin formation directly at the surface. Defocusing the laser causes a reduction of the effect, beginning at the center, up to the point at which twins are found only near the edge of the irradiated area [3.219]. Presumably, this is caused by the attenuation waves mentioned above.

[3.257] contributed to an understanding of the mechanism of deformation, by carrying out transmission electron microscopic examinations on copper with varying contents of aluminum (comp. Fig. 3.100). At a low 1 Ma.-% aluminum content, he observed the formation of a cell structure. Mean stacking fault energy and the cross slip processes enabled by it provide an explanation. It indicates good

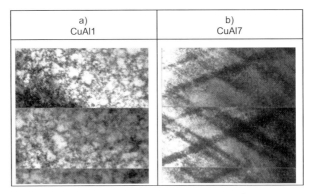

Fig. 3.100: Transmission electron microscopic view of the microstructures in CuAl1 (a) and CuAl7 (b) after laser shock treatment (XeCl-excimer-laser, glue band, water) [3.257]

deformability, allowing for high degrees of plastic deformation and thus, high residual stress values. By contrast, a planar dislocation arrangement and numerous twins are found at the high,7 Ma.-% aluminum content. The reason for this is the comparatively low level of stacking fault energy. It causes the splitting between the dislocations to be greater, while twin initiation stress, which is proportional to the square root of stacking fault energy, is reduced [3.258]. Thus, ductility is reduced and as a result, plastic strains and, therefore, the residual stress values, are smaller. For austenitic AISI316L steel, [3.253] reports that the microstructure is influenced also by power density. Low power densities result in threefold slip due to comparatively low deformation velocities, whereas high power densities cause single slip and twin formation, as strain rates are comparatively high.

f)   Influences on Phase Fractions

To date, there are hardly any studies which deal with the effects of laser shock treatments on the phase fractions which occur close to the surface. [3.253] showed that in contrast to shot peened states, α'-martensite is not detectable in austenitic X2CrNiMo18–12–2 steel after laser shock treatments. Evidently, the deformations created by laser shock treatment are not great enough to cause a deformation-induced transformation of retained austenite. However, [3.259] employed pure iron irradiated using a Nd:glass laser and a pressure greater than 13 GPa to demonstrate that bcc α-Fe can be transformed into hex ε-Fe high-pressure phase.

References

3.1 P. Martin: Beitrag zur Ermittlung der Einflussgrößen beim Kugelstrahlen durch Einzelkornversuche, Dissertation, Universität der Bundeswehr Hamburg, 1980.

3.2 F. Burgahn, V. Schulze, O. Vöhringer, E. Macherauch: Modellierung des Einflusses von Temperatur und Verformungsgeschwindigleit auf die Fließspannung von Ck45 im Tieftemperaturbereich T < 0,3 TS, Materialwissenschaft und Werkstofftechnik 27 (1996), pp. 521–530.

3.3 V. Schulze, O. Vöhringer: Influence of alloying elements on the strain rate and temperature dependence of the flow stress of steel, Metallurgical and Materials Transactions 31A (2000), pp. 825–830.

3.4 V. Schulze, O. Vöhringer: Constitutive description of plastic deformation, In: Encyclopedia Materialia: Science and Technology, Elsevier, 2001, pp. 7050–7064.

3.5 A. Seeger: Theorie der Kristallplastizität III: Temperatur- umd Geschwindigkeitsabhängigkeit der Kristallplastizität, Zeitschrift Naturforschung 8a (1954), pp. 870–881.

3.6 U. F. Kocks, A. S. Argon, M. F. Ashby: Thermodynamics and Kinetics of Slip, Progress in Materials Science 19 (1975).

3.7 K. Ono: Temperature dependence of dispersed barrier hardening, Journal of Applied Physics 39(1968), pp. 1803–1806.

3.8 O. Vöhringer: Temperatur und Geschwindigkeitsabhängigkeit der Streckgrenze von Kupferlegierungen, Zcitschrift für Metallkunde 65 (1974), pp. 32–36.

3.9 A. Seeger: The temperature and strain-rate dependence of the flow stress of body centered cubic metals: A theory based on kink-kink-interactions, Zeitschrift für Metallkunde 72 (1981), pp. 369–380.

3.10 B. Sestak, A. Seeger: Gleitung und Verfestigung in kubisch raumzentrierten Metallen und Legierungen I-III, Zeitschrift für Metallkunde 69 (1978), pp. 195–202, 355–363, 425–432.

3.11 J. Harding: Material behaviour at high rates of strain, In: C. Y. Chiem, H.-D. Kunze, L. W. Meyer (eds.), Impact loading and dynamic behaviour of materials, DGM-Informationsgesellschaft, Oberursel, 1988, pp. 23–42.

3.12 A. Kumar, F. E. Hauser: Viscous drag on dislocations in aluminium at high strain rates, Acta Metallurgica 16 (1968), pp. 1189–1197.

3.13 C. K. H. Dharan: Dislocation mobility in aluminium at high strain rates, In: Proc. Int. Conf. Metals and Alloys 2, ASM, Ohio, 1970, pp. 1006–1007.

3.14 G. P. Huffmann, N. Lonat: Interaction between electrons and moving dislocations, In: J. A. Simmons, R. de Witt, R. Bullough (eds.), Fundamental Aspects of Dislocation Theory, US National Bureau of Standards, 1970, pp. 1303–1323.

3.15 H. Wohlfahrt: Ein Modell zur Vorhersage kugelstrahlbedingter Eigenspannungszustände, In: E. Macherauch, V. Hauk (eds.), Eigenspannungen, DGM-Informationsgesellschaft, Oberursel, 1983, pp. 301–319.

3.16 R. Schreiber: Untersuchungen zum Dauerschwingverhalten des kugelgestrahlten Einsatzstähle 16 MnCr 5 in verschiedenen Wärmebehandlungszuständen, Dissertation, Universität Karlsruhe (TH), 1976.

3.17 J. Bauschinger: Über die Veränderung der Elastizitätsgrenze und der Festigkeit des Eisens und Stahls durch Strecken und Quetschen, durch Erwärmen und Abkühlen und durch oftmals wiederholte Beanspruchung, Mittheilungen aus dem Mechanisch-Technischen Laboratorium der K. Technischen Hochschule in München 13 (1886).

3.18 B. Scholtes: Die Auswirkungen des Bauschingereffekts auf das Verformungsverhalten technisch wichtiger Vielkristalle, Dissertation, Universität Karlsruhe (TH), 1980.

3.19 G. Masing: Zur Heyn'schen Theorie der Verfestigung der Metalle durch verborgene elastische Spannungen, Wiss. Ver. Siemens Konzern 3 (1924), pp. 231–239.

3.20 G. Masing: Berechnung von Dehnungs- und Stauchungslinien auf Grund von inneren Spannungen, Wiss. Ver. Siemens Konzern 5 (1926), pp. 135–141.

3.21 E. Heyn: Festschrift der Kaiser-Wilhelm-Gesellschaft, 1921.

3.22 E. Heyn: Internal strains in cold-wrought metals and some troubles caused thereby, The Journal of the Inst. of Metals (1914).

3.23 E. Orowan: Classification and nomenclature of internal stresses, In: Symposium on internal stresses, Inst. of Metals, 1947, pp. 47–59, 451–453.

3.24 H. Margolin, F. Hazaveh, H. Yaguchi: The grain boundary contribution to the Bauschinger effect, Scripta Metallurgica 12 (1978), pp. 1141–1145.

3.25 G. Sachs, H. Shoji: Zug-Druckversuche an Messingkristallen – Bauschingereffekt, Mitteilung aus dem Kaiser Wilhelm-Institut für Metallforschung (1927), pp. 776–796.

3.26 L. M. Brown: Orowans explanation of the Bauschinger Effect, Scripta Metallurgica 11 (1977), pp. 127–131.

3.27 E. Macherauch O. Vöhringer: Das Verhalten metallischer Werkstoffe unter mechanischer Beanspruchung, Zeitschrift für Werkstofftechnik 9 (1978), pp. 370–391.

3.28 J. C. Fisher, E. W. Hart, R. H. Pry: The hardening of metal crystals by precipitate particles, Acta Metallurgica 1 (1953), pp. 336–339.

3.29 L. M. Brown: Back-stresses, image stresses, and work-hardening, Acta Metallurgica 21 (1973), pp. 879–885.

3.30 W. M. Stobbs: Dispersion strengthening, In: AJME, Cincinnati, 1975, pp. 127–149.

3.31 L. M. Brown, W. M. Stobbs: The work hardening of copper-silica: II. The role of plastic relaxation, Phil. Mag. 23 (1971), pp. 1201–1233.

3.32 J. D. Atkinson, L. M. Brown, W. M. Stobbs: The work hardening of copper-silica: IV. The Bauschinger effect and plastic relaxation, Phil. Mag. 30 (1974), pp. 1247–1280.

3.33 L. M. Brown, W. M. Stobbs: The work hardening of copper-silica: V. Equilibrium plastic relaxation by secondary dislocations, Phil. Mag. 34 (1976), pp. 351–372.

3.34 A. Abel, H. Muir: The Bauschinger Effect and stacking fault energy, Phil. Mag. 27 (1973), pp. 585–594.

3.35 R. Milligan, H. W. Koo: The Bauschinger Effect in copper and copper-zinc alloys, Technical Reprot WVT-7022 Benet R+E Laboratories Watervliet Arsenal, 1970.

3.36 A. Abel, H. Muir: The Bauschinger Effect and discontinuous yielding, Phil. Mag. 26 (1973), pp. 489–504.

3.37 S. Kumakura: The Bauschinger effect in carbon steel, Bulletin of JSME 11 (1968), pp. 426–436.

3.38 A. Abel H. Muir: Mechnaical hysteresis and the initial stage of fatigue, Metal Science 9 (1975), pp. 459–463.

3.39 W. S. Owen, J. V. Carisella: The role of dislocation substructure in the monotonic and yield and strain hardening of steels, In: Proc. Int. Conf. Mech. Behavior of Materials 2, ASM, Boston, 1976, pp. 15–41.

3.40 S. N. Buckley, K. M. Entwistle: The Bauschinger Effect in super-pure aluminium single crystals and polycrystals, Acta Metallurgica 4 (1956), pp. 352–361.

3.41 T. C. Harrison, R. T. Weiner: Influence on the Bauschinger Effect in high-yield steel pipes, Journal of the Iron and Steel Institute (1972), pp. 334–336.

3.42 R. L. Woolley: The Bauschinger Effect in some face-centred and body-centred cubic metals, Phil. Mag. 44 (1953), pp. 597–618.

3.43 Y. Saleh H. Margolin: Bauschinger effect during cyclic straining of two ductile phase alloys, Acta Metallurgica 27 (1979), pp. 535–544.

3.44 J. A. delValle, R. Romero, A. C. Picasso: Bauschinger effect in age-hardened Inconel X-750 alloy, Materials Science and Engineering A311 (2001), pp. 100–107.

3.45 B. Scholtes, O. Vöhringer: Untersuchungen zum Bauschingereffekt von Ck 45 in unterschiedlichen Wärmebehandlungszuständen, Härterei Technische Mitteilungen 41 (1986), pp. 347–354.

3.46 H. Hertz: Über die Berührung fester elastischer Körper, Z. für reine und angewandte Mathematik 92 (1881), pp. 155.

3.47 K. L. Johnson: Contact mechanics, Cambridge University Press, Cambridge, 1985.

3.48 D. Tabor: Proc. Roy. Soc. A192 (1947), pp. 247–274.

3.49 D. Tabor: The Hardness of Metals, Clarendon Press, Oxford, 1951.

3.50 M. C. Shaw G. J. De Salvo: On the plastic flow beneath a blunt axisymmetric indenter, Journal of Engineering for Industry (1970), pp. 480–494.

3.51 J. Boussinesq: Application des Potentials, Comptes Rendus 114 (1892), pp. 1150.

3.52 S. T. S. Al-Hassani: Mechanical aspects of residual stress development in shot peening, In: A. Niku-Lari (ed.), Proc. Int. Conf. Shot Peening 1, Pergamon, Paris, 1981, pp. 583–602.

3.53 S. T. S. Al-Hassani: The Shot Peening of Metals – Mechanics and Structures, In: Shot Peening for Advanced Aerospace Design, Anaheim, 1981, pp. 1–14.

3.54 S. T. S. Al-Hassani: An engineering approach to shot peening machines, In: H. O. Fuchs (ed.), Proc. Int. Conf. Shot Peening 2, American Shot Peening Society, Paramus, 1984, pp. 275–282.

3.55 Y. F. Al-Obaid: A rudimentary analysis of improving fatigue life of metals by shot-peening, Journal of Applied Mechanics 57 (1990), pp. 307–312.

3.56 Y. F. Al-Obaid: Mechanical aspects of plasticity development in shot peening, In: Proc. Int. Conf. Technology of Plasticity 4, Bejing, 1993, pp. 2006–2013.

3.57 Y. F. Al-Obaid: Shot peening mechanics: experimental and theoretical analysis, Mechanics and Materials 19 (1995), pp. 251–260.

3.58 Y. Watanabe, N. Hasegawa: Simulation of residual stress distribution on shot peening, In: J. Champaigne (ed.), Proc. Int. Conf. Shot Peening 6, San Francisco, 1996, pp. 530–535.

3.59 D. A. Hills: Some aspects of post-yield contact problems, Wear 85 (1983), pp. 107–119.

3.60 D. A. Hills, R. B. Waterhouse, B. Noble: An analysis of shot peening, Journal of Strain Analysis 18 (1983) 2, pp. 95–100.

3.61 J. K. Li, M. Yao, D. Wang: Mechanical approach to the residual stress field induced by shot peening, Materials Science and Engineering A147 (1991) 2, pp. 167–173.

3.62 J. F. Flavenot, A. Niku-Lari: La Mesure des Contraintes Résiduelles: Methode de la Flêche – Methode de la Source des Contraintes, Les mémoires techniques du CETIM 31 (1977).

3.63 A. Niku-Lari: Methode de la flêche – methode de la source des contraintes résiduelles, pp. 237–247.

3.64 H. Guechichi, L. Castex, J. Frelat, G. Inglebert: Prevision des contraintes residuelles dués au grenaillage, CETIM, 1985.

3.65 H. Guechichi, L. Castex, J. Frelat, G. Inglebert: Predicting residual stresses due to shot peening, In: E. Macherauch, V. Hauk (eds.), Residual Stresses in Science and Technology, Proc. Int. Conf. Residual Stresses 1, DGM-Informationsgesellschaft, Oberursel, 1987, pp. 449–456.

3.66 J. Zarka, J. Frelat, G. Inglebert, P. Kasmai-Navidi: A new approach in inelastic analysis of structures, CADLM, France, 1990.

3.67 M. T. Khabou, L. Castex, G. Inglebert: A theoretical prediction of residual stresses introduced by shot peening, In: G. Beck, S. Denis, A. Simon (eds.), Proc. Int. Conf. Residual Stresses 2, Elsevier, London, 1988, pp. 619–624.

3.68 M. T. Khabou, L. Castex, G. Inglebert: The effects of material behavior law on the theoretical shot peening results, European Journal of Mechanics and Solids 9 (1990) 6, pp. 537–549.

3.69 S. Slim, T. Harm, G. Inglebert: Prediction of residual stresses due to shot peening using modelling of the material behavior, Materials and Manufacturing Processes 10 (1995) 3, pp. 579–588.

3.70 W. Cao, R. Fathallah, L. Castex: Correlation of almen arc height with residual stresses in shot peening process, Materials Science and Technology 11 (1995), pp. 967–973.

3.71 R. Fathallah, G. Inglebert, L. Castex: Modelling of shot peening residual stresses and plastic deformation induced in metallic parts, In: J. Champaigne (ed.), Proc. Int. Conf. Shot Peening 6, San Francisco, 1996, pp. 464–473.

3.72 R. Fathallah, G. Inglebert, L. Castex: Prediction of plastic deformation and residual stresses induced in metallic parts by shot peening, Materials Science and Technology 14 (1998), pp. 631–639.

3.73 W. Cao, R. Fathallah, J. Barralis, L. Castex: Residual stresses in shot peened metal components, Shotpeen: an interactive prediction sofware, In: M. R. James (ed.), Proc. Int. Conf. Residual Stresses 4, Baltimore, 1994, pp. 589–597.

3.74 R. Fathallah, S. Slim, W. Cao, G. Inglebert, L. Castex: Prediction of residual stresses due to shot peening, In: J. Flavenvot, B. Scholtes (eds.), Proc. French-German Shot Peening Meeting, Strasbourg, 1994.

3.75 Y. LeGuernic, J. S. Eckersley: Peenstress software selects shot peening parameters, In: J. Champaigne (ed.), Proc. Int. Conf. Shot Peening 6, San Francisco, 1996, pp. 481–492.

3.76 N. Hamdane: Modélisation des contraintes résiduelles introduites par martelage dans les parois d'un tube de générateur, Thèse PhD, Ensam Paris, 1990.

3.77 C. Hardy, C. N. Baronet, G. V. Tordion: The elasto-plastic indentation of a half-space by a rigid sphere, International Journal of Numerical Methods in Engineering 3 (1971), pp. 451–462.

3.78 G. Z. Voyiadjis, E. N. Buckner: Indentation of a half-space with a rigid indentor, International Journal of Numerical Methods in Engineering 19 (1983), pp. 1555–1578.

3.79 P. S. Follansbee G. B. Sinclair: Quasi-static normal indentation of an elasto-plastic half-space by a rigid sphere-I, International Journal of Solids and Strucutres 20 (1984), pp. 81–91.

3.80 G. B. Sinclair, P. S. Follansbee, K. L. Johnson: Quasi-static normal indentation of an elasto-plastic half-space by a rigid sphere-II: Results, International Journal of Solids and Strucutres 21 (1985), pp. 865–888.

3.81 E. R. Kral, K. Komvopoulos, D. B. Bogy: Elastic-plastic finite element analysis of repeated indentation of a half-space by a rigid sphere, Journal of Applied Mechanics 60 (1993), pp. 829–841.

3.82 D. S. Kyriacou: Shot-peening mechanics, a theoretical study, In: J. Champaigne (ed.), Proc. Int. Conf. Shot Peening 6, San Francisco, 1996, pp. 505–516.

3.83 Y. F. Al-Obaid: Three-dimensional dynamic finite element analysis for shot peening mechanics, Computers & Structures 36 (1990) 4, pp. 681–689.

3.84 Y. F. Al-Obaid: The automated three-dimensional analysis of steel plate to shot peening mechanics, In: K. Iida (ed.), Proc. Int. Conf. Shot Peening 4, Japan Society of Precision Engineering, Tokyo, 1990, pp. 595–606.

3.85 Y. F. Al-Obaid: The automated simulation of dynamic non-linearity to shot-peening mechanics, Computers & Structures 40 (1991) 6, pp. 1451–1460.

3.86 Y. F. Al-Obaid: Multiple shot analysis in shot-peening using finite elements, Computer Methods and Experimental Measurements for Surface Treatment Effects (1993), pp. 155–168.

3.87 K. Iida: Dent and Affected Layer Produced by Shot Peening, In: H. O. Fuchs (ed.), Proc. Int. Conf. Shot Peening 2, American Shot Peening Society, Paramus, 1984, pp. 283–292.

3.88 K. Han, D. Peric, A. J. L. Crook, D. R. J. Owen: A combined finite/discrete element simulation of shot peening processes. Part I: studies on 2D interaction laws, Engineering Computations 17 (2000) 5, pp. 593–619.

3.89 K. Han, D. Peric, D. R. J. Owen, J. Yu: A combined finite/discrete element simulation of shot peening processes. Part II: 3D interaction laws, Engineering Computations 17 (2000) 6, pp. 680–702.

3.90 F. Wüstefeld: Modelle zur quantitativen Abschätzung der Strahlmittelwirkung beim Kugelstrahlumformen, Dissertation, RWTH Aachen, 1993.

3.91 K. Mori, K. Osakada, N. Matsuoka: Finite element analysis of peening process with plastically deforming shot, Journal of Materials Processing Technology 45 (1994), pp. 607–612.

3.92 K. Mori, K. Osakada, N. Matsuoka: Rigid-plastic finite element simulation of peening process with plastically deforming workpiece, JSME Inernational Journal 39 (1996) 3, pp. 306–312.

3.93 J. Edberg, L. Lindgren, K. Mori: Shot peening simulated by two different finite element formulations, In: Shen & Dawson (eds.), Proc. Int. Conf. Numerical Methods in Industrial Forming Processes 5, Ithaca, New York, USA, 1995, pp. 425–430.

3.94 L. V. Grasty, C. Andrew: Shot peen forming sheet metal: finite element prediction of deformed shape, Journal of Engineering Manufacture, Part B 210 (1996), pp. 361–366.

3.95 A. Levers, A. Prior: Finite Element Simulation of Shot Peening, The Shot Peener 9 (1985) 3, pp. 14–16.

3.96 S. A. Meguid, G. Shagal, J. C. Stranart, J. J. Daly: Three-dimensional dynamic finite element analysis of shot-peening induced residual stresses, Finite Elements in Analysis and Design 31 (1999), pp. 179–191.

3.97 S. A. Meguid, G. Shagal, J. C. Stranart: Finite element modelling of shot-peening residual stresses, Journal of Materials Processing Technology 92–93 (1999), pp. 401–404.

3.98 S. T. S. Al-Hassani: Numerical simulation of multiple shot impact, In: A. Nakonieczny (ed.), Proc. Int. Conf. Shot Peening 7, Warschau, 1999, pp. 217–227.

3.99 M. Guagliano, L. Vergani, M. Bandini, F. Gili: An approach to relate the shot peening parameters to the induced residual stresses, In: A. Nakonieczny (ed.), Proc. Int. Conf. Shot Peening 7, Warschau, 1999, pp. 274–282.

3.100 C. Droste gen. Helling: Numerische Untersuchungen zum Kugelstrahlen, Dissertation, Universität-Gesamthochschule Siegen, 2000.

3.101 J. Schwarzer, V. Schulze, O. Vöhringer: Finite element simulation of shot peening – A Method to evaluate the influence of peening parameters on surface characteristics, In: L. Wagner (ed.), Shot Peening, Wiley-VCh, Weinheim, 2003, pp. 507–515.

3.102 B. Scholtes: Eigenspannungen in mechanisch randschichtverformten Werkstoffzuständen, Ursachen-Ermittlung-Bewertung, DGM-Informationsgesellschaft, Oberursel, 1990.

3.103 R. Kopp, H.-W. Ball: Recent Developments in Shot Peen Forming, In: H. Wohlfahrt, R. Kopp, O. Vöhringer (eds.), Shot Peening, Proc. Int. Conf. Shot Peening 3, DGM-Informationsgesellschaft, Oberursel, 1987, pp. 297–308.

3.104 K.-P. Hornauer, W. Köhler: Development of the Peen Forming Process for Spherical Shaped Components, In: K. Iida (ed.), Proc. Int. Conf. Shot Peening 4, Japan Society of Precision Engineering, Tokyo, 1990, pp. 585–594.

3.105 F. Wüstefeld, S. Kittel, R. Kopp, W. Linnemann, G. Werner, W. Dürr, W. Köhler: 1/4 Tank bulkhead segment for the European Ariane 5, In: J. Champaigne (ed.), Proc. Int. Conf. Shot Peening 6, San Francisco, 1996, pp. 87–94.

3.106 H. Holzapfel, A. Wick, V. Schulze, O. Vöhringer: Zum Einfluss der Kugelstrahlparameter auf die Randschichteigenschaften vom 42 CrMo 4 in verschiedenen Wärmebehandlungszuständen, Härterei Technische Mitteilung 53 (1998), pp. 155–163.

3.107 W. Zinn, D. Deiseroth, B. Scholtes: Randschichtzustand der Mg-Basislegierung AZ31 nach mechanischer Oberflächenbehandlung, Materialwissenschaft und Werkstofftechnik 29(1998), pp. 163–169.

3.108 J. M. Duchazeaubeneix: Stressonic shot peening (Ultrasonic Process), In: A. Nakonieczny (ed.), Proc. Int. Conf. Shot Peening 7, Warschau, 1999, pp. 444–452.

3.109 H. Wohlfahrt: Residual stress and stress relaxation, In: E. Kula, V. Weiss (eds.), Sagamore Army Mat. Res. Conference, 1982, pp. 71.

3.110 B. Hoffmann: Einfluss des Anlassens auf die Mikrostruktur und das Verformungverhalten gehärteter Stähle, Dissertation, Universität Karlsruhe (TH), 1996.

3.111 W. Köhler, B. Scholtes: Einfluss einer Kugelstrahlbehandlung auf die Eigenspannungen einer geschweißten AlZnMg-Legierung, In: E. Macherauch, V. Hauk (eds.), Eigenspannungen, DGM-Informationsgesellschaft, Oberursel, 1983, pp. 331–341.

3.112 W. Zinn: Eigenspannungen bei unterschiedlichen Strahlmittelgeschwindigkeiten, DGM-Fachausschuss Mechanische Oberflächenbehandlung, Cottbus, 2001.

3.113 R. Schreiber, H. Wohlfahrt, E. Macherauch: Der Einfluss des Kugelstrahlens auf das Biegewechselverhalten von blindgehärtetem 16 MnCr 5, Archiv für das Eisenhüttenwesen 48 (1977), pp. 653–657.

3.114 R. Schreiber, H. Wohlfahrt, E. Macherauch: Verbesserung des Biegewechselverhaltens eines kugelgestrahlten 16MnCr5 durch Oberflächennachbehandlung, Archiv für das Eisenhüttenwesen 49 (1978), pp. 207–210.

3.115 R. Schreiber, H. Wohlfahrt, E. Macherauch: Der Einfluss des Kugelstrahlens auf das Biegewechselverhalten von einsatzgehärtetem 16MnCr5, Archiv für das Eisenhüttenwesen 49(1978), pp. 37–41.

3.116 R. Schreiber, H. Wohlfahrt, E. Macherauch: Zum Einfluss von Kugelstrahlbehandlungen auf das Biegewechselverhalten von einsatzgehärtetem 16MnCr5 im angelassenen Zustand, Archiv für das Eisenhüttenwesen 49 (1978) 5, pp. 265–269.

3.117 R. Schreiber, H. Wohlfahrt, E. Macherauch: Der Einfluss des Kugelstrahlens auf das Biegewechsel-verhalten von normalgeglühtem 16MnCr5, Archiv für das Eisenhütten-wesen 48 (1977), pp. 649–652.

3.118 A. Ebenau, O. Vöhringer, E. Macherauch: Influence of the Shot Peening Angle on the Condition of Near Surface Layers in Materials, In: H. Wohlfahrt, R. Kopp, O. Vöhringer (eds.), Shot Peening, Proc. Int. Conf. on Shot Peening 3, DGM-Informa-tionsgesellschaft, Oberursel, 1987, pp. 253–260.

3.119 R. Clausen: Ermittlung von Einfluss-größen beim Kugelstrahlen durch Ein-zelkornversuche, In: A. Niku-Lari (ed.), Proc. Int. Conf. Shot Peening 1, Pergamon, Paris, 1981, pp. 279–288.

3.120 J. E. Hoffmann: Der Einfluss ferti-gungsbedingter Eigenspannungen auf das Biegewechselverhalten von glatten und gekerbten Proben aus Ck45 in verschiedenen Werkstoffzuständen, Dissertation, Universität Karlsruhe (TH), 1984.

3.121 J. Bergström, T. Ericsson: X-Ray Microstructure and Residual Stress Analysis of Shot Peened Surface Layers during Fatigue Loading, In: H. Wohlfahrt, R. Kopp, O. Vöhringer (eds.), Shot Peening, Proc. Int. Conf. Shot Peening 3, DGM-Informations-gesellschaft, Oberursel, 1987, pp. 221–230.

3.122 D. Zhang, K. Xu, X. Wang, N. Hu: Residual Stress Concentration and its Effect on Notch Fatigue Strength, In: H. Wohlfahrt, R. Kopp, O. Vöhringer (eds.), Shot Peening, Proc. Int. Conf. on Shot Peening 3, DGM-Informa-tionsgesellschaft, Oberursel, 1987, pp. 625–630.

3.123 A. Bignonnet: Fatigue Strength of Shot Peened Grade 35 NCD 16 Steel, Variation of Residual Stresses intro-duced by Shot Peening according to Type of Loading, In: H. Wohlfahrt, R. Kopp, O. Vöhringer (eds.), Shot Peening, Proc. Int. Conf. Shot Peen-ing 3, DGM Informationsgesellschaft, Oberursel, 1987, pp. 659–666.

3.124 K. Xu: unveröffentlicht, Universität Karlsruhe (TH), 1988.

3.125 H. Schilling, B. Scholtes: unveröffent-licht, Universität Karlsruhe (TH), 1984.

3.126 V. Hauk, B. Krüger: Residual stresses in Cu-FE-TiC sintered materials after grinding and after shot peening, Zeit-schrift für Metallkunde 93 (2002) 1, pp. 76–80.

3.127 O. Vöhringer: Changes in the state of the material by shot peening, In: H. Wohlfahrt, R. Kopp, O. Vöhringer (eds.), Shot Peening, Proc. Int. Conf. Shot Peening 3, DGM-Informations-gesellschaft, Oberursel, 1987, pp. 185–204.

3.128 D. E. Mikkola, J. B. Cohen: Examples of applications of line broadening, In: J. B. Cohen, J. E. Hilliard (eds.), Local atomic arrangements studied by X-ray diffraction, Met. Soc. Conf. 36, New York, 1965, pp. 289–333.

3.129 J. L. Snoek: Effect of small quantities of carbon and nitrogen on the elastic and plastic properties of iron, Physica 8 (1941) 7, pp. 711–733.

3.130 F. Burgahn, O. Vöhringer, E. Macherauch: Mikroeigenspan-nungszustände kugelgestrahlter Randschichten von 42 CrMo 4, Zeit-schrift für Metallkunde 84(1994), pp. 224–229.

3.131 F. Burgahn, O. Vöhringer, E. Macherauch: Microstructural inves-tigations of the shot peened steel 42 CrMo 4 in different heat treatment conditions by the aid of a X-ray profile analysis, In: K. Iida (ed.), Proc. Int. Conf. Shot Peening 4, Japan Society of Precision Engineering, Tokyo, 1990, pp. 199–207.

3.132 W. P. Evans, R. W. Buenneke: In: Trans. Met. Soc. AIME 227, 1963, pp. 447–451.

3.133 W. P. Evans, R. E. Rickleffs, R. E. Millian: In: J. B. Cohen, J. E. Hilliard (eds.), Met. Soc. Conf. 36, New York, 1965, pp. 351–377.

3.134 B. Hoffmann, O. Vöhringer, E. Macherauch: Effect of compressive plastic deformation on mean lattice strains, dislocation densities and flow stresses of martensitically hardened steels, Materials Science and Engineering A319–321 (2001), pp. 299–303.

3.135 H. H. Racké, T. Fett: Die Bestimmung biaxialer Eigenspannungen in Kunststoff-Oberflächen durch Knoop-Härtemessung, Materialprüfung 13 (1971), pp. 37–42.

3.136 U. Martin, I. Altenberger, K. Kremmer, B. Scholtes, H. Oettel: Characterization of the microsturcture-depth profile of shot peened steels, Praktische Metallographie 35 (1998), pp. 327–334.

3.137 F. Y. Hu, H.-J. Klaar: Querschnittspräparation für die Untersuchung von dünnen Schichten, Grenzflächen, Pulvern und Fasern im Transmissionselektronenmikroskop, Praktische Metallographie 32 (1995), pp. 603–615.

3.138 H. Lu, B. Scholtes, E. Macherauch: Randschichtzustände von normalisiertem und vergütetem 42 CrMo 4 nach konsekutiven Kugelstrahlen- und Festwalzbehandlungen, Materialwissenschaft und Werkstofftechnik 23 (1992), pp. 388–394.

3.139 B. Scholtes: unveröffentlicht, Universität Gesamthochschule Kassel, 1993.

3.140 B. Scholtes, O. Vöhringer: Ursachen, Ermittlung und Bewertung von Randschichtveränderungen durch Kugelstrahlen, Materialwissenschaft und Werkstofftechnik 24 (1993), pp. 421–432.

3.141 U. Martin, I. Altenberger, B. Scholtes, K. Kremmer, H. Oettel: Cyclic deformation and near surface microstructures of normalized shot peened steel SAE 1045, Materials Science and Engineering A246 (1998), pp. 69–80.

3.142 I. Altenberger: Mikrostrukturelle Untersuchungen mechanisch randschichtverfestigter Bereiche schwingend beanspruchter metallischer Werkstoffe, Dissertation, Universität Gesamthochschule Kassel, 2000.

3.143 I. Altenberger, U. Martin, B. Scholtes, H. Oettel: Near surface microstructures in mechanically surface treated materials and their consequences on cyclic deformation behaviour, In: A. Nakonieczny (ed.), Proc. Int. Conf. Shot Peening 7, Warschau, 1999, pp. 79–87.

3.144 H. Chardin, M. Jeandin: Ultrasonic shot-peening of a low-carbon porous sintered steel, In: Jian Lu (ed.), Mat-Tec 96: improvement of materials, Paris, 1996, pp. 195–204.

3.145 S. Pakrasi, J. Betzold: Effect of Shot Peening on Properties of Machined Surfaces, In: A. Niku-Lari (ed.), Proc. Int. Conf. Shot Peening 1, Pergamon, Paris, 1981, pp. 193–200.

3.146 D. Kirk: Residual Stresses and Retained Austenite in Shot Peened Steels, In: A. Niku-Lari (ed.), Proc. Int. Conf. Shot Peening 1, Pergamon, Paris, 1981, pp. 271–278.

3.147 M. Kikuchi, H. Ueda, K. Hanai, T. Naito: The Improvement of Fatigue Durability of Carburized Steels with Surface Structure Anomalies by Shot Peening, In: H. O. Fuchs (ed.), Proc. Int. Conf. Shot Peening 2, American Shot Peening Society, Paramus, 1984, pp. 208–214.

3.148 S. Hisamatsu, T. Kanazawa: Size Effect on Fatigue Strength of Shot Peened Carburized Steel, In: H. Wohlfahrt, R. Kopp, O. Vöhringer (eds.), Shot Peening, Proc. Int. Conf. Shot Peening 3, DGM-Informationsgesellschaft, Oberursel, 1987, pp. 517–524.

3.149 T. Hirsch, H. Wohlfahrt, E. Macherauch: Fatigue Strength of Case Hardened and Shot Peened Gears, In: H. Wohlfahrt, R. Kopp, O. Vöhringer (eds.), Shot Peening, Proc. Int. Conf. Shot Peening 3, DGM-Informationsgesellschaft, Oberursel, 1987, pp. 547–560.

3.150 D. Wehrle, A. Ebenau: unveröffentlicht, Universität Karlsruhe (TH), 1987.

3.151 H. Wohlfahrt: Festigkeitsstrahlen, In: D. Aurich (ed.), Bauteile 86 – Die Bauteiloberfläche, 1986, pp. 23–35.

3.152 R. Herzog, W. Zinn, B. Scholtes, H. Wohlfahrt: The significance of Almen intensity for the generation of shot peening residual stresses, In: J. Champaigne (ed.), Proc. Int. Conf. Shot Peening 6, San Francisco, 1996, pp. 270–281.

3.153 D. Kirk, N. J. Payne: Transformation induced in austenitic stainless steels by shot peening, In: A. Nakonieczny (ed.), Proc. Int. Conf. Shot Peening 7, Warschau, 1999, pp. 15–22.

3.154 A. Ebenau: Verhalten von kugel-gestrahltem 42 CrMo 4 im normali-sierten und vergüteten Zustand unter einachsig homogener und inhomoge-ner Wechselbeanspruchung, Disserta-tion, Universität Karlsruhe (TH), 1989.

3.155 G. Liu, J. Lu, K. Lu: Surface nanocrys-tallization of 316L stainless steel induced by ultrasonic shot peening, Materials Science and Engineering A286 (2000), pp. 91–95.

3.156 M. K. Tufft: Instrumented single parti-cle impact tests using production shot: the role of velocity, incidence angle and shot size on impact response, induced plastic strain and life behav-ior, In: J. Champaigne (ed.), Proc. Int. Conf. Shot Peening 6, San Francisco, 1996.

3.157 M. K. Tufft: Development of a fracture mechanics/Threshold behavior model to assess the effects of competing mechanisms induced by shot peening on cyclic life of a nickel-base superal-loy, René 88DT, PhD-Thesis, Univer-sity of Dayton, Ohio, 1997.

3.158 W. Renzhi, Z. Xuecong, S. Deyu, Y. Yuanfa: Shot-peening of superalloys and its fatigue properties at elevated temperature, In: A. Niku-Lari (ed.), Proc. Int. Conf. Shot Peening 1, Per-gamon, Paris, 1981, pp. 395–403.

3.159 G. Maurer, H. Neff, B. Scholtes, E. Macherauch: Texture and lattice deformation pole figures of machined surfaces, Textures and Microstructures 8 (1988) 9, pp. 639–678.

3.160 R. Zeller: Kugelstrahlen unter Zugvor-spannung, Materialprüfung 35 (1993), pp. 218–221.

3.161 R. Zeller, R. Prümmer: Erhöhung der Wechselfestigkeit von Teilen aus Stahl durch gezieltes Einbringen von Ober-flächen-Druckeigenspannungen, Bericht W 11/86, 1986.

3.162 E. Müller: Die Ausbildung von Eigen-spannungen an Torsionsproben beim Spannungsstrahlen, Materialwis-senschaft und Werkstofftechnik 27 (1996), pp. 354–358.

3.163 A. Wick: Randschichtzustand und Schwingfestigkeit von 42CrMo4 nach Kugelstrahlen unter Vorspannung und bei erhöhter Temperatur, Disser-tation, Universität Karlsruhe (TH), 1999.

3.164 R. Kopp, K.-P. Hornauer, H.-W. Ball: Kugelstrahlumformen, Neuere tech-nologische und theoretische Entwick-lungen, In: H. O. Fuchs (ed.), Proc. Int. Conf. Shot Peening 2, American Shot Peening Society, Paramus, 1984, pp. 6–14.

3.165 J. C. Xu, D. Zhang, B. Shen: The Fati-gue Strength and Fracture Morphol-ogy of Leaf Spring Steel After Pre-stressed Shot Peening, In: A. Niku-Lari (ed.), Proc. Int. Conf. Shot Peening 1, Pergamon, Paris, 1981, pp. 367–374.

3.166 J. Krobb, H. Weiß: Stress peening of ultra high strength steels, In: T. S. Sudarshan, M. Jeandin, K. A. Khor (ed.), Surface modification technolo-gies XI, London, 1997, pp. 423–429.

3.167 L. Bonus: Auswirkungen des Span-nungsstrahlens auf die Eigenschaften von hoch vergüteten Bremsspeicher- und Torsionsfedern, Dissertation, RWTH Aachen, 1994.

3.168 V. Schulze: Warm- und Spannungs-strahlen – Wege zur Erzeugung stabi-lisierter und erhöhter Druckeigen-spannungen, In: Deutscher Verband für Materialforschung und -prüfung e.V., DVM-Tag "Federn im Fahrzeug-bau", Berlin, 2002, pp. 45–54.

3.169 D. V. Wilson, B. Russell: The Contri-bution of Atmosphere Locking to the Strain-Ageing of Low Carbon Steels, Acta Metallurgica 8 (1960), pp. 36–45.

3.170 J. D. Baird: The effect of strain-ageing due to interstitial solutes of the me-chanical properties of metals, Metal-lurgical Review 16 (1971), pp. 1–18.

3.171 J. D. Baird: Strain Aging of steel – a critical review – Part I: Practical Aspects, Journal of Iron and Steel 5 (1963), pp. 186–192.

3.172 J. D. Baird: Strain Aging of steel – a critical review – Part I: Practical Aspects (continued), Journal of Iron and Steel 6 (1963), pp. 326–334.

3.173 J. D. Baird: Strain Aging of steel – a critical review – Part II: The Theory of Strain Aging, Journal of Iron and Steel 7 (1963), pp. 368–374.

3.174 J. D. Baird: Strain Aging of steel – a critical review – Part II: The Theory of Strain Aging (continued), Journal of Iron and Steel 8 (1963), pp. 400–405.

3.175 D. V. Wilson, B. Russell: Stress Induced Ordering and Strain-Ageing in Low Carbon Steels, Acta Metallurgica 7 (1959), pp. 628–631.

3.176 Y. Nakada, A. S. Keh: Kinetics of Snoek ordering and Cottrell atmosphere formation in Fe-N single crystals, Acta Metallurgica 15 (1967), pp. 879–883.

3.177 A. H. Cottrell B. A. Bilby: Proc. Phys. Soc. 62A (1949), pp. 49–62.

3.178 A. H. Cottrell B. A. Bilby: Dislocation Theory on Yielding and Strain Aging of Iron, Proc. Phys. Soc. 62 (1949) 1, pp. 49–62.

3.179 S. Harper: Phys. Rev. 83 (1951), pp. 709–712.

3.180 W. C. Leslie: The physical metallurgy of steels, New York, 1981.

3.181 R. Pradhan: Metallurgy of of vacuum-degassed steel products, TMS, Warrendale, 1990, pp. 309–325.

3.182 A. V. Snick, K. Lips, S. Vandeputte, B. C. DeCooman, J. Dilewijns: Modern LC and ULC sheet for cold forming: Processing and properties, In: W. Bleck (ed.), Verlag Mainz, Wissenschaftsverlag, Aachen, 1998, pp. 413–424.

3.183 P. Messien, V. Leroy: Scavening additions of boron in low C-low Al steels, Steel Research 60 (1989), pp. 320–328.

3.184 S. Hanai, N. Takemoto, Y. Tokunaga, Y. Mizuyama: Trans. Iron Steel Inst. Jpn. 24 (1984), pp. 17–23.

3.185 J. Z. Zhao, A. K. De, B. C. DeCooman: Formation of the Cottrell atmosphere during strain aging of bake-hardenable steels, Metallurgical And Materials Transaction 32A (2001), pp. 417–423.

3.186 J. D. Baird, C. R. Mackenzie: Journal of the Iron and Steel Institute 202 (1964), pp. 427.

3.187 W. C. Leslie, A. S. Keh: In: A.I.M.E. Met.Soc.Conf., 1965, pp. 337.

3.188 K. R. Carson, J. Weertman: Transaction of Met. Soc. A.I.M.E 242 (1968), pp. 1313.

3.189 A. S. Keh, Y. Nakada, W. C. Leslie: Dynamic strain ageing in iron and steel, In: A. R. Rosenfield et al. (eds.), Dislocation Dynamics, McGraw-Hill, New York, 1968, pp. 381.

3.190 D. J. Dingley, D. McLean: Components of the Flow Stress of Iron, Acta Metallurgica 15 (1967), pp. 885–901.

3.191 B. J. Brindley, J. T. Barnby: Dynamic Strain Ageing in Mild Steel, Acta Metallurgica 14 (1966), pp. 1765–1780.

3.192 J. D. Baird, A. Jamieson: Journal of the Iron and Steel Institute 204 (1966), pp. 793.

3.193 D. J. Lloyd: The deformation of commercial Aluminum-Magnesium Alloys, Metallurgical Transactions 11A (1980), pp. 1287–1294.

3.194 J. Glen: 128, ASTM, 1952, pp. 184.

3.195 J. Glen: Journal of the Iron and Steel Institute 186 (1957), pp. 21.

3.196 M. Schilling-Praetzel: Einfluss der Werkstücktemperatur beim Kugelstrahlen auf die Schwingfestigkeit von Drehstabfedern, Dissertation, RWTH Aachen, 1995.

3.197 M. Schilling-Praetzel, F. Hegemann, P. Gome, G. Gottstein: Influence of Temperature of Shot Peening on Fatigue Life, In: D. Kirk (ed.), Proc. Int. Conf. Shot Peening 5, Oxford, 1993, pp. 227–238.

3.198 A. Tange, H. Koyama, H. Tsuji: Study on warm shot peening for suspension coil spring, SAE Technical Paper Series 1999-01-0415, International Congress and Exposition, Detroit, 1999, pp. 1–5.

3.199 A. Tange, H. Koyama, H. Tsuji: The effect of warm shot peening on the fatigue strength of springs, Springs (2000), pp. 58–63.

3.200 Y. Harada, K. Mori, Y. Fukuoka, S. Maki: Effect of processing temperature on warm shot peening of spring steel, Journal of Japan Society for Technology of Plasticity 40 (2000), pp. 260–264.

3.201 A. Wick, V. Schulze, O. Vöhringer: Effects of warm peening on fatigue life and relaxation behaviour of residual stresses of AISI 4140, Materials Science and Engineering A293 (2000), pp. 191–197.

3.202 A. Wick, V. Schulze, O. Vöhringer: Effects of stress- and/or warm peening of AISI 4140 on fatigue life, Steel Research 71 (2000) 8, pp. 316–321.

3.203 K.-H. Kloos, E. Broszeit, B. Fuchsbauer, F. Schmidt: Optimierung von Schwingfestigkeitseigenschaften beim Oberflächendrücken gekerbter Umlaufbiegeproben unter Berücksichtigung der Probengröße, Zeitschrift für Werkstofftechnik 12 (1981), pp. 359–365.

3.204 E. Broszeit: Grundlagen der Schwingfestigkeitssteigerung durch Fest- und Glattwalzen, Zeitschrift für Werkstofftechnik 15 (1984), pp. 416–420.

3.205 A. Liebisch: Bestimmung der Festwalzeigenspannungen in einsatzgehärteten Zahnrädern mit der Finite-Elemente-Methode, Verein Deutscher Ingenieure Zeitschrift 133 (1991) 12, pp. 73–79.

3.206 U. Jung: FEM-Simulation und experimentelle Optimierung des Festwalzens bauteilähnlicher Proben unterschiedlicher Größen, Dissertation, TH Darmstadt, 1996.

3.207 U. Jung, B. Kaiser, K.-H. Kloos, C. Berger: Festwalz-Eigenspannungen per Computer-Simulation bestimmen, Materialwissenschaft und Werkstofftechnik 27 (1996), pp. 159–164.

3.208 U. Jung, B. Kaiser, C. Berger: Computer-Simulation und experimentelle Optimierung des Festwalzens, In: H. Zenner (ed.), DVM-Bericht 122 "Leichtbau durch innovative Fertigungsverfahren", 1996, pp. 119–125.

3.209 K. Lange: Umformtechnik: Handbuch für Industrie und Wissenschaft. Band 1: Grundlagen, Springer-Verlag, Berlin, 1984.

3.210 C. Achmus, J. Betzold, H. Wohlfahrt: Messung von Festwalzeigenspannungsverteilungen an Bauteilen, Materialwissenschaft und Werkstofftechnik 28 (1997), pp. 153–157.

3.211 C. Achmus: Messung und Berechnung des Randschichtzustands komplexer Bauteile nach dem Festwalzen, Dissertation, TU Braunschweig, 1998.

3.212 C. Achmus, U. Jung, B. Kaiser, H. Wohlfahrt: FEM-Simulation des Festwalzens von Kurbelwellen, Konstruktion 10 (1997), pp. 31–34.

3.213 C. Achmus, W. Reimers, H. Wohlfahrt: Eigenspannungen in festgewalzten Bauteilen, Materialprüfung 40 (1998), pp. 88–91.

3.214 U. Jung, R. Schaal, C. Berger, H.-W. Reinig, H. Traiser: Berechnung der Schwingfestigkeit festgewalzter Kurbelwellen, Materialwissenschaft und Werkstofftechnik 29 (1998) 10, pp. 569–572.

3.215 K.-H. Kloos J. Adelmann: Schwingfestigkeitssteigerung durch Festwalzen, Materialwissenschaft und Werkstofftechnik 19 (1988), pp. 15–23.

3.216 S. Gruber, G. Holzheimer, H. Naundorf: Glatt- und Festwalzen an PKW Fahrgestell- und Antriebsbauteilen, Zeitschrift für Werkstofftechnik 15 (1984), pp. 41–45.

3.217 E. Broszeit, H. Steindorf, V. Hauk: State of residual stresses and texture after surface rolling of sinter materials, In: V. Hauk, H. Hourgardy, E. Macherauch (eds.), Resiudal stresses, measurement, calculation, eavluation, DGM-Informationsgesellschaft Verlag, Oberursel, 1991, pp. 245–252.

3.218 L. Berthe, R. Fabbro, P. Peyre, E. Bartnicki: Laser shock processing of materials: experimental study of breakdown plasma effects at the surface of confining water, Laboratoire pour l'Application des Laser de Puissance, 2002, .

3.219 K. Eisner: Prozesstechnologische Grundlagen zur Schockverfestigung von metallischen Werkstoffen mit einem kommerziellen Excimerlaser, Dissertation, Universität Erlangen-Nürnberg, 1998.

3.220 A. N. Pirri: Theory for momentum transfer to a surface with a high-power laser, Physical Fluids 16 (1973) 9, pp. 1435.

3.221 R. Fabbrro et al.: Physical study of laser-produced plasma in confined geometry, Journal of Applied Physics 50 (1990) 3, pp. 775.

3.222 P. Ballard: Contraintes résiduelles induites par impact rapide. Application au choc-laser, Thèse PhD, Ecole Polytéchnique, 1991.

3.223 B. Schmidt: Einordnung des Schock-härtens mit Laserstrahlung in die Verfahren zur Erzeugung mechanisch bedingter Eigenspannungen in Randschichten von Festkörpern, Dissertation, RWTH Aachen, 1994.

3.224 G. T. Gray: High pressure shock compression of solids, Springer-Verlag, New York, 1993.

3.225 K. A. Johnson. K. P. Staudhammer: In: L. E. Murr et al. (eds.), Metallurgical applications of shock-wave and high-strain-rate phenomena, M. Dekker, New York, 1986, pp. 525.

3.226 M. A. Mogilevsky. L. A. Teplyakova: In: L. E. Murr et al. (eds.), Metallurgical applications of shock-wave and high-strain-rate phenomena, New York, 1986, pp. 419.

3.227 M. A. Meyers, O. Vöhringer, V. A. Lubarda: The Onset of Twinning in Metals: A Constitutive Description, Acta Metallurgica 49 (2001), pp. 4025–4039.

3.228 K. P. Staudhammer, L. E. Murr, S. S. Hecker: Nucleation and Evolution of Strain-Induced Martensitic (B.C.C.) Embryos and Substructure in Stainless Steel: A Transmission Electron Microscope Study, Acta Metallurgica 31 (1983) 2, pp. 267–274.

3.229 M. A. Meyers. L. E. Murr: Shock waves and high-strain-rate phenomena, Plenum Press, New York, 1980.

3.230 L. E. Murr et al.: Metallurgical applications of shock-wave and high-strain-rate phenomena, M. Dekker, New York, 1986.

3.231 L. E. Murr: In: L. E. Murr, M. A. Meyers (eds.), Shock waves and high-strain-rate phenomena, Plenum Press, New York, 1980, pp. 607.

3.232 M. A. Meyers: Dynamic Behavior of Materials, John Wiley & Sons, New York, 1994.

3.233 K. P. Staudhammer: Shock Wave effects and metallurgical parameters, In: C. Y. Chiem, H.-D. Kunze, L. W. Meyer (eds.), Impact loading and dynamic behaviour of materials, Vol.II, DGM Informationsgesellschaft, Oberursel, 1988, pp. 93–110.

3.234 J. R. Klepaczko: Plastic shearing at high and very high strain rates, Journal de Physique IV 4 (1994), pp. 35–40.

3.235 R. Dormeval: Adiabatic shear phenomena, In: C. Y. Chiem, H.-D. Kunze, L. W. Meyer (eds.), Impact loading and dynamic behaviour of materials, DGM Informationsgesellschaft, Oberursel, 1988, pp. 43–56.

3.236 R. F. Recht: Journal of Applied Mechanics 31(1964), pp. 189.

3.237 D. A. Shockey: Materials aspects of the adiabatic shear phenomenon, In: L. E. Murr, K. E. Staudhammer, M. A. Meyers (eds.), Metallurgical applications of shock-wave and high-strain-rate phenomena, M. Dekker, New York, 1986, pp. 633–656.

3.238 M. Stelly, R. Dormeval: Adiabatic shearing, In: L. E. Murr, K. E. Staudhammer, M. A. Meyers (eds.), Metallurgical applications of shock-wave and high-strain-rate phenomena, M. Dekker, New York, 1986, pp. 607–632.

3.239 S. LaRouche, D. E. Mikkola: Shock Hardening Behaviour of Cu-8.7Ge at very short pulse durations, Scripta Metallurgica 12 (1978), pp. 543–547.

3.240 K. P. Staudhammer, K. A. Johnson, B. W. Olinger: In: 41st Annual Meeting of the Elect. Microscopy Soc. of America, Elect. Microscopy Soc. of America, San Francisco, 1983, pp. 272.

3.241 C. Dubouchet: Traitements thermo-méchaniques de surfaces métalliques à l'aide de lasers CO2 continues et de laser impulsionnel, Thèse PhD, Université de Paris-Sud Centre d'Orsay, 1993.

3.242 V. Schulze, R. Menig, O. Vöhringer: Comparison of surface characteristics and thermal residual stress relaxation of laser peened and shot peened AISI 4140, In: L. Wagner (ed.), Shot Peening, Wiley-VCH, Weinheim, 2002, pp. 145–160.

3.243 K. Eisner, A. Lang, K. Schutte, H. W. Bergmann: Shock hardening as a novel technique for materials processing, In: G. Sayegh, M. R. Osborne (eds.), EUROPTO '96, Besancon, 1996, pp. 274.

3.244 S. D. Thompson, D. W. See, C. D. Lykins, P. G. Sampson: Laser shock peening vs shot peening a damage tolerance investigation, In: J. K. Gregory, H. J. Rack, D. Eylon (eds.), Proc. Symp. Surface Performance of Titanium, Cincinnati, 1997, pp. 239–251.

3.245 Z. Hong, Y. Chengye: Laser shock processing of 2024-T62 aluminium alloy, Materials Science and Engineering A257 (1998) 322–327.

3.246 T. R. Tucker, A. H. Clauer: Laser processing of materials, MCIC-83–48, Metals and Ceramics Information Center, 1983.

3.247 J. E. Masse, G. Barreau: Surface modification by laser induced shock waves, Surface Engineering 11 (1995) 2, pp. 131–132.

3.248 P. S. Prevey, D. J. Hornbach, P. W. Mason: Thermal residual stress relaxation and distortion in surface enhanced gas turbine engine components, In: D. L. Milam, D. A. Poteet, Jr. , G. D. Pfaffmann, V. Rudnev, A. Muehlbauer, W. B. Albert (eds.), 17th ASM Heat Treating Society Conference, ASM, Metals Park, 1997, pp. 3–12.

3.249 P. S. Prevey: The effect of cold work on the thermal stability of residual compression in surface enhanced IN718, In: K. Funatani, G. E. Totten (eds.), 20th ASM Heat Treating Society Conference, ASM, Metals Park, 2000, pp. 426–434.

3.250 M. Obata, Y. Sano, N. Mukai, M. Yoda, S. Shima, M. Kanno: Effect of laser peening on residual stress and stress corrosion cracking for typ. 304 stainless steel, In: A. Nakonieczny (ed.), Proc. Int. Conf. Shot Peening 7, Warschau, 1999, pp. 387–394.

3.251 P. Peyre, R. Fabbro, L. Berthe: Laser shock processing to improve mechanical properties of metallic materials, In: Proc. Int. Conf. Computer Methods and Experimental Measurements for Surface Treatment Effects 3, Oxford, 1997, pp. 267–276.

3.252 P. Peyre, R. Fabbro, P. Merrien, H. P. Lieurade: Laser shock processing of aluminium alloys. Application

to high cycle fatigue behaviour, Materials Science and Engineering A210 (1995), pp. 102–113.

3.253 P. Peyre, X. Scherpereel, L. Berthe, C. Carboni, R. Fabbro, G. Béranger, C. Lemaitre: Surface modifications induced in 316L steel by laser peening and shot-peening. Influence on pitting corrosion resistance, Materials Science and Engineering A280 (1999), pp. 294–302.

3.254 J. Kaspar, A. Luft: Beeinflussung der Mikrostruktur kubisch-raumzentrierter Metalle durch die Einwirkung kurzer Laserpulse hoher Intensität, In: J. Tobolski H.-W. Bergmann (eds.), Kurzzeitmetallurgie, Lehrstuhl für Metallische Werkstoffe der Universität, Bayreuth, 1999, pp. 173–184.

3.255 A. H. Clauer, J. H. Holbrook, B. P. Fairand: Shock waves and high-strain-rate phenomena in metals, In: M. A. Meyers L. E. Murr (eds.), Shock waves and high-strain-rate phenomena, Plenum Press, New York, 1980, pp. 675–702.

3.256 J. Kaspar, A. Luft: Untersuchung der Mikrostruktur kubisch-raumzentrierter Metalle nach Laserschockbehandlung, Praktische Metallographie 37 (2000) 4, pp. 181–193.

3.257 S. Reichstein: Mikrostruktur und Eigenspannungen in CuAl-Legierungen nach Randschichtverfestigung durch hochenergetische ns-Laserpulse, In: J. Tobolski H.-W. Bergmann (eds.), Kurzzeitmetallurgie, Lehrstuhl für Metallische Werkstoffe der Universität, Bayreuth, 1999.

3.258 M. A. Meyers, O. Vöhringer, Y. J. Chen: A constitutive description of the slip-twinning transition in metals, In: S. Ankem C. S. Pande (eds.), Advances in twinning, TMS-AIME, 1999, pp. 43–65.

3.259 M. Hallouin, F. Cottet, J. P. Romain, L. Marty: Microstructural transformations in iron induced by intense laser shock loading, In: C. Y. Chiem, H.-D. Kunze, L. W. Meyer (eds.), Impact loading and dynamic behaviour of materials, Vol.II, DGM Informationsgesellschaft, Oberursel, 1988, pp. 1051–1056.

3.260 B. Scholtes, E. Macherauch: Auswirkungen mechanischer Randschichtverformungen auf das Festigkeitsverhalten metallischer Werkstoffe, Zeitschrift für Metallkunde 77 (1986) 5, pp. 322–337.

3.261 J. Schwarzer, V. Schulze: unveröffentlicht, Universität Karlsruhe (TH), 2003.

3.262 R. Menig V. Schulze: unveröffentlicht, Universität Karlsruhe (TH), 2002.

3.263 H. Steindorf, E. Broszeit, K- H. Kloos: Festwalzen – Ein Verfahren zur Verbesserung der Eigenschaften von Sinterstählen, In: E. Broszeit, H. Steindorf (eds.), Mechanische Oberflächenbehandlung: Festwalzen, Kugelstrahlen, Sonderverfahren, DGM Informationsgesellschaft, Oberursel, 1989, pp. 55.

4
Changes of Surface States due to Thermal Loading

4.1
Process Models

4.1.1
Elementary Processes

Depending on temperature, thermal loading of mechanically surface-treated material states causes fundamentally different processes, which may be divided as follows. When material states that have received mechanical surface layer deformation are exposed to temperatures $T > 0.5\ T_M$ (T_M being melting- or solidus temperature) for several hours, re-crystallization occurs. This causes grain re-formation, which results in a removal of all macro residual stresses and of the vast majority of micro residual stresses. The reason why micro residual stresses are not removed completely is that lattice distortions, such as dislocations, remain. In multi-phase materials, thermally induced micro-residual stresses are inevitably created when the material cools down from processing temperature, due to the different thermal expansion coefficients of the individual phases. By contrast, thermally activated creep- and relaxation processes, in which temperature and time are largely exchangeable, are expected to occur at temperatures $T < 0.5\ T_M$. They are based on a diffusion-controlled dislocation movement, changing from volume diffusion to dislocation core diffusion as temperature drops. This is associated with a reduction of the processes' apparent activation enthalpy. Differences in the relaxation of micro- and macro residual stresses appear in this temperature range, as well. While the removal of macro residual stresses only requires dislocation movement, a significant relaxation of micro residual stresses requires a decrease of the number of lattice defects, resulting in a slower relaxation of the micro residual stresses compared to that of the macro residual stresses. Fig. 4.1 schematically shows which processes dominate the surface layer changes at a given temperature, time remaining constant.

In addition to the relaxation processes described, strain aging effects may occur in steels and other strain aging materials during thermal loading at annealing temperatures that are relatively low. These effects counteract the relaxation of micro- and macro residual stresses (comp. Sect. 3.3.1a). The residual stress-reducing

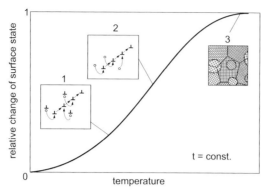

Fig. 4.1: Influence of annealing temperature on the processes controlling changes of the surface state (1 = dislocation core diffusion-controlled dislocation creep, 2 = volume diffusion-controlled dislocation creep. 3 = recrystallization)

dislocation movement is controlled by volume- or dislocation core diffusion and has an activation enthalpy $\Delta H_A \approx \Delta H_S$ or $\Delta H_A \approx \frac{1}{2} \cdot H_S$ where ΔH_S is the activation enthalpy for self-diffusion – about 2.8 eV in α-iron. At the same time, carbon clouds form around the dislocations. At an activation enthalpy of about 1 eV, this occurs much faster than diffusion-controlled dislocation movement. Thus, the movement of carbon-clouded slip dislocations is obstructed. If, after a certain incubation period, strain aging causes carbides to form in the vicinity of the dislocations, the result may be that further dislocation movement is impeded more effectively and ultimately, the dislocations are pinned. However, this may be undone by a coarsening of the carbides at a later stage. Fig. 4.2 is a schematic summary of the results to be expected, focusing on the temperature-dependency of the surface layer changes, while time remains constant.

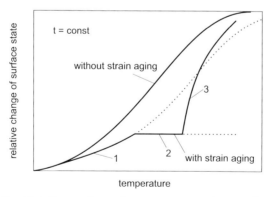

Fig. 4.2: Influence of annealing temperature on changes of the surface state for strain aging materials (schematically, 1 = decelerated movement of mobile dislocations bearing interstitial clouds, 2 = pinning of dislocations due to finest carbides, 3 = accelerated movement of dislocations due to coarsening of carbides)

By contrast, there are quenching-induced, over-saturated states in light metals in which the plastic deformation occurring during mechanical surface treatment changes the kinetics of precipitation hardening. Resulting time-temperature-precipitation diagrams, e.g. for titanium alloys, show that the curves describing the end of precipitation in the surface layer appear before those describing the start of precipitation in the core region. Accordingly, a selective surface hardening which occurs without a hardening of the core region is possible. The precipitation curves for aluminum alloys, on the other hand, are shifted forward to a lesser extent. These alloys permit only preferred surface hardening, the reason being that the precipitations which form in the surface layer are smaller and much more finely distributed than they are in the core region. Fig. 4.3 is a schematic representation of the changes which mechanical surface treatments cause in the time-temperature-precipitation diagrams for selective and preferred surface hardening.

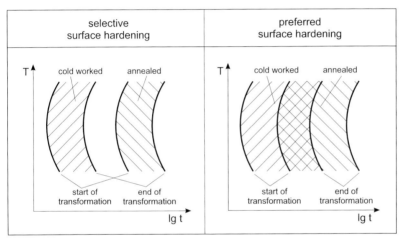

Fig. 4.3: Schematical time-temperature-precipitation diagrams for selective and preferred surface hardening

4.1.2
Quantitative Description of Processes

According to [4.1], the residual stress changes during thermal loading may be described using a Zener-Wert-Avrami equation

$$\frac{\sigma^{rs}(T, t)}{\sigma_0^{rs}} = \exp\left\{-\left[C\exp\left(-\frac{\Delta H_A}{kT}\right)t\right]^m\right\}. \tag{4.1}$$

$\sigma^{rs}(T,t)$ is the residual stress value which remains after annealing at temperature T lasting time t, σ_0^{rs} is the initial residual stress value, ΔH_A is the activation enthalpy of the rate-controlling process, m is an exponent, C is a velocity constant and k is the Boltzmann constant. Several methods may be used to determine the

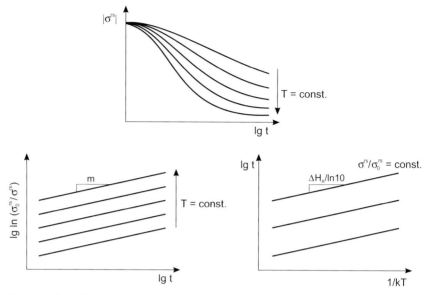

Fig. 4.4: Scheme for the common determination of the parameters of the Avrami approach for the description of thermal residual stress relaxation [4.3]

material's parameters ΔH_A, m and C on the basis of Equ. 4.1. The conventional method of analysis [4.2] is shown schematically in Fig. 4.4. The measurement results are first plotted as $\lg \ln(\sigma_0^{rs}/\sigma^{rs})$ versus $\lg t$. Since

$$\lg \ln\left(\frac{\sigma_0^{rs}}{\sigma^{rs}(T,t)}\right) = \left(m\lg C - \frac{m}{\ln 10}\frac{\Delta H_A}{kT}\right) + m\lg t \tag{4.2}$$

follows by converting Equ. 4.1, straight lines are expected for constant annealing temperatures in this graph. Their slopes are given by exponent m and should be independent of annealing temperature. Corresponding pairs of values of annealing time and -temperature may be determined for a constant relaxation rate $\sigma^{rs}/\sigma_0^{rs} = $ const. and applied in a diagram $\lg t$ versus $1/kT$. A conversion of Eq. 4.2 results in

$$\lg t = \left[\frac{1}{m}\lg \ln\left(\frac{\sigma_0^{rs}}{\sigma^{rs}(T,t)}\right) - \lg C\right] + \frac{\Delta H_A}{\ln 10}\frac{1}{kT}. \tag{4.3}$$

Therefore, straight lines should appear in this graph as well. In this case, their slopes should be $\Delta H_A/\ln 10$, permitting activation enthalpy ΔH_A to be determined. Finally, by inserting this into Eq. 4.2, velocity constant C can be determined. This method usually forces the use of interpolations and daring extrapolations of the measured values to arrive at high or low degrees of relaxation, which inevitably leads to uncertainties regarding the factors determined. For this reason, [4.3] introduced a method based on an algorithm for solving non-linear fitting problems [4.4]. It allows for an iterative calculation of the factors ΔH_A, m and C by way of the least squares method. Thus it is possible to determine these parameters without graphically linearizing the measurement values. It is no longer necessary to extrapolate very long or very short times from the measured values, as required by the method described before. Thus, all of them are weighted equally and are included in the analysis of the experiment. Replacing the ratio $\sigma^{rs}(T,t)/\sigma_o^{rs}$ in Eq. 4.1 by $<\varepsilon^2(T,t)>^{1/2}/<\varepsilon_o^2>^{1/2}$ or by $\Delta HWB(T,t)/\Delta HBW_o$ permits a representation of the relaxation of the mean strains and half widths which describe the micro residual stress state. In order to minimize the influence of any effects due to the technical devices used, the model does not employ the absolute values of the half widths, $HWB(T,t)$, but only the changes $\Delta HBW(T,t) = HBW(T,t) - HBW_N$ versus the value HBW_N measured in the material states with the lowest possible deformation. In steels, this value is determined in the normalized state.

According to [4.5], the changes in residual stress caused by supplying thermal energy can also be described by using the Norton approach, known from high temperature creep deformation, in the form of

$$\dot{\varepsilon}_p = A\left(\sigma_m^{rs}\right)^n \exp\left(\frac{-\Delta H_N}{kT}\right). \tag{4.4}$$

$\dot{\varepsilon}_p$ is the plastic strain rate, σ_m^{rs} is the mean amount of residual stress, ΔH_N is the activation enthalpy, n is an exponent and A is a velocity constant. For an analysis of measured values $\sigma^{rs}(T,t)$, these must first be converted, under the precondition that

$$\dot{\varepsilon}_t = \dot{\varepsilon}_e + \dot{\varepsilon}_p = 0 \tag{4.5}$$

is valid for macro residual stress relaxation, i.e. that $\dot{\varepsilon}_p$ equals negative elastic strain rate $\dot{\varepsilon}_e$.

The measured values $\sigma_i^{rs}(T,t_i)$ and $\sigma_j^{rs}(T,t_j)$, which remain after different annealing times, t_i and t_j ($t_j > t_i$) with T = const., together with the temperature-dependent Young's modulus $E(T)$ result in

$$\dot{\varepsilon}_{p,i} = \dot{\varepsilon}_{e,i} = -\frac{\sigma_j^{rs}(T,t_j) - \sigma_i^{rs}(T,t_i)}{E(T)(t_j - t_i)} \tag{4.6}$$

and the mean amount of residual stress

$$\sigma_{m,i}^{rs} = \frac{1}{2}\left|\sigma_j^{rs}(T,t_j) + \sigma_i^{rs}(T,t_i)\right|. \tag{4.7}$$

As they are strongly dependent on temperature and stress, factors ΔH_N, n and A are determined using the linearizations shown schematically in Fig. 4.5 as graphs lg $\dot{\varepsilon}_p$ versus lg σ_m^{rs} – and lg $\dot{\varepsilon}_p$ versus $1/kT$.

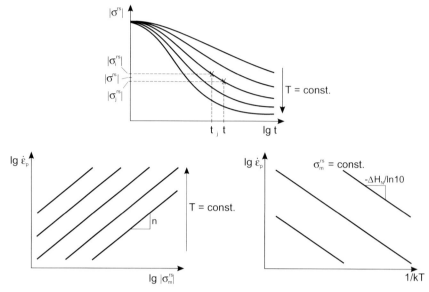

Fig. 4.5: Scheme for the determination of the parameters of the Norton approach for the description of thermal residual stress relaxation [4.3]

4.2
Experimental Results and their Descriptions

4.2.1
Influences on Shape and Topography

There are hardly any known examinations which focus on the influence of thermal loading on the shape and topography of mechanically surface-treated components or specimens. The most pronounced distortions caused by thermal loading are expected to occur in thin-walled components, due to their low stiffness and the great depth of the residual stresses in relation to specimen thickness. [4.6] reported on annealing experiments involving X22CrMoV12–1 steel strips the size of Almen test strips (76.2x19x1.295 mm³), which had been shot peened on one side and therefore showed a deflection of about 0.6 mm. During annealing, the strips were clamped between two flat plates, in order to eliminate the effect of deflection during annealing. Remaining deflections were measured after annealing treatments using different times and temperatures. The resulting correlations with annealing temperatures and -times are shown in Fig. 4.6. At all temperatures, the greatest decrease of the deflections occurs at the beginning of the annealing period. While deflections remain constant after about 5 hours at 150 °C and 300 °C, they continue to decrease at 450 °C and 600 °C. This ultimately results in deflections of less than 0.1 mm after 36 hours. [4.7] reported on dilatometric

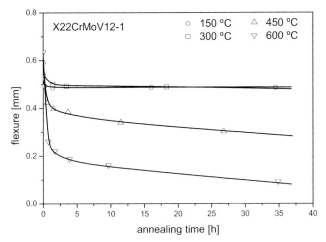

Fig. 4.6: Changes of deflection of single-sidedly shot peened plates of X22CrMoV12–1 for different annealing treatments [4.6]

measurements carried out on 10 x 10 x 20 mm^3 specimens of 38CrMo4 steel cut from shot-peened discs with a height of 20 mm. The shot peened areas formed the face sides of the dilatometer specimens, which made it possible to analyze changes of the strain state at an angle perpendicular to the shot peened surface. The specimens were heated to 500 °C, kept at this temperature for one hour and then cooled down to room temperature. This resulted in a permanent elongation of the specimens, which was compared to that of stress-relieved specimens. The differences in specimen elongation are a measure of the dimensional changes perpendicular to the shot peened surface, amounting to roughly 2.4 µm.

Using finite-element simulations, [4.8] estimated the possible extent of thermally-induced distortions. A thin component modeled on a turbine blade was heated on one side, to the point at which most shot peening-induced residual stresses were removed on that side, with no relaxation occurring on the other. Assuming that the component, measuring about 350 mm in length and up to 5 mm in thickness, was clamped down on one end and able to move freely at the other, displacements were estimated at the free end. I. TiAl6V4 alloy these can measure up to roughly 5 mm. By contrast, smaller shifts are predicted for Inconel 718 nickel-base alloy, which is much stiffer. on the whole, the few published studies confirm the expectation that thin-walled components, in particular, will experience changes of shape.

There are no known examinations regarding changes in the topography of mechanically surface-treated material states due to thermal loading.

4.2.2
Influences on Residual Stress State

a) **Basic Results**

Fig. 4.7 shows the amounts of macro residual stresses in shot peened quenched and tempered AISI4140 for different annealing temperatures, plotted versus the logarithm of annealing time [4.3]. Residual stresses decrease as time and temperature are increased. Analyzing the measured data by using the Avrami function, following the iterative least squares method, results in parameters $\Delta H_A = 3.29$ eV, $m = 0.122$ and $C = 1.22 \times 10^{21}$ l/min, which were used to calculate the curves in Fig. 4.7. These serve exceptionally well to describe the dependence of the macro residual stress values – marked by different symbols – on annealing time at the individual temperature settings. By means of a conventional analysis of the same data, [4.3] arrived at parameters $\Delta H_{A,50\%} = 4.47$ eV, $m_{mean} = 0.122$ and $C_{50\%} = 3.4 \cdot 10^{29}$ l/min. Here, the 250 °C setting necessitated an extrapolation of the experimental durations to a questionably high $3 \cdot 10^6$ years. The advantages of the iterative method become evident in the variance of each measured value, which is reduced from 8.9 MPa to 5.3 MPa by changing from the common method to the iterative method. For the same measurement data, Fig. 4.8 shows the correlations between plastic strain rates and mean residual stresses, which were determined using the approach presented in Equations 4.4–4.7. At all annealing temperatures, the data can be approximated entirely or sectionally by straight lines, the slopes of which provide the temperature-dependent exponent n. This exponent is reduced from 120 at 250 °C to 20 at 350 °C. At 400 °C and 450 °C, great stresses and thus,

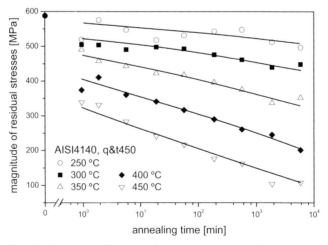

Fig. 4.7: Magnitudes of macro residual stresses vs. annealing time for different annealing temperatures in shot peened AISI4140 in a quenched and tempered state and their description by the Avrami-equation (Eq. 4.1) [4.3]

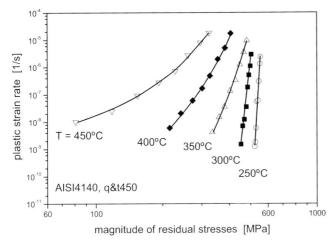

Fig. 4.8: Plastic strain rate vs. magnitude of mean residual stresses for shot peened AISI4140 in a quenched and tempered state [4.3]

short times, cause the values of n to be greater than for smaller stresses. For example, $n = 9$ and $n = 3$ are results found at 450 °C. Thermally activated dislocation movement across short-range obstacles is assumed to be the deformation process which determines velocity at values of n greater than 4. This is termed 'power law breakdown'. Only stress exponent $n = 3$, occuring at 450 °C and for small stresses, is compatible with dislocation creep, which is controlled by volume diffusion. The velocity of dislocation creep is determined by climb processes. The strong temperature-dependence of the n-values prohibits a determination of the activation enthalpy. Therefore, fit quality cannot be compared with the fits made using the Avrami equation. While the Avrami function allows for a consistent description of residual stress relaxation behavior, this is not possible using the Norton approach. Yet the latter permits a clear observation of the temperature- or stress-dependent mechanism changes which are inevitably neglected when the Avrami function and only one set of parameters are used.

Fig. 4.9 shows the residual stress values after shot peening and subsequent annealing of X22CrMoV12–1 steel, which used for turbine blades. The values are depicted as lg t versus $1/kT$, required for determining parameters in the conventional Avrami-analysis [4.6]. There is a noticeable change of mechanisms which occurs at about 300 °C and causes activation enthalpy to rise from about 1.4 eV to about 3.5 eV. This points at the fact that volume diffusion, rather than dislocation core diffusion, determines the velocity of residual stress relaxation above 300 °C.

There are studies which focus on the thermal stability of shot peening-induced residual stress states in nickel-base alloys, as well [see e.g. 4.5,4.8–4.11]. Using the Norton approach, [4.5] analyzed NiCr22Co12Mo9 sheet alloy. His findings on the correlations between strain rate and residual stresses at 600 °C and 850 °C are represented in Fig. 4.10. Given short annealing times, and thus, great residual stress

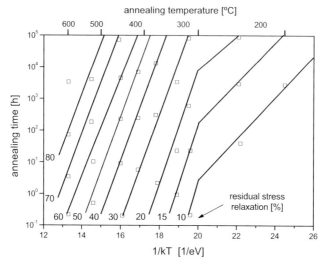

Fig. 4.9: Annealing times required for achieving certain values of residual stress relaxation vs. 1/kT for shot peened X22CrMoV12–1 [4.6]

values, very large Norton exponents are found at 600 °C due to power-law-break-down and the effects of dynamic strain aging processes. Long annealing times create exponents of around 6, indicating viscous slip, i.e. initially pinned edge dislocations experiencing volume diffusion-controlled climb processes. By contrast, re-crystallization processes close to the surface (comp. Sect. 4.2.4b) occuring at 850 °C result in very small residual stress values, while the exponent value 6 found in deeper regions again indicates viscous slip.

Only long annealing times and the greatest depth examined show exponents measuring 11, which are caused by dislocations that were pinned due to carbide formation. The example of shot peened Inconel 718 [4.11] in Fig. 4.11 shows that a 10-minute annealing at 670 °C decreases the remaining fractions of the original residual stress values as the degree of surface layer deformation grows. The latter was determined from the widths of the interference lines obtained from the X-ray measurements of the residual stress analysis. Accordingly, great dislocation densities, brought about by strong surface layer deformations, counteract residual stress stability, as they make it easier for dislocation climb processes to occur.

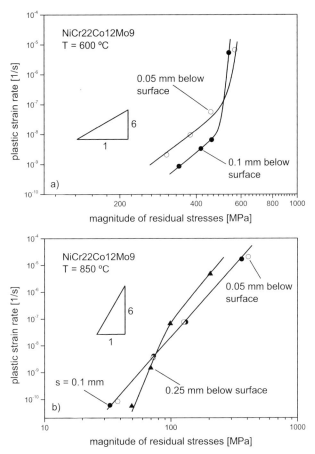

Fig. 4.10: Plastic strain rate vs. residual stresses induced by shot peening for NiCr22Co12Mo9 at 600 (a) and 850 °C (b) at different depths [4.5]

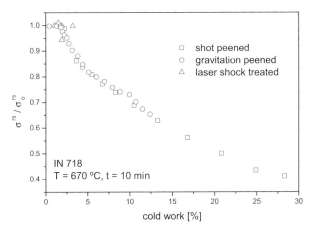

Fig. 4.11: Residual stresses related to their initial values in shot peened or laser shock treated IN718 vs. degree of cold work after a 10-minute annealing treatment at 670 °C [4.11]

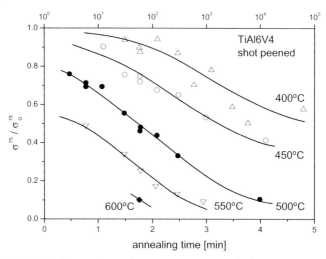

Fig. 4.12: Influence of annealing temperature on residual stresses vs. annealing time for shot peened TiAl6V4 [4.12]

Examinations on TiAl6V4 titanium alloy regarding the thermal stability of shot peening-induced residual stress states were published e.g. in [4.11,4.12]. I. Fig. 4.12, [4.12] shows that the values measured in the 400–500 °C range can be described by the Avrami function. Using the conventional method of determining the Avrami function's parameters, an activation enthalpy of 2.8 eV was derived, which is slightly higher than the activation enthalpy for self-diffusion of 2.51 eV in α-titanium.

Fig. 4.13 shows the thermal relaxation of residual stresses in quenched and tempered SAE5155 shot peened under a torsional prestress of 700 MPa [4.13]. Compared to the values observed in quenched and tempered AISI4140 after conventional shot peening – depicted in Fig. 4.7 –, the thermal relaxation of residual stress is clearly more pronounced. This may be due to the fact that residual stress values are significantly greater after stress peening. [4.13] stated an activation enthalpy value of 4.25 eV, which is high in comparison to the activation enthalpy for self-diffusion found in α-iron.

In warm peened AISI4140 quenched and tempered at 450 °C, a thermal relaxation of residual stresses is found at 290 °C, as shown Fig. 4.14. This relaxation occurs slighty slower at 300 °C and a little faster at 450 °C than after conventional shot peening. The effect of dislocation pinning due to strain aging seems to be predominant at the lower temperature, stabilizing residual stresses. However, at 450 °C, it is probable that the pinning effect of strain aging serves to stabilize residual stresses only when annealing times are very short. One must assume that the dislocations remain mobilized due to the dislocation movement occurring together with the initial relaxation of residual stresses, and that an elevated density of mobile dislocations also contributes to residual stress relaxation, due to the diffuse dislocation structure present after warm peening [4.14].

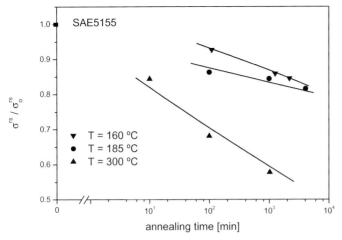

Fig. 4.13: Residual stress relaxation due to annealing of quenched and tempered SAE5155 spring steel, previously stress peened using a torsional prestress of 700 MPa [4.13]

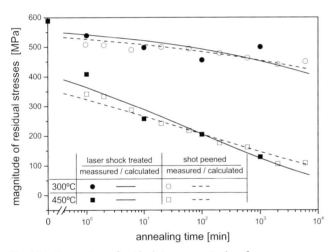

Fig. 4.14: Comparison of residual stress magnitudes after shot peening or laser shock treatment vs. annealing time at different annealing temperatures and their description by the Avrami-equation (Eq. 4.1), for quenched and tempered AISI4140 [4.3, 4.15]

Fig. 4.14 also shows the residual stress stability under thermal loading at 300 °C and 450 °C for laser shock-treated quenched and tempered AISI4140. The values are described well by the Avrami function, which is also depicted and for which [4.15] stated parameters ΔH_A = 2.49 eV, m = 0.17 and C = 2.91 10^{15} 1/min. Thus, the value for activation enthalpy is close to the activation enthalpy of 2.8 eV found in α-iron. The residual stresses compare to the values stated in [4.3] for shot pee-ned specimens of the same material state, which were already presented in Fig. 4.7 and are also depicted here.

It is hardly possible to arrive at a systematic, self-contained assessment of resid-ual stress relaxation using the Avrami function's parameters. However, there are some tendencies that are generally applicable with only a few exceptions. These tendencies have been derived in the past from the measurements presented above, as well as from other measurements which are not shown here as their

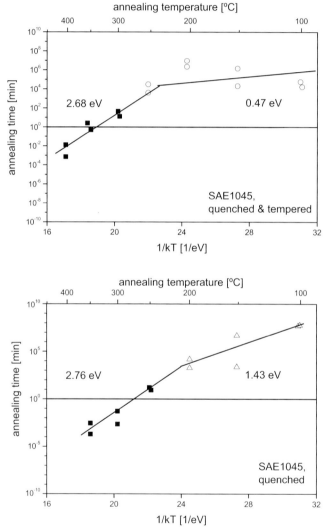

Fig. 4.15: Annealing time for 50% residual stress relaxation vs. annealing temperature for shot peened SAE1045 in a quenched and tempered (a) and a quenched (b) state, acc. [4.16]

representation would require a great amount of space. Firstly, it should be noted that the nature of the surface treatment process itself is, in most cases, irrelevant to residual stress stability during thermal loading. Regarding all the examined materials, the relaxation of residual stresses can be described using the Avrami function or the Norton approach. An integral evaluation of all examined temperatures shows that mean activation enthalpy can be determined for each one. When high and low temperatures are examined separately, small activation enthalpies are found for low temperatures and larger ones are found for high temperatures. The probable cause is the change of mechanisms from dislocation core diffusion-controlled dislocation climb – its activation enthalpy being equivalent to half the value of the activation enthalpy for self-diffusion – to volume diffusion-controlled dislocation climb, which occurs at higher temperatures, with an activation enthalpy equivalent to the full amount of the activation enthalpy for self-diffusion. This is demonstrated impressively by Fig. 4.15 [4.6], in which the logarithms of the annealing times required for a 50% relaxation of residual stresses are plotted versus $1/kT$ for quenched and tempered and for quenched SAE1045 [4.16]. As evidenced by the change of the slope at medium temperatures, residual stress relaxation is controlled by dislocation core diffusion at low temperatures and by volume diffusion at high temperatures. In the analysis of residual stress relaxation using the Norton approach, the change of mechanisms is reflected by the change from power-law-breakdown, which occurs at low temperatures and great stresses, to a classic creep process with $n = 3$ at high temperatures and at rather low stresses (comp. Fig. 4.8, [4.3]).

In steels, the heat treatment state shows typical effects on residual stress relaxation. For shot peened AISI4140, Fig. 4.16 shows that at 450 °C, residual stress relaxation takes place much more slowly in the normalized state than in the states quenched and tempered at 600 °C and 450 °C [4.17]. Similar findings were reported by [4.2], comparing results for shot peened SAE1045 in normalized, quenched and quenched and tempered states. He found residual stress relaxation was similarly accelerated in both the quenched and quenched and tempered states versus the normalized state.

Fig. 4.17 focuses on different materials and material states bearing shot peening-induced residual stresses. It shows activation enthalpies related to the activation enthalpies for self-diffusion as functions of temperature $T_{1/2}$ related to the melting point –, which lead to a 50% relaxation of residual stresses within one hour. This shows that related activation enthalpies grow as related temperature rises. In the case of AISI4140, the activation enthalpy and the required temperature appear to be greater than for SAE1045. In both steels, the quenched states show lower activation enthalpies and temperatures, while their nomalized states exhibit higher activation enthalpies and temperatures. After laser shock treatment, on the other hand, activation enthalpy is reduced while the required temperature remains the same. In the titanium alloy, and particularly, in the aluminum alloys and brass, activation enthalpies are reduced and temperatures increased in comparison to the steels. The differences between the individual materials and material states may, in part, be due to the fact that temperatures too

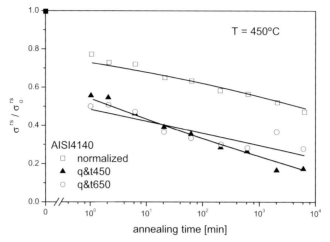

Fig. 4.16: Influence of material state on residual stresses related to their initial values vs. annealing time for shot peened AISI4140 [4.17]

few in number and very close to each other were included in the analysis using the Avrami function. Thus, activation enthalpy ΔH_A and velocity constant C may have been interdependently variable, within certain limits, while the result for residual stress relaxation remained the same. An additional reason for the different positions of the individual materials or material states in Fig. 4.17 may be found in the different mechanisms of residual stress relaxation that occur in the individual

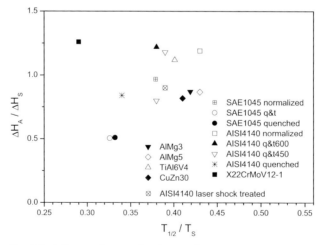

Fig. 4.17: Influence of material and material state on activation enthalpy for thermal residual stress relaxation related to self-diffusion enthalpy vs. annealing temperature for 50% residual stress relaxation within 1 h related to solidus temperature, acc. [4.19]

materials or material states. In evenly distributed dislocation structures, e.g. quenched steels, residual stress relaxation is more likely to be controlled by dislocation core diffusion, resulting in a pronounced relaxation already at lower temperatures. On the other hand, planar dislocation arrangements, cell walls, subgrain boundaries or small angle grain boundaries show a greater resistance to dislocation climb movements, as the latter tend to be controlled by volume diffusion. This often applies to softer material states, such as normalized steels. The categorization of quenched and tempered states falls somewhere inbetween, depending on hardness or annealing temperature [4.18].

b) Behavior Beneath the Surface

The depth-dependency of thermal residual stress relaxation behavior was examined by [4.3,4.19] regarding steels and by [4.5,4.8] regarding nickel-base alloys. Their corresponding findings show that thermal stress relaxation is slightly slower beneath the surface than it is directly at the surface. Fig. 4.18d shows an example of this [4.3], and additionally depicts the relaxation behavior beneath the surface as described by the Avrami function. The Avrami function's parameters, plotted in Fig. 4.18a-c as a function of distance to surface, were obtained using 45 measurement points for each one. Starting from the surface, no significant tendencies are found, apart from a small reduction of ΔH_A and C and an increase of m for surface distances greater than 0.04 mm. Therefore, two sets of parameters are fully sufficient for describing distributions at any annealing temperature or -time. The Avrami parameters mentioned in the context of Fig. 4.7 are to be applied in respect to macro residual stress relaxation directly at the surface. On the other hand, the mean parameters $\Delta H_{A,m} = 2.99$ eV, $C = 6.09 \cdot 10^{17}$ l/min and $m = 0.172$ – shown as plateau levels of the curves in Fig. 4.18a-c – must be used in respect to residual stress relaxation beneath the surface. Starting from the distribution prior to annealing, the distributions may be calculated for any given set of annealing conditions. As an example, Fig. 4.18d shows predictions for 450 °C and times of 6, 60 and 6000 minutes, which are compared with experimental results. The measurements and the calculation correspond well. Modeling results which incorporate the same parameters for the entire surface layer, including the surface itself, arrive at inferior results. This again indicates that the relaxation behavior of macro residual stresses directly at the surface is significantly different from that found in regions close to the surface.

As mentioned within the context of Fig. 4.10, [4.5] showed that in nickel-base alloys, re-crystallization processes, occurring exclusively in surface proximity, may result in residual stress relaxation being particularly pronounced in these regions. It was also shown that at particularly great depths, the slip movements, which are usually viscous, may be prevented by carbide formation and the pinning of dislocations.

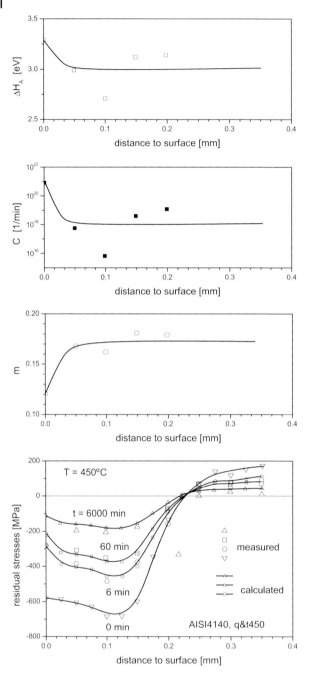

Fig. 4.18: Distributions of properties of the Avrami-equation
(Eq. 4.1) (a-c) and comparison of measured and calculated
distributions of residual stresses at 450 °C (d) for shot peened
AISI4140 in a quenched and tempered state [4.3]

c) Transient Behavior During Heating

By means of measurements made on quenched and tempered and shot peened AISI4140, Fig. 4.19 proves that residual stress relaxation behavior is particularly pronounced within the first minute of annealing [4.3]. As this is also the period in which the component or specimen is heated up, it is essential to discuss in detail the transient residual stress relaxation processes which occur during the heating phase. This shall be done using the Avrami function, which is taken to be isothermal, and by using model assumptions first described by [4.3]. The so-called stress transient method was developed to this end, and it is depicted schematically in Fig. 4.20. It assumes that the residual stress relaxation occurring in non-isothermal processes within a small time increment $\Delta t = t_{i+1} - t_i$ is equivalent to the isothermal relaxation which occurs at a mean temperature $T_{m,i} = 1/2\ (T_{i+1} + T_i)$ within the same time increment, starting from the same initial residual stress value σ_i^{rs}, i.e. that the development of residual stress relaxation can be described by the current residual stress value and by the temperature currently reached. Residual stress relaxation calculated in this manner is shown in Fig. 4.21, regarding specimens which had been immersed in a slat bath with a temperature of 450 °C for up to 90 seconds. Temperature development at the surface of the specimen is represented by the $T(t)$ curve. Residual stress relaxation calculated using the stress transient method is indicated by the broken-line curve, and it is compared to the residual stress values measured after the interruption of the experiment and rapid cooling to room temperature. These values are indicated by triangles. The arithmetic description of the experimental data works very well. In addition, the isothermally calculated course of residual stress versus time is indicated in Fig. 4.21 by a dotted line. The results for the heating period, lasting about 20 seconds, show clear differences to the curve calculated using the stress transient method. The influence of the non-isothermal heating phase is negligible only when annealing times are sufficiently long, in which case the curves approximate each other. The efficacy of estimating transient residual stress relaxation by means of the stress transient method is emphasized by Fig. 4.22. It compares the measured values with the arithmetic descriptions of the residual stress relaxation which occurs during the initial 90 seconds of annealing in shot peened quenched and tempered AISI4140 at 300, 350 and 450 °C [4.14]. Transient residual stress relaxation is described almost ideally for 450 °C. While the stress transient method still achieves a good representation of the curves for both of the lower temperatures, the values of the remaining residual stresses are slightly over-estimated.

d) Annealing Aimed at Surface State Optimization

A number of variant methods of mechanical surface treatment have been developed recently which are able to achieve improved component properties in certain applications. They are based on annealing treatments which follow conventional shot peening treatments. A fundamental distinction must be made between a short-term annealing of steels [see e.g. 4.20–4.25], which, like the warm peening treatments described in detail in Sect. 3, cause strain aging effects, and an

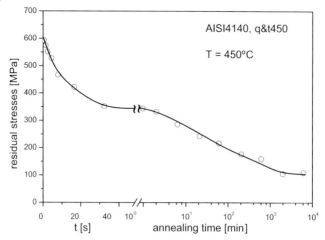

Fig. 4.19: Reduction of the magnitudes of residual stresses for short- and long term annealing at 450 °C for shot peened AISI4140 in a quenched and tempered state [4.3]

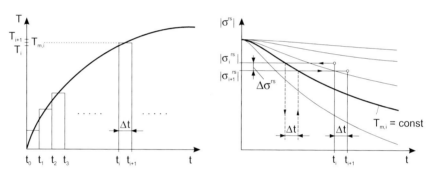

Fig. 4.20: Scheme differentiating temperature-time-course (a) and stress-transient method (b) for the description of transient residual stress relaxation, acc. [4.3]

annealing of titanium or aluminum alloys [see e.g. 4.26–4.29], which essentially results in local re-crystallization or hardening effects. Residual stresses introduced by the initial step of mechanical surface treatment – usually shot peening – are reduced to a greater or lesser extent, as described above. Fig. 4.23 shows the thermally-induced relaxation of residual stress observed by [4.30] in quenched and tempered AISI4140 when the shot peening treatment was optimized by subsequent annealing. This relaxation can be described by the Avrami function, using the parameters $\Delta H_A = 2.23$ eV, $m = 0.138$ and $C = 4.96 \ 10^{13}$ 1/min. The deviations from the parameters found in Chap. 4.2.2a are presumably due to the small number of temperatures contained in the data used in this fit. None of the reasons postulated in Sect. 4.1.1 – such as a restricted movement of clouded slip disloca-

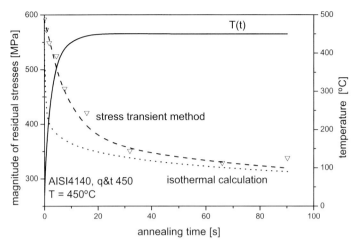

Fig. 4.21: Magnitudes of residual stresses after short-term immersion in a 450 °C-salt bath and comparison with courses measured, calculated by the stress-transient method and calculated assuming isothermal annealing for shot peened AISI4140 in a quenched and tempered state [4.3]

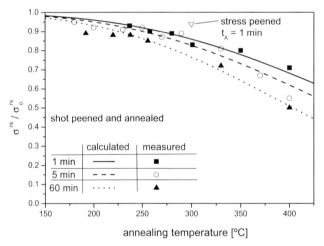

Fig. 4.22: Magnitudes of residual stresses after short-term immersion in a salt bath at 300, 350 and 450 °C, compared to courses measured and calculated by the stress-transient method for shot peened AISI4140 in a quenched and tempered state [4.3]

tions, obstructions of dislocation movement, or dislocations being pinned by carbides and released again later due to a coarsening of the carbides – appear to cause any deviations from the expected course of the Avrami function. After the annealing treatment, which lead to the optimal alternating bending strength

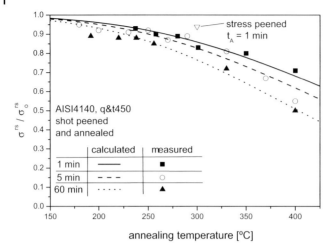

Fig. 4.23: Residual stresses related to their initial values vs. annealing time and their description by the Avrami-equation (Eq. 4.1) for AISI4140 in a quenched and tempered state conventionally peened and stress peened using a prestress of 500 MPa [4.30]

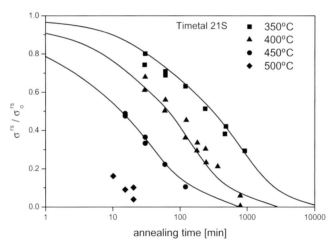

Fig. 4.24: Residual stresses related to their initial values vs. annealing time for different annealing temperatures and their description by the Avrami-equation (Eq. 4.1) for shot peened Timetal 21S [4.28]

achievable by such combined shot peening and annealing treatments, the residual stresses remained at more than 80 % of their initial values [4.30]. The residual stresses which remained after stress peening and subsequent annealing for 1 minute at 300 °C are also indicated. Their initial amount of 791 MPa before annealing is markedly higher than the 660 MPa amount present after conventional shot peening [4.31]. Despite the higher stress value, the remaining residual stresses are slightly greater, related to the initial value, than they are after conventional shot peening. [4.25] and [4.23] also indicated residual stress values for shot peened and subsequently annealed states of ZStE 220 BH bake-hardening steel and normalized SAE1045. They demonstrated that the residual stresses differed only marginally from those found immediately after shot peening, when fatigue strength was optimized. Contrastingly, [4.28] measured the residual stresses in Timetal 21S titanium alloy created by shot peening and subsequent annealing, as shown in Fig. 4.24. Optimum hardening occurred after 2 hours at 400 °C, or after 1 hour at 450 °C, residual stresses being reduced by at least 50 %. Thus, one factor which would contribute significantly to increased fatigue strength was available only in a substantially weakened form. Residual stress relaxation is even more extreme in titanium alloys. Here, optimized properties are achieved by annealing treatments which utilize local fine grain formation by re-crystallization. In these cases, residual stresses are removed completely.

4.2.3
Influences on Workhardening State

a) Basic Results

Fig. 4.25 shows the changes of surface half widths in shot peened quenched and tempered AISI4140 caused by annealing lasting longer than one minute, at temperatures between 250 and 450 °C [4.30]. Evidently, similar laws apply here as they do for the macro residual stresses depicted in Fig. 4.7. This is also indicated by the curves calculated using the Avrami function's parameters $\Delta H_{A,HBW}$ = 2.48 eV, m_{HBW} = 0.116 and C_{HBW} = 1.09 · 10^{13} 1/min, which were determined by the iterative method. The analysis did not involve the absolute values of the half widths, but used their respective differences to the initial value HBW_N = 1.65 °2θ of a normalized specimen. This reduced the influence of the technical devices, while taking account of the fact that, unlike macro residual stresses, half widths are not reduced to a marginal level. Fig. 4.26 shows the relaxation versus time of mean strains and micro residual stresses for the same temperatures, measured on the same specimens by analyzing the x-ray interference lines. Again, the depiction of the measured values is supplemented by the curves calculated according to the Avrami function using the iteratively determined parameters $\Delta H_{A,\varepsilon}$ = 2.64 eV, m_ε = 0.096 and C_ε = 5.31 · 10^{12} 1/min. Thus, the Avrami function is also very effective in describing the thermal relaxation of mean strains and micro residual stresses, as well as the half widths. The same method, however, is not suitable for describing the changes – shown in Fig. 4.27 – of the domain sizes measured at the surface after annealing treatments of various lengths and at temperatures between 250 and 450 °C. This is owed to the relatively strong scattering of the measured values. Their tendency to become greater may prompt the use of, for instance, the reciprocal values, yet even this does not render the method usable.

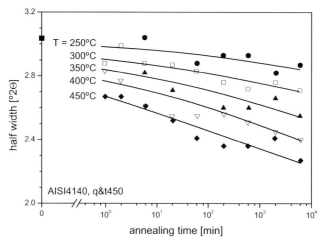

Fig. 4.25: Half widths vs. annealing time for different annealing temperatures in shot peened AISI4140 in a quenched and tempered state and their description by the Avrami-equation (Eq. 4.1) [4.3]

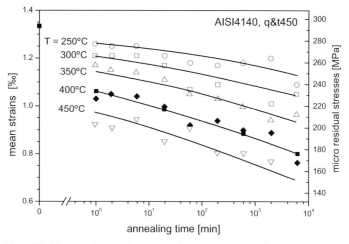

Fig. 4.26: Mean strains or micro residual stresses vs. annealing time for different annealing temperatures in shot peened AISI4140 in a quenched and tempered state and their description by the Avrami-equation (Eq. 4.1) [4.3]

For powder-metallurgically produced γ'-hardened Rene'95 nickel-base alloy, [4.9] showed that the shot peening-induced increases of micro residual stresses – determined using x-ray line profile analyses – were reduced by three-hour annealing treatments at elevated temperatures, and that they were completely removed at about 950 °C (comp. Fig. 4.28). For TiAl6V4, Fig. 4.29 shows that the Avrami function may also be used to describe how the shot peening-induced increases of

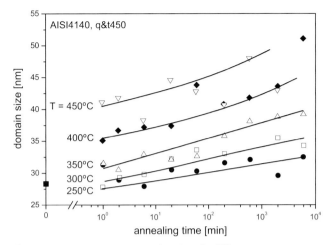

Fig. 4.27: Domain sizes vs. annealing time for different annealing temperatures in shot peened AISI4140 in a quenched and tempered state [4.3]

the half widths are altered by annealing. This is possible when activation enthalpy $\Delta H_{A,HBW} = 2.8$ eV, velocity constant $C_{HBW} = 2.19 \cdot 10^{15}$ l/min and exponent $m_{HBW} = 0.36$ are used [4.12]. Fig. 4.30 shows that the thermally-induced relaxation of the half widths in laser shock-treated quenched and tempered AISI4140 occurs in a similar way as it does after shot peening [4.15].

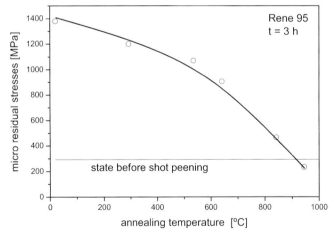

Fig. 4.28: Micro residual stresses vs. annealing temperature for an annealing time of 3 h in shot peened René 95 [4.9]

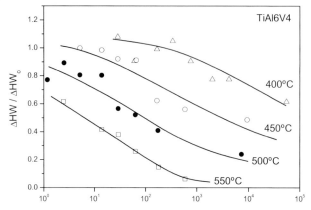

Fig. 4.29: Influence of annealing temperature on half width changes related to their initial values vs. annealing time for shot peened TiAl6V4 [4.12]

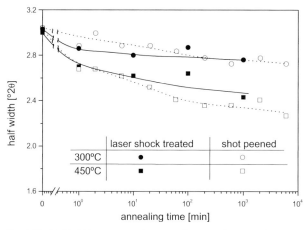

Fig. 4.30: Half widths vs. annealing time for annealing at 300 and 450 °C after laser shock treatment or shot peening of quenched and tempered AISI4140 [4.3, 4.15]

An obvious step to take is to compare the relaxation behavior of the micro residual stresses to that of the macro residual stresses. Fig. 4.31 shows results for the quenched and tempered AISI4140 discussed above, and a temperature of 350 °C [4.3]. Compared to the half widths or the mean strains, the macro residual stresses are reduced a little more quickly. This is primarily due to the difference in the velocity constant C of nine orders of magnitude, as the exponents m do not show any significant differences and the activation enthalpies are close to the enthalpy for self-diffusion in all cases, those describing the relaxation of micro residual stresses being smaller than those describing macro residual stress relaxation. These findings agree entirely with results obtained by [4.2] for shot peened nor-

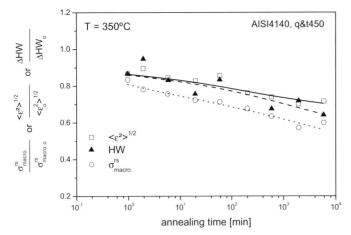

Fig. 4.31: Related macro residual stresses, related mean strains or micro residual stresses and related changes of half widths vs. annealing time for an annealing temperature of 350 °C in shot peened AISI4140 in a quenched and tempered state and their description by the Avrami-equation (Eq. 4.1) [4.3]

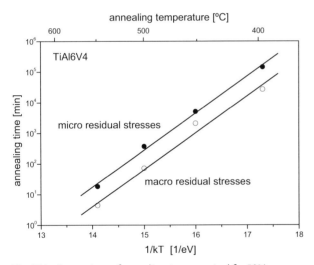

Fig. 4.32: Comparison of annealing times required for 50% relaxation of macro and micro residual stresses vs. annealing temperature for shot peened TiAl6V4 [4.12]

malized SAE1045. The correlation shown in Fig. 4.32 applies to TiAl6V4. While confirming that micro residual stress relaxation occurs more slowly than macro residual stress relaxation, it also shows equal activation enthalpies for macro- and micro residual stress relaxation, which are equivalent to the slopes of the lines indicated in the plots shown. Fig. 4.33 shows a comparison of the parameters of

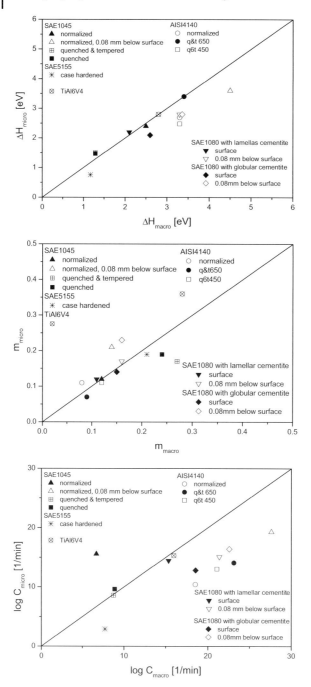

Fig. 4.33: Comparison of the activation enthalpies (a), exponents (b) and rate constants (c) of the Avrami-equation (Eq. 4.1) for the relaxation of macro and micro residual stresses in different materials

the Avrami function, as far as they are available, for different materials and material states. Most of the activation enthalpies for micro residual stress relaxation are slightly smaller than they are for macro residual stress relaxation. Exponents *m* show similar values for macro- and micro residual stress relaxation, exhibiting non-systematic deviations, while the velocity constants are generally smaller for micro residual stress relaxation than they are for macro residual stress relaxation, as already demonstrated by the examples selected. Accordingly, volume diffusion-controlled dislocation creep dominated by climbing of edge dislocations should determine the rate of macro- and micro residual stress relaxation in all cases. Nevertheless, differences in relaxation rates occur. These are explained by the fact that dislocation movements may suffice for a relaxation of macro residual stresses, while a pronounced relaxation of micro residual stresses additionally requires dislocation annihilations.

b) Behavior Beneath the Surface

Fig. 4.34 shows the depth distributions of mean strains and micro residual stresses in shot peened quenched and tempered AISI4140 for 400 °C and different annealing times. They are compared to the state present immediately after shot peening. After the shot peening treatment, an increase of mean strains is found within immediate proximity of the surface, yet depths of 0.05 to 0.15 mm show a relaxation of mean strains, which is caused by dislocations being rearranged into energetically preferable configurations, i.e. by microstructural softening effects. After two minutes of annealing, the shot peening-influenced surface layer up to 0.2 mm from the surface already shows a relaxation of mean strains, in comparison to the initial, un-peened state. This effect continues as annealing times grow

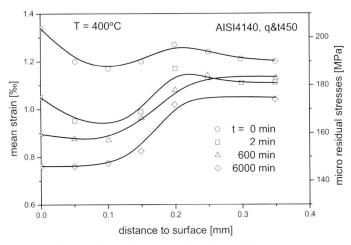

Fig. 4.34: Influence of annealing time on distribution of mean strains or micro residual stresses at an annealing temperature of 400 °C for shot peened AISI4140 in a quenched and tempered state [4.3]

longer, resulting in a pronounced decrease of mean strains in the surface layer after 6000 minutes. This is caused by the instability of the micro residual stress state of the shot peened surface layer which comes with microstructural softeing. The bulk material, on the other hand, shows only small changes of the mean strains, which are caused by tempering effects. Fig. 4.35 shows the depth distributions of the domain sizes for the same annealing conditions. It indicates that the domain size changes run counter to those of the mean strains. Dislocation densities are calculated from the mean strains and domain sizes according to [4.32]. Their depth distributions are represented in Fig. 4.36, confirming that annealing treatments cause the dislocation density in the surface layer to fall below the level

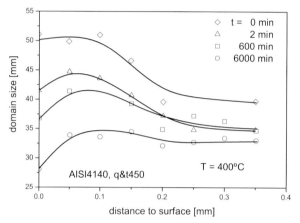

Fig. 4.35: Influence of annealing time on distribution of domain sizes at an annealing temperature of 400 °C for shot peened AISI4140 in a quenched and tempered state [4.3]

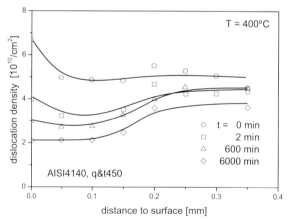

Fig. 4.36: Influence of annealing time on distribution of dislocation densities acc. Cohen at an annealing temperature of 400 °C for shot peened AISI4140 in a quenched and tempered state [4.3]

present in the uninfluenced bulk material. This is due to fact that the dislocation arrangements are unstable when thermal energy is introduced.

As shown in Fig. 4.37, [4.5] found that in NiCr22Co12Mo9 nickel base alloy, the shot peening-induced increases of the half widths close to the surface were changed only marginally by annealing at 600 °C, while the residual stresses were almost completely removed. Similar results were obtained in respect to the dislocation densities close to the surface.

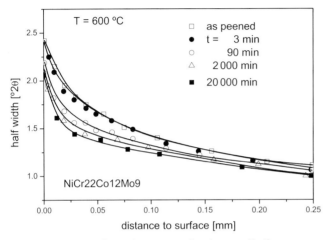

Fig. 4.37: Influence of annealing time on distribution of half widths at an annealing temperature of 600 °C for shot peened NiCr22Co12Mo9 [4.5]

c) Annealing Aimed at Surface State Optimization

The various methods of mechanical surface treatment incorporating subsequent high-temperature annealing do not only cause inevitable relaxations of the macro residual stress state – as mentioned in Sect. 4.2.2 – but also bring about relaxations of micro residual stresses. Fig. 4.38 shows this for shot peened quenched and tempered AISI4140. This is an example of steels which permit utilizing strain aging effects. 1-, 5- and 60-minute annealing treatments at temperatures between 180 and 400 °C result in significant relaxations of the half widths, related to their values after shot peening. These relaxations progress in a way similar to the one discussed above in respect to longer times [4.30]. The relaxation of half widths created by stress peening and subsequent short-term annealing at 300 °C – also shown in Fig. 4.38 – does not deviate from the behavior of conventionally peened specimens. Concurrently to half width relaxation, increases of hardness occur within the regions which were previously plastically deformed. This is due to static strain aging effects, which result in the pinning of slip dislocations by Cottrell clouds and ultra-fine carbides. Fig. 4.39 compares these hardness increases to those found in conventionally peened specimens. [4.23] observed similar, and even more impressive, effects in normalized SAE1045 after conventional shot

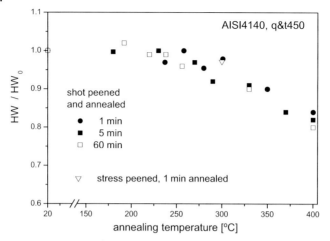

Fig. 4.38: Influence of annealing temperature on related half widths during short-term annealing (1, 5, and 60 min) of shot peened and stress peened (σ_{pre} = 500 MPa) AISI4140 in a quenched and tempered state [4.31]

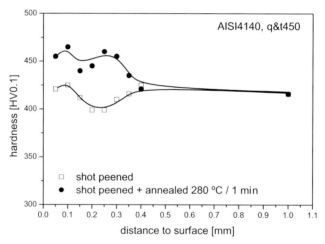

Fig. 4.39: Comparison of the distributions of microhardness after shot peening and shot peening plus subsequent annealing for 1 min at 280 °C for AISI4140 in a quenched and tempered state, acc. [4.31]

peening and subsequent short-term annealing at 350 °C. Fig. 4.40 shows pronounced hardness increases close to the surface and compares them with the distribution of hardness found after shot peening. Concurrently, the half widths decrease during annealing, as is the case for quenched and tempered AISI4140. The reason why half widths and micro hardness behave differently is found in the fact already discussed in Sect. 3.1.2d. Half widths merely represent the distortion

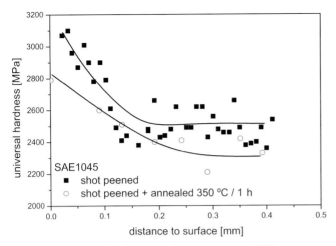

Fig. 4.40: Influence of a 1-hour annealing treatment at 350 °C
on the distribution of microhardness for shot peened
SAE1045 in a normalized state [4.23]

states induced by the microstructure, whereas plastic deformation needs to have
occurred if micro hardness is to be measured. This is why static strain aging
affects only measurements of micro hardness. Fig. 4.41 clearly illustrates the
hardening effect which annealing has on shot peened normalized SAE1045. Tensile workhardening-curves are shown for specimens trepanned after shot peening
and annealing at 350 °C using different times. Already strongly workhardened in
comparison to its pre-treatment state, the shot peened trepanned specimen shows
additional increases of flow stress. These are caused by strain aging effects and
are particularly pronounced in regions of small deformations. A contrary example
to the steels described above was found by [4.25], examining ZStE220 BH bake-
hardening steel. Here, 20-minute annealing treatments at temperatures up to
250 °C did not change the half widths significantly.

[4.28] examined Timetal 21S β-Ti alloy and found that shot peening treatments
followed by annealing treatments with different times and temperatures increased
micro hardness in the surface layer. This is shown in Fig. 4.42. The longer the
annealing times and the higher annealing temperatures were, the more hardness
increased. After passing through a maximum, however, micro hardness decreased
again at the highest temperatures. These changes of hardness were caused by
selective surface hardening. In β-Ti alloys, increases of hardness are created in the
previously deformed surface layer by the selective formation of precipitations, discussed above in the context of Fig. 4.3. Due to over-aging effects, hardness
decreases again at longer annealing times and higher temperatures. Fig. 4.43
shows similar results for BetaC Ti-alloy after shot peening or deep rolling and subsequent annealing [4.33]. An issue of particular interest here is that hardening creates a high density of nuclei and thus causes precipitations which are particularly
fine. This makes it possible to achieve hardness values close to the surface which

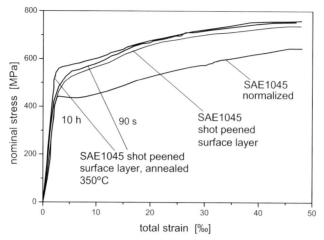

Fig. 4.41: Deformation behaviour of trepanned specimens of shot peened SAE1045 in a normalized state compared to the non-deformed core material [4.24]

an optimized hardening treatment of the bulk material alone cannot achieve. This is illustrated by the broken line which indicates the maximum hardness of the bulk material. Examining the same material, [4.34] showed that increasing deep rolling force permits the depth effect of the hardness increases to be extended noticeably. Preferred surface hardening of AlCuMg2 (AA 2024) was examined by [4.29] (comp. Fig. 4.44). Shot peening is followed by warm aging, creating hardness values in the core region typically achievable by hardening. As the surface layers, however, contain higher densities of nuclei, more and finer precipitations are created there. They cause the increases of hardness observed in these regions.

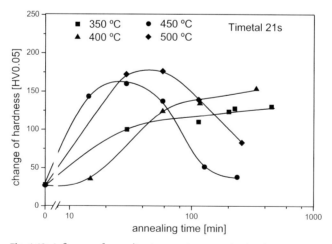

Fig. 4.42: Influence of annealing temperature on microhardness changes vs. annealing time for shot peened Timetal 21S [4.28]

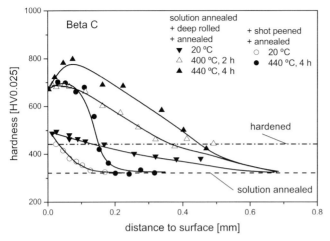

Fig. 4.43: Influence of annealing on the distributions of microhardness for solution-annealed Beta C after shot peening or deep rolling compared to microhardness values of a peak aged state [4.33, 4.38]

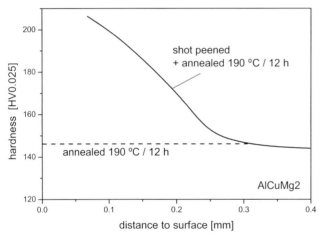

Fig. 4.44: Influence of shot peening on the distribution of microhardness in a state of AlCuMg2 warm aged for 12 h at 190 °C [4.29]

4.2.4
Influences on Microstructure

a)   **Basic Results**

There is only a small number of published studies which examine the influence of thermal treatments on the surface layer microstructure of mechanically surface-treated material states. [4.5] carried out extensive transmission electron miscroscopic studies on shot peened NiCo22Cr12Mo9, which also covered the influence of annealing time and -temperature on the depth distributions of the microstructure. Fig. 4.45 shows the distributions of the half widths and the corresponding dislocation structures, found at certain depths after shot peening, after 2000 minutes of annealing at 600 °C and after 90 minutes of annealing at 850 °C. The dislocation structures in the shot peened state and in the state anealed at 600 °C correspond, showing homogenous dislocation arrangements with high dislocation densities in the surface layer, along with numerous thin dislocation twins. Slip band arrangements and a low dislocation density are found directly at the surface. At 850 °C, however, large re-crystallized zones with a very low dislocation density appear in the region close to the surface where the half widths show a local minimum. By contrast, only a small re-crystallized zone is found in the local half width maximum area, while there is none at all in deeper regions. The carbide structure is affected by the annealing treatment, as well. While the shot peened state and the state annealed at 600 °C contain only a few large grain boundary carbides, a great number of precipitations of secondary carbides is found in the state annealed at 850 °C, occurring in the region of the former dislocation twins and on the grain boundaries. Again, a rather homogeneous distribution of carbides is observed in the deeper regions.

[4.9] determined sub-grain sizes and γ'-phase fractions in powder-metallurgically produced γ'-hardened René 95 nickel-base alloy. Shot peening causes sub-grain sizes to decrease from 180 to about 15 nm and γ'-phase fractions to decrease from 45 vol.-% to 25 vol.-%. As shown in Fig. 4.46, they increase again as annealing temperature is increased, the γ'-phase fraction returning to its initial value at around 900 °C. Sub-grain sizes grow a little more slowly, reaching 40 nm at 1000 °C. This is three times more than the value found after shot peening, yet still far smaller than the value before shot peening. The reason why shot peening reduces the γ'-phase fraction is that hydrostatic pressure abets the volume reduction which contributes to the creation of the denser γ'-phase. During annealing, the microstructure approximates its original initial γ'-phase value.

TiAl6V4 and AISI304 exhibit nano-crystalline surface layers after deep rolling. [4.35] examined their thermal stability by means of (in-situ-TEM) studies. No significant microstructural changes and in particular, no re-crystallization, were found in TiAl6V4 up to 900 °C and in AISI304 for temperatures up to 600 °C. Therefore, fatigue properties are expected to be particularly good at elevated temperatures.

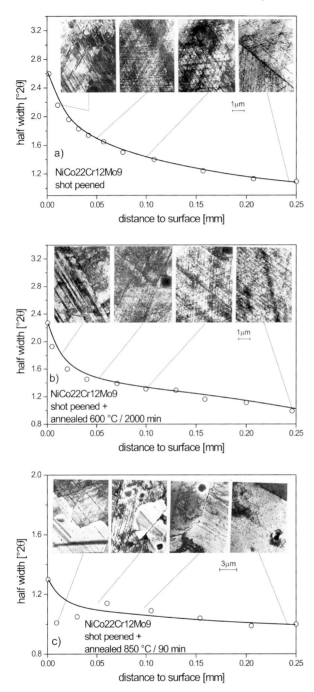

Fig. 4.45: Distributions of half widths and transmission electron microscopically-observed dislocation structures at different distances to surface for shot peened NiCo22Cr12Mo9 after shot peening (a), after annealing for 2000 min at 600 °C (b) and after annealing for 90 min at 850 °C (c) [4.5]

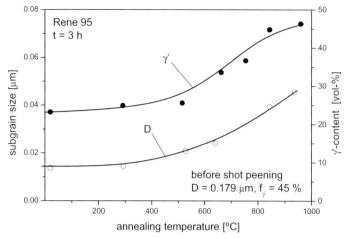

Fig. 4.46: Subgrain size and γ'-particle content vs. annealing temperature for a 3-hour annealing treatment of shot peened René 95 [4.9]

b) Annealing Aimed at Surface State Optimization

Short-term annealing of steels after conventional shot peening treatments is aimed at achieving static strain aging effects. It is almost impossible to deliver experimental evidence of these effects on the microstructural level. Only [4.23,4.24] is known to have carried out transmission electron microscopic examinations, documenting the microstructural changes in the surface layer state of shot peened normalized SAE1045 created by a 90-second annealing treatment at 350 °C. Fig. 4.47a shows the tangled dislocation structure typically found in the surface layers of shot peened specimens. Fig. 4.47b shows the results of the 90-second annealing treatment at 350 °C. While the dislocation structure remains the same, there are distortion contrasts which [4.23] attributed to coherency stresses around semi-coherent finest carbides 10 to 20 nm in size.

Annealing mechanically surface-treated α-Ti alloys, such as Ti-8Al, can be utilized specifically for the creation of fine-grain zones by re-crystallization which occurs close to the surface. Deformation causes a high density of nuclei, which aids re-crystallization and creates extremely fine grains in the deformed surface layer. A comparatively coarse grain structure is maintained in the core microstructure. Accordingly, the crack propagation rate is low during cyclic loading, and it is possible to increase surface layer strength under quasi-static and cyclic loading (comp. Sect. 6.3.5). Fig. 4.48 compares the surface microstructure of shot peened Ti-8Al and the surface microstructure of a specimen of the same material which was shot peened and annealed at 820 °C for one hour. Clearly discernable, there are traces of deformation close to the surface and grain sizes of around 100 μm in the shot peened state, while the specimen which was also annealed shows finely grained zones with grain sizes in the 20 μm range within a distance to surface of about 80 μm [4.26,4.27].

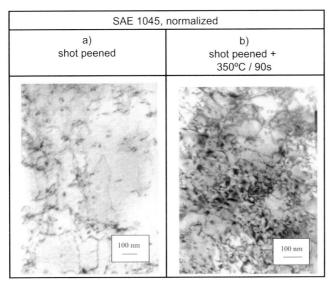

Fig. 4.47: Microstructure observed using a transmission electron microscope at a distance to surface of about 0.05 mm in
normalized AISI4140, shot peened (a) and shot peened and
subsequently annealed for 90 s at 350 °C (b) [4.23, 4.24]

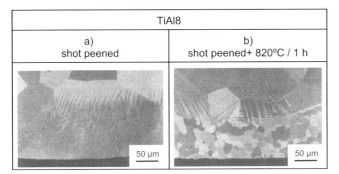

Fig. 4.48: Microstructures close to the surface in Ti8Al after
shot peening (a) and after shot peening plus subsequent
annealing for 1 h at 820 °C (b) [4.26,4.27]

In ($\alpha+\beta$)-Ti alloys, such as Ti6Al4V or Ti6Al2Sn4Zr2Mo, thermal annealing
treatments lead to re-crystallization an therefore alter the grain morphologies
close to the surface. Fig. 4.49 verifies this, showing a polished cross section of the
surface layer of a specimen which was shot peened and subsequently annealed at
850 °C for one hour. A finely grained, lamellar core microstructure is found within
a fine- and equiaxially-grained surface layer with a thickness of about 200 μm
[4.27]. Thus, the microstructure is optimized for creep-fatigue applications,

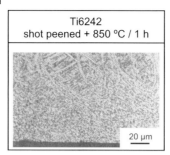

Fig. 4.49: Microstructure observed by light microscopy close to the surface of shot peened and annealed Ti6242 [4.27]

because the lamellar basic structure is a barrier against creep effects and the equiaxial surface layer impedes the initiation and growth of fatigue cracks (comp. Sect. 6.3.5).

There is only limited photographic documentation available regarding the microstructure of β-Ti alloys showing selective surface hardening. Deep rolled BetaC Ti alloy is depicted in [4.34,4.36]. Selective surface hardening is observable up to a distance of about 300 µm to the surface. It becomes visible in the altered etching behavior of the regions which were hardened at 440 °C for 4 hours (comp. Fig. 4.50). As mentioned several times above, the surface layer exhibits an increased dislocation density and thus, an increased density of nuclei, when elevated temperatures cause precipitations of semi-coherent or incoherent hexagonal α-particles within the body-centered cubic β-matrix. In a time-temperature-precipitation diagram, transformation times are shifted forward to such an extent that hardening is already concluded within the surface layer before it has begun in the uninfluenced core region. Therefore, core hardness and microstructure are not affected by the annealing treatment. Contrastingly, Fig. 4.51 shows the surface layer microstructure which is achievable by selective surface hardening in a shot peened AlCuMg2 (AA2024) specimen warm-aged at 190 °C for 12 hours. It is compared to the microstructure of the same material which underwent a similar annealing treatment but no prior cold work [4.29]. As is the case in selective surface hardening, dislocation density and thus, the density of nuclei, are increased here, as well. Therefore, annealing prematurely creates an increased number of precipitations which are refined in comparison to the uninfluenced bulk material. However, time shift in the time-temperature-precipitation diagram is only marginal. Thus, only the number and the size of the precipitations are increased in comparison to the base material state, and the bulk material hardens, as well. In the AlCuMg2 alloy depicted, θ-phase Al_2Cu precipitations occur in the shot peening-affected surface layer, which are significantly finer than after conventional warm aging. Fig. 4.51 offers impressive evidence of this, explaining the noticeable increase of hardness in the surface layer which is shown in Fig. 4.44.

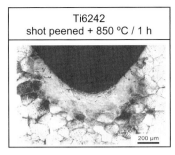

Ti6242
shot peened + 850 °C / 1 h

200 μm

Fig. 4.50: Microstructures close to the surface in Beta C after deep rolling and subsequent annealing for 4 h at 440 °C [4.34,4.36]

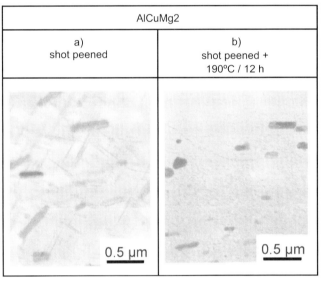

AlCuMg2	
a) shot peened	b) shot peened + 190°C / 12 h

0.5 μm | 0.5 μm

Fig. 4.51: Microstructures of AlCuMg2 after conventional warm aging for 12 h at 190 °C (a) and after the same warm aging treatment following a shot peening treatment (b) [4.29]

References

4.1 O. Vöhringer: Abbau von Eigenspannungen, In: V. Hauk, E. Macherauch (eds.), Eigenspannungen, DGM-Informationsgesellschaft, Oberursel, 1983, pp. 49–83.

4.2 J. Hoffmann: Entwicklung schneller röntgenographischer Spannungsmessverfahren und ihre Anwendung bei Untersuchungen zum thermischen Eigenspannungsabbau, Dissertation, Universität Karlsruhe (TH), 1985.

4.3 V. Schulze, F. Burgahn, O. Vöhringer, E. Macherauch: Zum thermischen Abbau von Kugelstrahl-Eigenspannungen bei vergütetem 42 CrMo 4, Materialwissenschaft und Werkstofftechnik 24 (1993), pp. 258–267.

4.4 J. Stoer: Einführung in die numerische Mathematik I. Springer Verlag, Berlin, 1983.

4.5 D. Viereck, D. Löhe, O. Vöhringer, E. Macherauch: Relaxation of residual stresses in a nickel-base superalloy due to dislocation creep, In: D. G. Brandon, R. Chaim, A. Rosen (eds.), Proc. Int. Conf. Strength of Metals and Alloys 9, Freund Publishing, London, 1991, pp. 14–19.

4.6 M. Roth: Die thermische Stabilität von Eigenspannungen in kugelgestrahlten Oberflächen, Zeitschrift für Werkstofftechnik 18 (1987), pp. 225–228.

4.7 I. Lillamand, L. Barrallier, J. M. Sprauel, R. Chiron: Macroscopic and microscopic evolutions of a shot-peened layer during isothermal recovery, Metallurgical and Materials Transactiona 31A (2000), pp. 213–224.

4.8 P. S. Prevey, D. J. Hornbach, P. W. Mason: Thermal residual stress relaxation and distortion in surface enhanced gas turbine engine components, In: D. L. Milam, D. A. Poteet Jr., G. D. Pfaffmann, V. Rudnev, A. Muehlbauer, W. B. Albert (eds.), 17th AS. Heat Treating Society Conference, ASM. Metals Park, 1997, pp. 3–12.

4.9 R. Jilai, Wang Renzhi, L. Xiangbin: Investigation on shot peening strengthening of Rene'95 powder alloy, In: J. Champaigne (ed.), Proc. Int. Conf.

Shot Peening 6, San Francisco, 1996, pp. 338–347.

4.10 Wang Renzhi, R. Jilai: Investigation on the microstructure changes of NiCrAlY coating due to shot peening plastic deformation and aging, In: A. Nakonieczny (ed.), Proc. Int. Conf. Shot Peening 7, Warschau, 1999, pp. 33–41.

4.11 P. S. Prevey: The effect of cold work on the thermal stability of residual compression in surface enhanced IN718, In: K. Funatani G. E. Totten (eds.), 20th AS. Heat Treating Society Conference, ASM. Metals Park, 2000, pp. 426–434.

4.12 T. Hirsch, O. Vöhringer, E. Macherauch: Der thermische Abbau von Stahleigenspannungen bei TiAl 6 V 4, Härterei Technische Mitteilung (1983) 38, pp. 229–233.

4.13 L. Bonus: Auswirkungen des Spannungsstrahlens auf die Eigenschaften von hoch vergüteten Bremsspeicher- und Torsionsfedern, Dissertation, RWT. Aachen, 1994.

4.14 R. Menig, V. Schulze: unveröffentlicht, Universität Karlsruhe (TH), 2002.

4.15 V. Schulze, R. Menig, O. Vöhringer: Comparison of surface characteristics and thermal residual stress relaxation of laser peened and shot peened AISI 4140, In: L. Wagner (ed.), Shot Peening, Wiley-VCH, Weinheim, 2002, pp. 145–160.

4.16 U. Schlaak: Röntgenographische Ermittlung der Eigenspannungsumlagerung bei erhöhter Temperatur, Universität Bremen, 1988.

4.17 H. Holzapfel: unveröffentlicht, Universität Karlsruhe (TH), 1994.

4.18 D. Löhe, O. Vöhringer: Stability of residual stresses, In: G. Totten, M. Howes, T. Inoue (eds.), Handbook of residual stress and deformation of steel, AS. International, Metals Park, 2002, pp. 54–69.

4.19 J. Hoffmann, B. Scholtes, O. Vöhringer, E. Macherauch: Relaxation of residual stresses of various sources by annealing, In: E. Macherauch, V. Hauk (eds.), Residual stresses in science and technology, Proc. Int. Conf. Residual Stresses 1, DG. Informationsgesellschaft, Oberursel, 1987, pp. 695–702.

4.20 A. Wick, V. Schulze, O. Vöhringer: Effects of warm peening on fatigue life and relaxation behaviour of residual stresses of AISI 4140, Materials Science and Engineering A293 (2000), pp. 191–197.

4.21 R. Menig, V. Schulze, O. Vöhringer: Optimized warm peening of the quenched and tempered steel AISI 4140, Materials Science and Engineering A335 (2002), pp. 198–206.

4.22 A. Tange, K. Ando: Study on the shot peening processes of coil spring, In: Proc. Int. Conf. Residual Stresses 6, IO. Communications, Oxford, 2000, pp. 897–904.

4.23 I. Altenberger, B. Scholtes: Improvement of fatigue behaviour of mechanically surface treated materials by annealing, Scripta Materialia 42 (1999), pp. 873–881.

4.24 I. Altenberger: Mikrostrukturelle Untersuchungen mechanisch randschichtverfestigter Bereiche schwingend beanspruchter metallischer Werkstoffe, Dissertation, Universität Gesamthochschule Kassel, 2000.

4.25 A. Rössler, J. K. Gregory: Thermal decay of shot-peening induced residual stresses during annealing of a bake-hardening steel, In: C. A. Brebbia J. M. Kenny (eds.), Surface Treatment IV – Computer Methods and Experimental Measurements, Southampton, 1999, pp. 311–320.

4.26 L. Wagner, J. K. Gregory: Thermomechanical surface treatment of titanium alloy, In: Second European AS. Heat Treatment and Surface Engineering Conference, ASM. Metals Park, 1993, pp. 1–24.

4.27 H. Gray, L. Wagner, G. Lütjering: Effect of modified surface layer microstructures through shot peening and subsequent heat treatment on the elevated temperature fatigue behavior of Ti alloys, In: H. Wohlfahrt, R. Kopp, O. Vöhringer (eds.), Shot Peening, Proc. 3rd Int. Conf. on Shot Peening, DGM-Informationsgesellschaft, Oberursel, 1987, pp. 467–475.

4.28 M.-C. Berger, J. K. Gregory: Selective hardening and residual stress relaxation in shot peened Timetal 21s, In: C. A. Brebbia J. M. Kenny (eds.), Surface Treatment IV – Computer Methods and Experimental Measurements, Southampton, 1999, pp. 341–348.

4.29 J. K. Gregory, C. Müller, L. Wagner: Bevorzugte Randschichtaushärtung: Neue Verfahren zur Verbesserung des Dauerschwingverhaltens mechanisch belasteter Bauteile, Metall 47 (1993), pp. 915–919.

4.30 R. Menig, V. Schulze, O. Vöhringer: Effects of Static Strain Aging on Residual Stress Stability and Alternating Bending Strength of Shot Peened AISI 4140, Zeitschrift für Metallkunde 93 (2002) 7, pp. 635–640.

4.31 R. Menig: Randschichtzustand, Eigenspannungsstabilität und Schwingfestigkeit von unterschiedlich wärmebehandeltem 42 CrMo 4 nach modifizierten Kugelstrahlbehandlungen, Dissertation, Universität Karlsruhe (TH), 2002.

4.32 D. E. Mikkola, J. B. Cohen: Examples of applications of line broadening, In: J. B. Cohen, J. E. Hilliard (eds.), Local atomic arrangements studied by X-ray diffraction, Met.Soc.Conf. 36, New York, 1965, pp. 289–333.

4.33 J. K. Gregory, L. Wagner, C. Müller: Selective Surface Aging in the High-Strength Beta Titanium Alloy Beta-C. In: P. Mayr (ed.), Proc. Surface Engineering, DGM-Informationsgesellschaft, Oberursel, 1985, pp. 435–440.

4.34 A. Berg, L. Wagner: Near surface gradient microstructures in metastable beta-titanium alloys for improved fatigue performance, In: W. A. Kaysser (ed.), Proc. Functionally Graded Materials 1998, Trans Tech Publications, 1999, pp. 307–312.

4.35 I. Altenberger: Alternative surface treatments: microstructures, residual stresses and fatigue behavior, In: L. Wagner (ed.), Shot Peening, Wiley-VCH, Weinheim, 2003, pp. 421–434.

4.36 A. Berg: Gradientengefüge und mechanische Eigenschaften hochfester Titanlegierungen, Dissertation, BT. Cottbus, 2000.

4.37 V. Schulze O. Vöhringer: unveröffentlicht, 1993.

4.38 W. Walz, T. Dörr, J. Kiese, L. Wagner:
Schwingfestigkeit höchstfester Leicht-
bauwerkstoffe auf Titanbasis nach Fest-
walzen und Ausscheidungshärtung im
Kerbgrund, In: H. Zenner (ed.), DVM-
Bericht 122 "Leichtbau durch innovative
Fertigungsverfahren", 1996, pp. 51–59.

5
Changes of Surface Layer States due to Quasi-static Loading

5.1
Process Models

5.1.1
Elementary Processes

The surface layer states created by mechanical surface treatments influence behavior during mechanical loading. This is due, firstly, to pre-deformation, which changes deformation behavior in the surface layer, and secondly, to the residual stress state present. The way in which mechanical surface treatments modify deformation behavior is determined, in particular, by the changes of the workhardening state close to the surface and by the microstructural changes discussed at length in Sect. 3. These can be characterized in terms of measured values, such as the half widths or integral widths of the interference lines – determined using x-rays –, or in terms of characteristics, such as mean strains, domain sizes, or dislocation densities, – derived from the analysis of the x-ray interference line profiles, or by studying the dislocation structure using electron microscopy. As described in detail in Sect. 3 in respect to the various methods of mechanical surface treatment, the phase fractions present in the surface layer, and its textural state, may change. The result is that the deformation behavior of mechanically surface-treated components or specimens changes, as well. The plastic deformations introduced into the surface layer during mechanical surface treatments can be viewed as pre-deformations. Therefore, this section and the discussion of quasi-static loading of mechanically surface-treated material states always includes possible repercussions of the Bauschinger effect, as presented in Sect. 3.1.1a.

When mechanical loading takes place, the residual stress states induced close to the surface by mechanical surface treatments are added tensorially to the loading stresses introduced from the outside. Therefore, they are of particular interest when assessing the onset of plastic deformation. Conversely, such calculations of superpositions also permit estimates for different loading types of the loading stress ranges in which the residual stresses remain stable. As component geometries may contain superpositions that are quite complex, the following discussion is restricted to simple loading cases. Uniaxial homogeneous loading appears in

Modern Mechanical Surface Treatment. Volker Schulze
Copyright © 2006 WILEY-VCH Verlag GmbH & Co. KGaA, Weinheim
ISBN: 3-527-31371-0

the form of tensile and compressive loadings, uniaxial inhomogeneous loading is represented by bending loadings, and biaxial inhomogeneous loading is constituted by torsional loadings.

5.1.2
Quantitative Description of Processes

Assuming uniaxial residual stress states, the changes of surface layer- and core residual stresses created by quasi-static tensile, compressive or bending deformation can be examined by means of the surface-core model developed in [5.1–5.3]. The model is based on a prismatic rod, viewed as a compound, and divides the rod's total cross-sectional area A_o into a surface layer region A_{surf} and a core region A_{core}. It is assumed that longitudinal residual stresses σ_{surf}^{rs} and σ_{core}^{rs} are effective within these regions and that both regions always exhibit equal total strains. The next step is to assign state-specific surface layer- and core workhardening curves $\sigma_{surf}(\varepsilon_t)$ and $\sigma_{core}(\varepsilon_t)$, permitting an assessment of the deformation behavior of the compound and of the residual stress changes due to loading stress σ^{ls}.

$$\sigma^{ls}(\varepsilon_t) = \frac{A_{surf}}{A_o}\sigma_{surf}(\varepsilon_t) + \frac{A_{core}}{A_o}\sigma_{core}(\varepsilon_t) \tag{5.1}$$

is valid for every total strain. This allows for a quantitative assessment of residual stress relaxation according to

$$\begin{aligned} \sigma_{surf}^{rs} &= \sigma_{surf} - \sigma^{ls}(\varepsilon_t) \\ \sigma_{core}^{rs} &= \sigma_{core} - \sigma^{ls}(\varepsilon_t) \end{aligned}, \tag{5.2}$$

assuming linear-elastic unloading processes. Fig. 5.1 is a schematic representation of the model's implications regarding the deformation behavior of the compound and the dependencies of the remaining residual stresses on total strain [based on 5.4]. This is shown by using different, idealized deformation behaviors for the surface layer and the core, as well as different ratios of yield strength for the surface layer and the core. In each case it is assumed that compressive residual stresses are present in the surface layer, as is the case for all of the mechanical surface treatments examined here. Given ideal elasto-plastic behavior and a yield strength of the surface layer which is unchanged in comparison to the core (a), the deformation behavior of the compound initially shows a region of workhardening upon reaching yield strength. Beyond this point the compound, too, is deformed without workhardening. After the yield srength of the compound is exceeded, the residual stresses show a linear decrease to zero and remain there. If, however, the yield strength of the surface layer is assumed to be higher than that of the core (b), tensile loading may cause the compressive residual stresses initially present in the surface layer to become tensile and remain that way. By contrast, compressive loading effects only an incomplete relaxation of the compressive residual stresses in the surface layer. When the surface layer is worksoftened (c), tensile loading causes an incomplete relaxation of residual stresses, while compressive loading can result in residual stresses changing mathematical signs. Given elasto-plastic deformation behavior and linear workhardening in the surface layer and the core (d), the workhardening curve of the compound is sectionally linear. Depending on the relations of yield strengths and workhardening rates in the surface layer and the core, the residual stresses may also show multiple shifts from compressive to tensile values and vice versa.

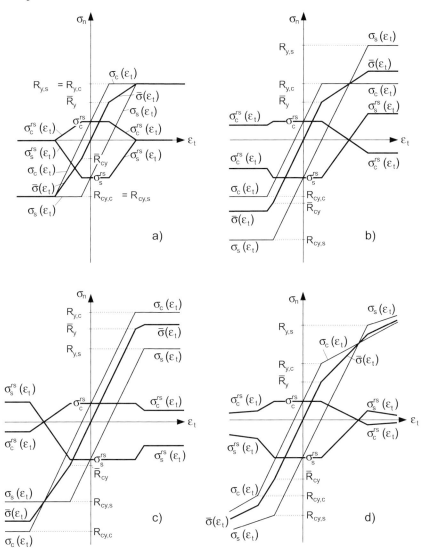

Fig. 5.1: Scheme showing deformation behavior and changes of residual stresses according to the surface-core-model for ideal elasto-plastic material behavior without change of surface yield strength (a), with increased surface yield strength (b), with decreased surface yield strength (c) and for linearly workhardening material behavior (d), acc. [5.4]

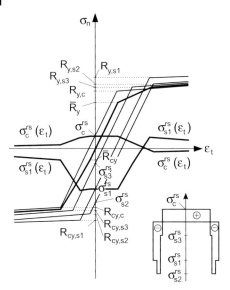

Fig. 5.2: Scheme showing deformation behavior and changes of residual stresses according to the three-surface-layers-core model, acc. [5.3]

[5.3] was the first to extend the model to include several surface layers, showing that this decreases the relaxation rates of the residual stresses (Fig. 5.2). In addition, [5.3,5.5] first incorporated ideas on the multi-axiality of deformations in the assessment, thereby showing that workhardening of the surface layer due to quasi-static loading results in multi-axial stress states even if there was a prior absence of residual stresses. [5.6,5.7] augmented this model with an algorithm, shown schematically in Fig. 5.3, by which deformation behavior and residual stress relaxation can be calculated for a compound body which is assumed to be multi-layered. Longitudinal and transversal residual stresses are applied to the individual layers and each layer receives designated nominal stress-total strain curves to represent deformation behavior. Employing the v. Mises hypothesis [5.8], the flow rule according to Prandtl-Reuss [5.9] and the method of initial stresses [5.10] permits calculating the deformation behavior of the compound, as well as the dependency of the residual stresses on loading stress and total deformation.

[5.11] presented the finite-element model (FEM model), which is even more precise and can also be used for calculating deformation behavior and residual stress relaxation. This model allows for an exact representation of the initial depth distribution of the residual stresses and for assigning different workhardening curves to the individual layers of the specimens. Sect. 5.2.2b describes applications of the model approaches listed here.

Contrary to the course of action described previously, all of the model approaches described above permit a determination of the compressive yield strengths of the surface layer, $R_{cy,surf}$. This is achieved by examining the initial residual stresses

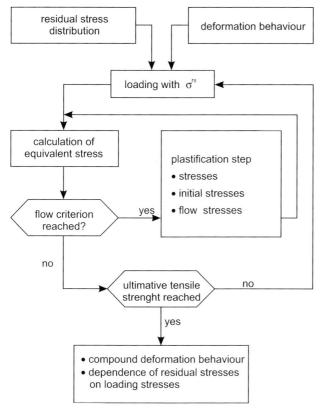

Fig. 5.3: Scheme showing the algorithm for the multi-layer-model, acc. [5.6]

and the critical loading stress σ_{crit}, which is equivalent to the compound's compressive yield strength $\bar{R}_{cy}$ required for residual stress relaxation to set in. In the uniaxial surface-core model presented first, the compound compressive yield strength and the initial residual stresses σ_{surf}^{rs} can be used to determine the compressive yield strength of the surface according to

$$R_{cy,surf} = \bar{R}_{cy} + \sigma_{surf}^{rs}. \tag{5.3}$$

The v. Mises hypothesis can be used to take into account a biaxial residual stress state. It calls for the use of the improved method introduced by [5.5] to assess the compressive yield strength of the surface

$$R_{cy,surf} = -\sqrt{\left(\bar{R}_{cy} + \sigma_{ax,surf}^{rs}\right)^2 + \left(\sigma_{tan,surf}^{rs}\right) - \left(\bar{R}_{cy} + \sigma_{ax,surf}^{rs}\right)\sigma_{tan,s}^{rs}}. \tag{5.4}$$

Within certain limits, the models additionally allow for determining the deformation behavior of the surface layer regions, as they incorporate a precise analysis of the residual stress relaxation and the deformation behavior of the compound and the core region. Applications of the equations presented here are covered in Sect. 5.2.2d.

5.2
Experimental Results and their Descriptions

5.2.1
Influences on Shape and Deformation Behavior

There have been no examinations to date in respect to how quasi-static loadings influence the topography of components whose surface layer was affected by mechanical workhardening. Neither are there clear statements regarding changes of shape caused by quasi-static loadings. Yet it is to be expected that the inhomogeneous deformation behavior of the individual surface layer zones will have an influence on the changes of shape which occur, as inhomogeneously distributed residual stresses are caused by inhomogeneous plastic deformations and thus lead to inhomogeneous workhardening states. Therefore, this study will focus only on changes in deformation behavior in comparison to equivalent states without mechanical surface treatment. These changes, in turn, are usually quite small, as the surface layer regions which are influenced by mechanical workhardening are small compared to the total cross-section. Therefore, deformation behavior is mostly unaffected, aside from the region of the onset of plastic deformation.

This is shown in Figs. 5.4–5.6 which compare the deformation behavior of shot peened and of untreated AISI4140 specimens in different heat treatment states [5.6,5.12].

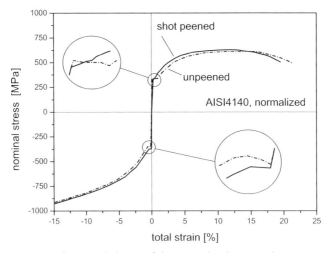

Fig. 5.4: Deformation behavior of shot peened and unpeened round specimens of normalized AISI4140 with a 5 mm gauge diameter during tensile or compressive loading [5.12]

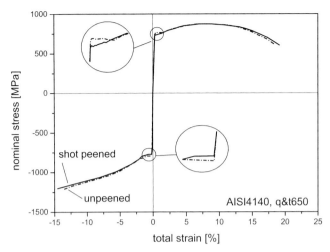

Fig. 5.5: Deformation behavior of shot peened and unpeened round specimens of AISI4140 quenched and tempered at 650 °C with a 5 mm gauge diameter during tensile or compressive loading [5.6]

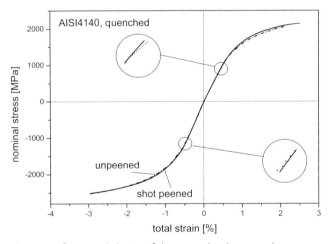

Fig. 5.6: Deformation behavior of shot peened and unpeened round specimens of quenched AISI4140 with a 5 mm gauge diameter during tensile or compressive loading [5.12]

5.2.2
Influences on Residual Stress State

a) **Basic Results**

The influence of the residual stresses has fundamentally different effects on behavior during homogeneous uniaxial tensile or compressive loading, inhomogeneous uniaxial bending loading or biaxial torsional loading. Accordingly, the basic
results regarding the changes of residual stress distribution and of residual stresses at the surface will be discussed separately for each of the loading types mentioned above. Notch loadings cannot be covered due to a lack of data.
    The depth distributions of residual stresses after tensile and compressive load-
ings are influenced by the initial residual stress state and by changes in the workhardening behavior of the surface layer versus the core. Fig. 5.7 shows results for
tensile deformation of shot peened copper [5.13]. Residual stress distributions are
found to change mathematical signs between 0.25 % and 0.31 % in regions close
to the surface, as the simple surface-core model predicts, qualitatively, for workhardened surface layers (Fig. 5.1b). Shot peened nickel also shows residual stresses becoming tensile in regions close to the surface, following a critical plastic
deformation [5.13]. Fig. 5.8 shows contrasting results for AISI4140 quenched and
tempered at 650 °C. Here, shot peening and loading using sufficiently strong tensile stresses result in small longitudinal tensile residual stresses, which occur only
immediately at the surface [5.7]. However, considerable longitudinal compressive
residual stresses are found beneath the surface. The transversal residual stresses
remain strong and within the compressive range in the entire surface layer. Residual
stresses show the greatest changes within immediate proximity to the surface,
because deformation behavior here differs from that of the rest of the surface layer.

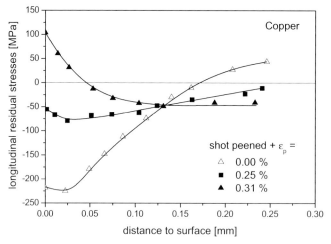

Fig. 5.7: Influence of plastic deformation on distribution of
longitudinal residual stresses after tensile loading of shot peened copper [5.13]

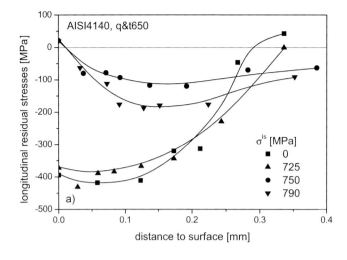

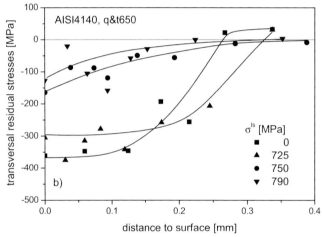

Fig. 5.8: Influence of loading stress on distribution of longitudinal (a) and transversal (b) residual stresses after tensile loading of shot peened AISI4140 in a state quenched and tempered at 650 °C [5.7]

As shown in Fig. 5.9, the findings are similar for normalized AISI4140. There is a smaller reduction in residual stresses in the transversal direction than in the longitudinal direction. Within immediate proximity to the surface, the longitudinal residual stresses become tensile, given sufficiently strong loadings [5.11]. By contrast, the changes of the longitudinal residual stresses in quenched AISI4140 are more evenly distributed throughout the surface layer, while the transversal residual stresses remain essentially unchanged [5.12]. This is depicted in Fig. 5.10.

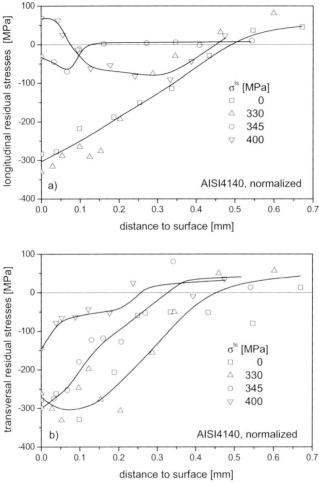

Fig. 5.9: Influence of loading stress on distribution of longitudinal (a) and transversal (b) residual stresses after tensile loading of shot peened AISI4140 in a normalized state [5.12]

The example of shot peened normalized AISI4140, depicted in Fig. 5.11, shows that residual stresses at the surface are dependent on the loading stresses applied [5.12]. When loading is compressive, the relaxation of longitudinal residual stresses sets in at a critical loading stress of -200 MPa. When tensile loading is applied, this relaxation starts at a noticeably higher level of 300 MPa. Both tensile and compressive loadings cause the longitudinal residual stresses at the surface to reach the tensile range. However, tensile loading causes small compressive residual stresses to re-develop quite rapidly. As expected, and in accordance with the schematic representation in Fig. 5.1, a workhardened surface layer is created, showing a decreased workhardening rate compared to the core. The transversal residual

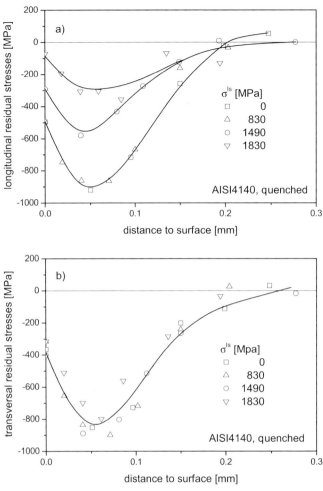

Fig. 5.10: Influence of loading stress on distribution of longitudinal (a) and transversal (b) residual stresses after tensile loading of shot peened AISI4140 in a quenched state [5.12]

stresses, on the other hand, change mathematical signs only when compressive loading is applied, and they will not be removed completely by tensile loading. The same residual stress values are depicted in Fig. 5.12, shown versus the total strains achieved. It is evident that major changes of the residual stresses occur even when the total strains are still very small.

Fig. 5.13 shows the corresponding values of shot peened AISI4140, quenched and tempered at 650 °C [5.6]. At 685 and -400 MPa, the loading stresses required for initiating residual stress relaxation are substantially higher than they are for the normalized state, and the loading stress interval is significantly smaller for tensile longitudinal stresses. Besides this, the effects which occur are similar.

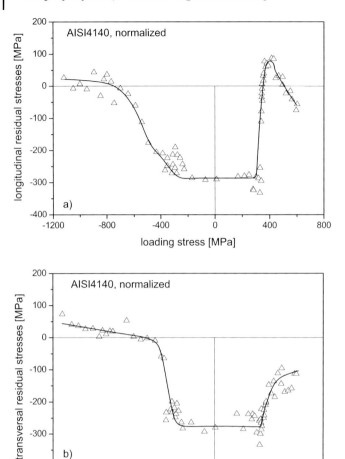

Fig. 5.11: Longitudinal (a) and transversal (b) residual stresses at the surface vs. tensile or compressive loading stress for shot peened AISI4140 in a normalized state [5.12]

Both longitudinal and transversal residual stresses are found to become tensile when compressive loading is applied. The relaxation of transversal residual stresses is incomplete, albeit more extensive, compared to the normalized state. Behavior during tensile loading indicates the presence of a workhardened surface layer, its workhardening rate reduced in comparison to the core. In contrast to this, the surface layer should, according to compressive loading behavior, show at least a decreased workhardening rate, if not even the additional appearance of a pronounced Bauschinger effect. Fig. 5.14 shows the dependency of the longitudinal surface residual stresses on loading stress for shot peened AISI4140, quenched and tempered at 450 °C [5.14]. In this higher strength material state, there are

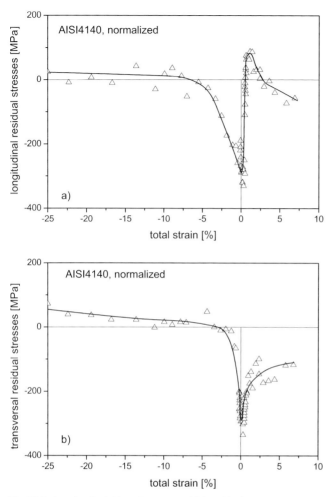

Fig. 5.12: Longitudinal (a) and transversal (b) residual stresses at the surface vs. total strain after tensile or compressive loading for shot peened AISI4140 in a normalized state [5.12]

further increases of the critical loading stresses, at which residual stress relaxation sets in, to 1150 and -600 MPa. Tensile loading effects only an incomplete relaxation of residual stresses, also in the longitudinal direction. Great compressive loading stresses fail to create an approximately constant level of tensile residual stresses. Instead, the occurring tensile residual stresses increase continuously as compressive loading stresses rise. Therefore it can be assumed that tensile loading does not cause a pronounced workhardening of the surface layer compared to the core, and/or that the workhardening rate of the surface layer remains essentially unchanged, which corresponds to Fig. 5.1. By contrast, compressive loading behavior shows that the surface layer must have, at least, a reduced workhardening

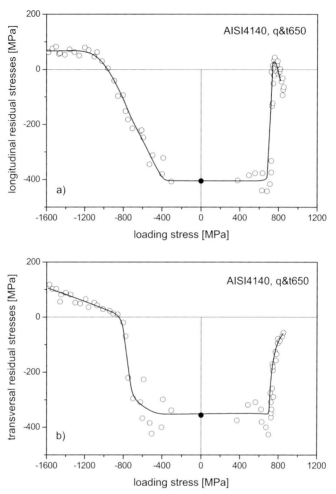

Fig. 5.13: Longitudinal (a) and transversal (b) residual stresses at the surface vs. tensile or compressive loading stress for shot peened AISI4140 in a state quenched and tempered at 650 °C [5.6]

rate compared to the core, or that there must be the additional appearance of the Bauschinger effect. This resembles the behavior of the state quenched and tempered at 650 °C. Fig. 5.15 shows results for shot peened quenched AISI4140. During tensile loading, such high strength material states show only an incomplete relaxation of longitudinal residual stresses, whereas compressive loading gives rise to small tensile residual stresses [5.12]. Transversal residual stresses remain nearly unaffected by tensile loading and are essentially removed completely by compressive loading. The decisive factor regarding tensile behavior is probably found in the low ductility of this state, permitting only small changes of residual

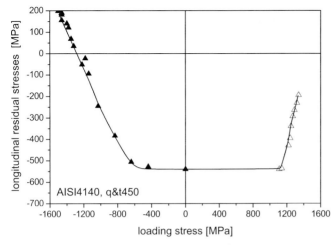

Fig. 5.14: Longitudinal residual stresses at the surface vs. tensile or compressive loading stress for shot peened AISI4140 in a state quenched and tempered at 450 °C [5.14]

stresses before failing. Compressive behavior, on the other hand, is in accordance with Fig. 5.1 in that the surface layer must show at least a reduced workhardening rate compared to the core, as is the case for the state quenched and tempered at 450 °C.

Fig. 5.16 shows an overview of the longitudinal surface residual stresses found in shot peened TiAl6V4. As in the low strength states of AISI4140, tensile loading causes a small loading stress interval in which tensile residual stresses are created. Therefore, the surface layer should be workhardened in comparison to the core and have a reduced workhardening rate [5.15]. The relaxation of residual stresses during compressive loading, however, remains incomplete, due to buckling problems in the chosen sample geometry. By example of shot peened AlCu5Mg2 aluminum alloy, Fig. 5.17 once again sums up the influence of the heat treatment state and thus, of strength and ductility, on longitudinal residual stresses during tensile or compressive loading [5.5]. The as-delivered state and the cold aged state experience a complete removal of residual stresses during tensile loading, along with an interval of tensile residual stresses, i.e. the surface layer is workhardened and has a lower workhardening rate than the warm aged state. The latter shows only a small relaxation, followed by a re-formation, of residual stresses. Therefore, the surface layer is assumed to worksoften. Compressive loading, on the other hand, always yields a complete removal of residual stresses, with the exception of the as-delivered state. The reason is that surface layer worksoftening due to Bauschinger effect, but at least a reduced workhardening rate, are expected to occur.

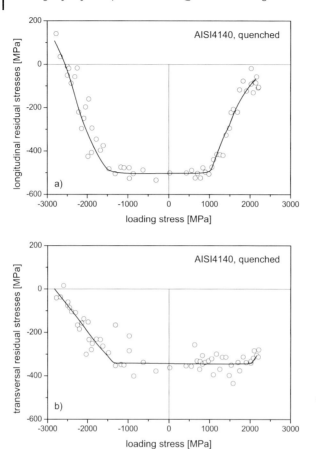

Fig. 5.15: Longitudinal (a) and transversal (b) residual stresses at the surface vs. tensile or compressive loading stress for shot peened AISI4140 in a quenched state [5.12]

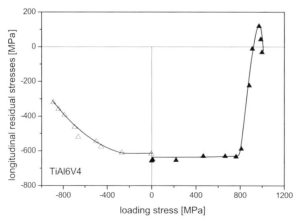

Fig. 5.16: Longitudinal residual stresses at the surface vs. tensile or compressive loading stress for shot peened TiAl6V4 [5.15]

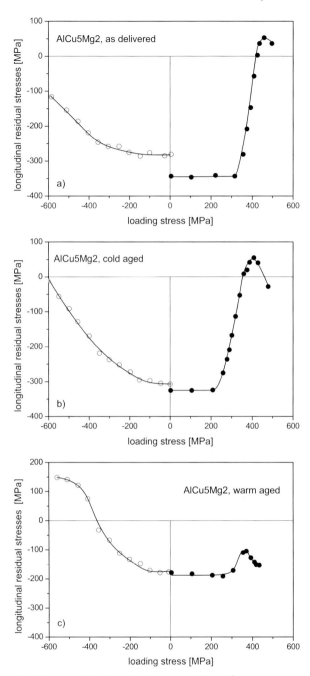

Fig. 5.17: Longitudinal residual stresses at the surface vs. tensile or compressive loading stress for shot peened AlCu5Mg2 in the as-delivered (a), cold aged (b) and warm aged state (c) [5.5]

The results for the different material states can be summarized based on the knowledge gained from the simple surface-core model: Tensile loading removes longitudinal stresses completely, if the surface layer is workhardened compared to the bulk material. In addition, the residual stresses repeatedly change mathematical signs if the workhardening rate in the workhardened surface layer is lower. The ductility of the material state is also crucial, as the changes of residual stresses cannot be completed while minor elongations to fracture are occurring. Regarding compressive loading, ductility is only a secondary aspect, and the region accessible to loading is restricted only by buckling phenomena. Therefore, compressive loading results in a complete removal of longitudinal residual stresses in nearly every case. This requires at least a reduced workhardening rate in the surface layer, and possibly worksoftening due to Bauschinger effect.

Residual stress distributions after bending loading were examined for AISI4140in the state quenched and tempered at 450 °C, primarily for analyzing the differences of residual stress stability after modified peening treatments [5.16,5.17]. This was carried out by applying a single loading with a fictitious loading stress at the surface of 1000 MPa. Unloading was followed by the application of an equal loading in the inverse direction. Combined, this is equivalent to the complete first load cycle of alternating bending loading. As Fig. 5.18 shows for the conventionally shot peened state, the longitudinal residual stresses decrease noticeably on the compressively loaded top side, while still remaining nearly unchanged on the bottom side which is experiencing tensile loading. After inverse loading, the residual stresses on the top and bottom sides again approximate each other, having undegone a marked relaxation on the bottom side, as well. Overall, residual stresses are reduced to about half their initial values. Fig. 5.19 shows that

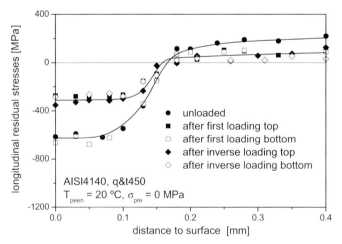

Fig. 5.18: Distribution of longitudinal residual stresses for conventionally shot peened specimens made of AISI4140 in a state quenched and tempered at 450 °C after single bending loading with a fictitious surface stress of 1000 MPa and subsequent inverse loading [5.16]

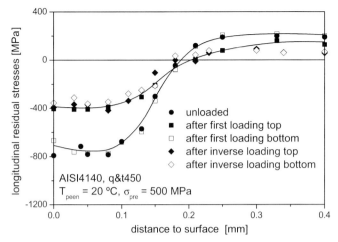

Fig. 5.19: Distribution of longitudinal residual stresses for specimens stress peened with $\sigma_{pre} = 500$ MPa made of AISI4140 in a state quenched and tempered at 450 °C after single bending loading with a fictitious surface stress of 1000 MPa and subsequent inverse loading [5.16]

the residual stresses after stress peening using a tensile prestress of 500 MPa are clearly greater initially. However, applying the same loading as to a conventionally peened state reduces the residual stresses to levels that are only insignificantly higher. By contrast, Fig. 5.20 shows that warm peening at 290 °C creates residual

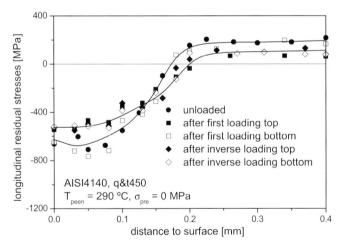

Fig. 5.20: Distribution of longitudinal residual stresses for specimens warm peened at a temperature of $T_{peen} = 290$ °C made of AISI4140 in a state quenched and tempered at 450 °C after single bending loading with a fictitious surface stress of 1000 MPa and subsequent inverse loading [5.17]

stresses which are only marginally greater than after conventional peening. However, they are far more stable during bending loading. Only stress peening at elevated temperatures, using a prestress of 500 MPa and a temperature of 290 °C, results in residual stress values being slightly increased over the warm peened state after bending loading. This is depicted in Fig. 5.21. Yet again, there is the penalty of residual stresses being reduced substantially during the initial loading.

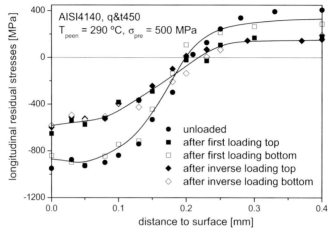

Fig. 5.21: Distribution of longitudinal residual stresses for specimens stress peened with $\sigma_{pre} = 500$ MPa at a temperature of $T_{peen} = 290$ °C made of AISI4140 in a state quenched and tempered at 450 °C after single bending loading with a fictitious surface stress of 1000 MPa and subsequent inverse loading [5.16]

Fig. 5.22 contains data regarding the loading stress-dependency of residual stress relaxation during bending loading for shot peened AISI4140 in different heat treatment states [according to 5.14,5.18]. Behavior during initial loading largely corresponds to the expectations raised by the tensile and compressive loadings described above. The loading stress at which residual stress relaxation begins is noticeably smaller on the compressively loaded side than it is on the side experiencing tensile loading. By contrast, during bending loading the relaxation rate on the compressively loaded side of a normalized specimen roughly equals the rate found during homogeneous compressive loading, while showing marked relaxations in quenched and tempered specimens. [5.14] attributes this to a more pronounced supporting effect of the deeper regions in normalized material, which is not found in this way in quenched and tempered states due to the plateau of residual stresses. As shown in Fig. 5.23, bending- and compressive loading of normalized specimens cause approximately equal gradients of the sum of loading- and residual stresses. In a material quenched and tempered at 450 °C, the sum of loading- and residual stresses decreases far more rapidly during bending loading than during purely compressive loading. This points toward a reduced residual stress

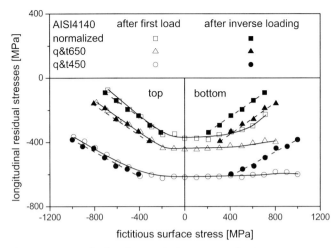

Fig. 5.22: Longitudinal surface residual stresses vs. fictitious surface stress at bending loading for shot peened AISI4140 in a normalized state and states quenched and tempered at 450 °C or 650 °C after single bending loading with a fictitious surface stress of 1000 MPa and subsequent inverse loading [5.14, 5.18]

relaxation rate. On the side experiencing tensile loading, however, relaxation behavior during bending is clearly changed in comparison to the behavior found for homogeneous tensile loading. The loading stresses at which residual stress relaxation starts are significantly smaller here than they are for homogeneous tensile loading. The reason is that instead of plastic deformation occurring in the core area, changes of the residual stresses take place first, caused by the relaxation of residual stresses on the compressively loaded side. Given greater loading stresses – which would cause a greater relaxation of residual stresses during tensile loading – bending loading results in a much smaller relaxation of residual stresses, as the supporting effects of deeper material regions again play a major part. This is particularly evident regarding the normalized state. Subsequent inverse loading causes a strong relaxation of residual stresses on the side which initially experienced tensile loading, while hardly any further changes of residual stresses occur on the initially compressively loaded side. This results in residual stress distributions that are approximately equal on both sides. These findings are confirmed by the results for shot peened TiAl6V4 shown in Fig. 5.24 [5.15]. The residual stress relaxation occurring on the compressively loaded side is comparable to the rate found for compressive loading alone, as is the case for normalized AISI4140. By contrast, residual stress relaxation on the side of tensile loading is strongly decelerated, as the supporting effect of deeper regions takes on an important role.

Residual stresses after torsional loading were examined in [5.19] for AISI4140 which was either conventionally peened, stress peened using 500 MPa of prestress, or warm peened at 290 °C. Fig. 5.25 shows how the principal residual

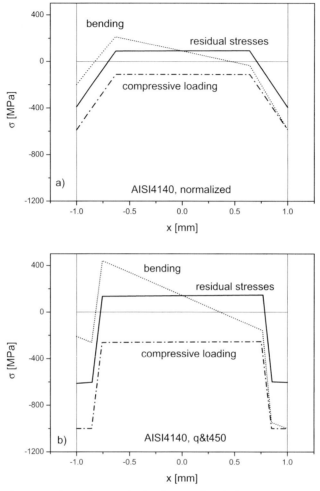

Fig. 5.23: Scheme showing the distribution of residual stresses and sum of loading and residual stresses at initiation of plastic deformation during compressive and bending loading for normalized (a) and quenched and tempered (b) AISI4140 [5.14]

stresses change in the direction of tensile loading at 45° and in the direction of compressive loading at 135° after loading with a fictitious surface shear stress of 800 MPa. Due to the overlap of compressive residual stresses and the compressive loading stress component, a much more pronounced relaxation of residual stresses is found in the 135° direction than in the 45° direction, in which residual stresses are essentially stable. As there is only limited data available, interactions between the individual stress components and supporting effects, such as those proven for bending loading, cannot be discussed.

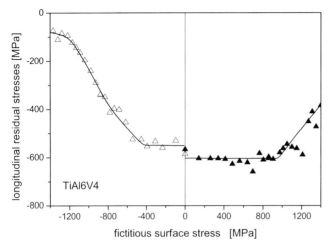

Fig. 5.24: Longitudinal residual stresses at the surface vs. fictitious surface stress after bending loading for shot peened TiAl6V4 [5.15]

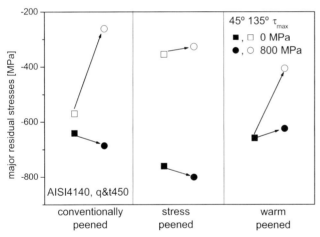

Fig. 5.25: Changes of principal residual stresses in the tensilely loaded 45°- and in the compressively loaded 135° direction after torsional loading with a fictitious surface shear stress of 800 MPa after conventional shot peening, stress peening with a prestress of 500 MPa and warm peening at 290 °C for AISI4140 in a state quenched and tempered at 450 °C [5.19]

The relaxation of residual stresses during tensile and compressive loading at elevated temperatures was examined by [5.14] for normalized AISI4140 and for shot peened AISI4140 quenched and tempered at 450 °C. The following will focus exemplarily on the results for the quenched and tempered state, deformed at 25, 250 and 400 °C. Due to thermal relaxation of residual stresses (comp.

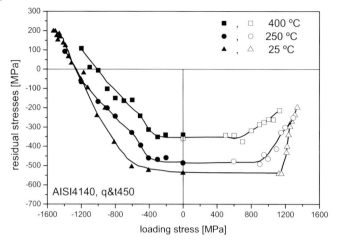

Fig. 5.26: Influence of temperature on longitudinal residual stresses at the surface vs. tensile or compressive loading stress for shot peened AISI4140 in a state quenched and tempered at 450 °C [5.14]

Sect. 4.2.2a), the residual stresses decrease already during the heating phase. As Fig. 5.26 shows, the initial relaxation of the residual stresses is the more pronounced the higher deformation temperature is. In respect to both tensile and compressive loading, higher temperatures see residual stress relaxation set in at lower values of loading stress. During tensile loading, the rates of residual stress relaxation decrease as temperatures increase, while remaining approximately constant during compressive loading. The findings for the normalized state are qualitatively equivalent [5.11,5.20].

b) Modeling of Residual Stress Relaxation

The model approaches presented in Sect. 5.1.2 have been used in a variety of ways to provide a quantitative description of the deformation behavior and the loading stress-dependency of residual stress relaxation during quasi-static loading. [5.3,5.5] developed the uniaxial surface-core model and applied it to the three heat treatment states of AlCu5Mg2 aluminum alloy presented above. Figs. 5.27–5.29 show that the model is capable of representing basic results for tensile and compressive loading, such as the onset of residual stress relaxation, any changes of the residual stresses' mathematical signs and the residual stress values which remain after great strains. However, significant deviations between the model and the experiment must be accepted, particularly in the area of mean strains. In agreement with the assumptions made in Sect. 5.2.1a, it becomes evident that the as-delivered state and the cold aged state correspond best, when applying surface layer workhardening and the concurrent reduction of the workhardening rate in the model. For the warm aged state (Fig. 5.29), however, a pronounced work-

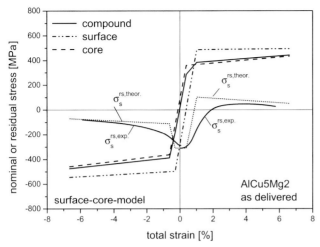

Fig. 5.27: Deformation behavior of surface and core assumed in the surface-core-model, calculated deformation behavior of the compound and calculated residual stress relaxation during tensile and compressive loading for shot peened AlCu5Mg2 in the as-delivered state [5.3]

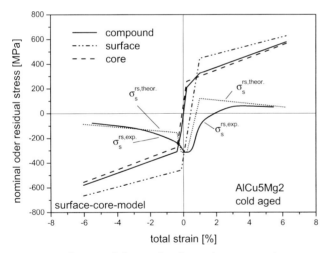

Fig. 5.28: Deformation behavior of surface and core assumed in the surface-core-model, calculated deformation behavior of the compound and calculated residual stress relaxation during tensile and compressive loading for shot peened AlCu5Mg2 in the cold aged state [5.3]

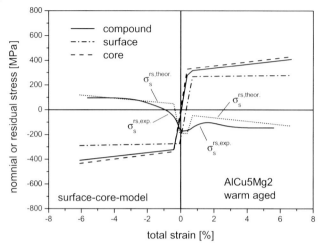

Fig. 5.29: Deformation behavior of surface and core assumed in the surface-core-model, calculated deformation behavior of the compound and calculated residual stress relaxation during tensile and compressive loading for shot peened AlCu5Mg2 in the warm aged state [5.3]

softening of the surface layer and, concurrently, ideal elasto-plastic material behavior must be assumed in order to explain the onset of residual stress relaxation during compressive loading, in particular, as well as the residual stress values that remain after great tensile and compressive strains. This is plausible regarding the microstructural level, as the state in question is a highest strength state which cannot be deformed much further by mechanical surface treatment, but will rather worksoften during the cyclic plastic deformation which occurs during shot peening.

[5.3] applied the same surface-core model to shot peened TiAl6V4. The assumption of surface layer workhardening occurring alongside a reduction of the workhardening rate again yields good descriptions, yet the problems outlined for the aluminum alloy in respect to describing mean strains also appear here. In the case of compressive loading, these deviations also extend to greater strains. Additionally, [5.3] applied the model to TiAl6V4 during bending loading. The results, depicted in Fig. 5.30, correspond well with the experimental values, when yield strength was reduced linearly down to the core's values – starting from elevated levels at the surface – and ideal elasto-plastic deformation behavior was assumed for each layer. As shown in Fig. 5.31, given these assumptions and a specimen geometry measuring 10 mm in height, residual stress relaxation on the side experiencing compressive loading sets in at a fictitious loading stress at surface of about -400 MPa in the compressive residual stress maximum. On the side receiving tensile loading, however, around 750 MPa of loading stress are required to initiate residual stress relaxation in the transitional region between the core and the workhardened surface layer.

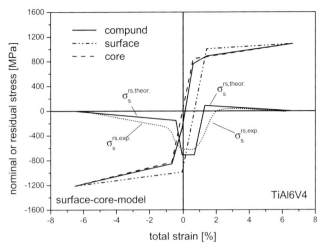

Fig. 5.30: Deformation behavior of surface and core assumed in the surface-core-model, calculated deformation behavior of the compound and calculated residual stress relaxation during tensile and compressive loading for shot peened TiAl6V4 [5.3]

[5.21] also used the surface-core model to describe the changes of residual stress caused by quasi-static loading of shot peened AISI4140 in a quenched and tempered state. Using a surface layer strength of about 30 % less compared to the core, it is possible to describe the initiation of residual stress relaxation during tensile and compressive loading, as shown in Fig. 5.32. While the residual stress relaxation caused by tensile loading is described quite well, the residual stress relaxation which occurs during compressive loading is clearly over-estimated for major regions of the loading stress. The model predicts that tensile loading creates an incomplete relaxation of residual stresses and compressive loading results in residual stresses becoming tensile. Thus, it may be stated that the model corresponds well on a qualitative level. Although a small increase of dislocation density is observed (Fig. 5.32), the surface layer strength is reduced compared to the core. This is a manifestation of the Bauschinger effect, which is particularly pronounced in quenchend and tempered steels. The Bauschinger effect occurs because compressive loading takes the reverse direction of the stretching created during shot peening. Residual stress stability can be described much more precisely using the FEM model introduced by [5.18]. The initial distribution of residual stresses is represented by 6 surface layer regions, as depicted in Fig. 5.33. Presumed deformation behavior, shown in Fig. 5.34, is independent of distance to surface in the case of tensile loading, while it is divided into three areas in the case of compressive loading. While the outer three surface layers are classified as a distinct, worksoftened 'surface' region, the three following surface layers show far less worksoftening compared to the core region. At the same time, it is assumed that the worksoftening incorporates a reduction of flow stress which is constant and independent of strain. As Fig. 5.32 shows, the relaxation of residual

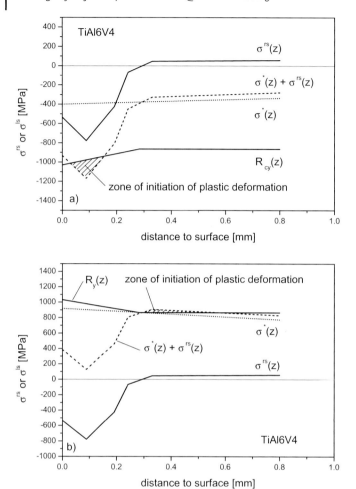

Fig. 5.31: Distributions of residual stresses, of yield strength assumed in the surface-core-model, of loading stresses and calculated sums of loading- and residual stresses at initiation of residual stress relaxation after bending loading with a fictitious surface stress of 920 MPa on the tensilely loaded side (a) and with a fictitious surface stress of –920 MPa on the compressively loaded side (b) of 10 mm-high bending bars made of shot peened TiAl6V4, acc. [5.3]

stresses is described very well for both tensile and compressive loading. This is because residual stress distribution and deformation behavior are represented at greater resolution and the multiple loading axes are taken into account. Fig. 5.35 shows a parameter study which indicates that the starting point of residual stress relaxation during compressive loading is determined by the compressive yield strength at the surface, while core strength affects residual stresses only at the highest values of loading stress.

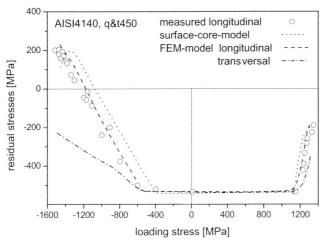

Fig. 5.32: Comparison of longitudinal and transversal residual stresses vs. loading stress after tensile and compressive loading, calculated in the surface-core-model or the FEM-model and experimentally determined for shot peened AISI4140 in a state quenched and tempered at 450 °C [5.11,5.21]

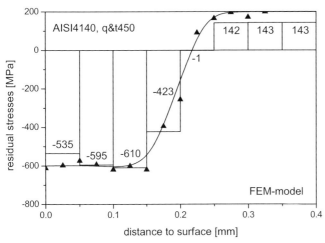

Fig. 5.33: Distributions of residual stresses as assumed in the FEM-calculations for shot peened AISI4140 in a state quenched and tempered at 450 °C [5.11]

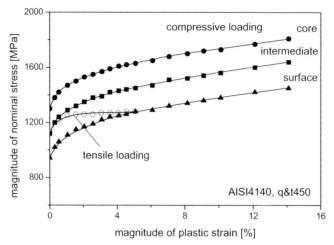

Fig. 5.34: Deformation behavior of surface layers and core region during tensile and compressive loading as assumed in the FEM-calculations for shot peened AISI4140 in a state quenched and tempered at 450 °C [5.11]

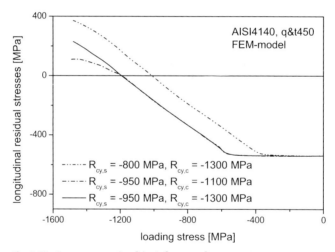

Fig. 5.35: Parameter study of the influence of compressive yield strength at the surface and the core on the residual stresses relaxation behavior during compressive loading performed in the FEM-calculations for shot peened AISI4140 in a state quenched and tempered at 450 °C [5.11]

Model calculations following the multi-layer model delineated by [5.6,5.7] are available for shot peened, AISI4140 quenched and tempered at 650 °C. The calculations are based on 6 surface layers and a core region. The initial depth distributions of longitudinal and transversal residual stresses were induced as indicated in Fig. 5.36. Fig. 5.37 depicts assumed deformation behavior, which shows incre-

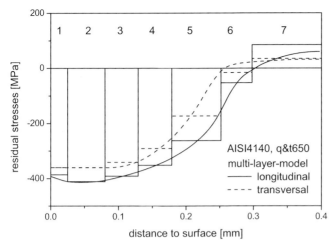

Fig. 5.36: Distributions of longitudinal and transversal residual stresses as assumed for the 7 layers in the multi-layer-model for shot peened AISI4140 in a state quenched and tempered at 650 °C and comparison with the measured values [5.6]

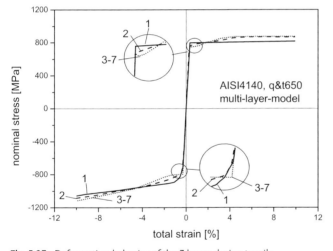

Fig. 5.37: Deformation behavior of the 7 layers during tensile and compressive loading as assumed in the multi-layer-model for shot peened AISI4140 in a state quenched and tempered at 650 °C [5.6]

ments for the two outer surface layers in comparison to the other layers. Deformation behavior under both tensile and compressive deformation shows a worksoftening of the surface layers at small strains, caused by the Bauschinger effect. Workhardening occurs at greater strains and is caused by pre-deformation. Fig. 5.38 demonstrates that using these assumptions allows for an excellent repre-

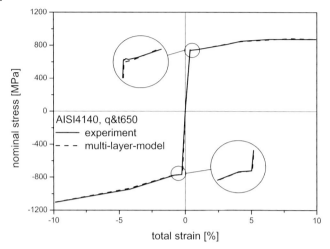

Fig. 5.38: Deformation behavior of a rod calculated using the multi-layer-model for shot peened AISI4140 in a state quenched and tempered at 650 °C and comparison with the measured values [5.6]

sentation of the deformation behavior of the shot peened rod for both tensile and compressive deformation. The calculated values and the experimental values of flow stress deviate by 1.5 % at most, even in the Lüders range, which is particularly difficult to describe. The residual stress changes are shown in Fig. 5.39 as a function of tensile or compressive loading stress. They, too, correspond very well with the experimental results. The start of residual stress relaxation, as well as the rates of relaxation – showing marked differences for tensile versus compressive loading – and the mathematical sign changes of the longitudinal residual stresses, are represented very well. Merely the distribution of the transversal residual stresses is described poorly for great loading stresses. This is due to residual stresses being reduced too far during tensile loading and, secondly, to the fact that the creation of tensile residual stresses is not represented for compressive loading.

Figs. 5.40 and 5.41 demonstrate that the multi-layer model is also quite capable of describing the deformation behavior and the dependency of residual stresses on loading stress for normalized AISI4140 [5.12]. All essential effects are described precisely, aside from the transversal residual stress values, which become minutely small in the model calculations for high loading stresses. By contrast, the FEM model – included in Fig. 5.41 – yields a description of residual stress relaxation which is less precise. This is particularly evident in that longitudinal residual stresses are misrepresented for high compressive loading stresses and in the fact that the tensile stresses at which the longitudinal residual stresses become tensile are a little too high in this model calculation [5.18].

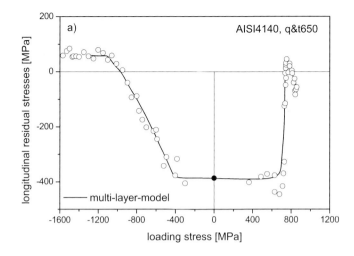

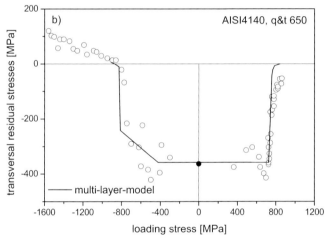

Fig. 5.39: Longitudinal (a) and transversal (b) residual stresses at the surface measured and calculated using the multi-layer-model vs. loading stress after tensile and compressive loading for shot peened AISI4140 in a state quenched and tempered at 650 °C [5.6]

Figs. 5.42 and 5.43 show that the multi-layer model also yields a very precise description of the behavior of shot peened quenched AISI4140 [5.12]. Both deformation behavior and the dependency of the longitudinal residual stresses on loading stress are reflected well in the model. Only the relaxation of the transversal residual stresses is clearly overestimated for tensile loading, while being slightly underestimated for compressive loading in the mid-range of the stress values.

The FEM model was used to describe the residual stress relaxation caused by isothermal tensile or compressive loading at 250 and 400 °C for shot peened

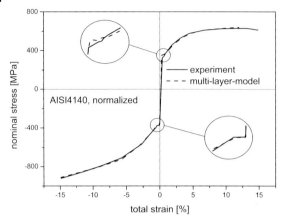

Fig. 5.40: Deformation behavior of a rod calculated using the multi-layer-model for shot peened AISI4140 in a normalized state and comparison with the measured values [5.12]

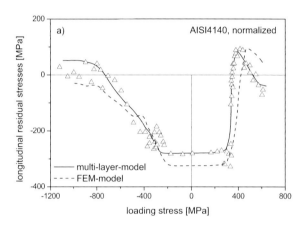

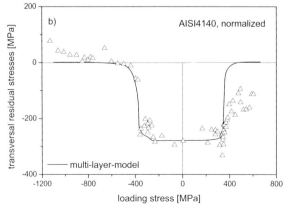

Fig. 5.41: Longitudinal (a) and transversal (b) residual stresses at the surface measured and calculated using the multi-layer-model or the FEM-model vs. loading stress after tensile and compressive loading for shot peened AISI4140 in a normalized state [5.11,5.12]

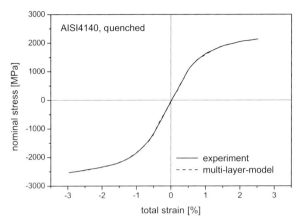

Fig. 5.42: Deformation behavior of a rod calculated using the multi-layer-model for shot peened AISI4140 in a quenched state and comparison with the measured values [5.12]

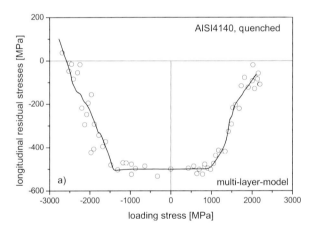

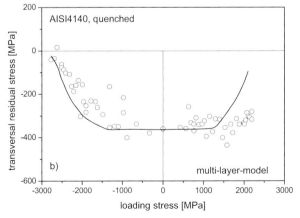

Fig. 5.43: Longitudinal (a) and transversal (b) residual stresses at the surface measured and calculated using the multi-layer-model vs. loading stress after tensile and compressive loading for shot peened AISI4140 in a quenched state [5.12]

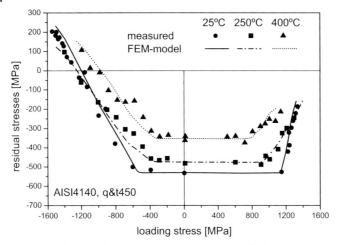

Fig. 5.44: Influence of temperature on longitudinal residual stresses at the surface measured and calculated using the FEM-model vs. tensile or compressive loading stress for shot peened AISI4140 in a state quenched and tempered at 450 °C [5.11,5.20]

AISI4140 in a normalized state and in a state quenched and tempered at 450 °C [5.11,5.20]. As depicted in Fig. 5.44, residual stress relaxation is described very well at all temperatures for the quenched and tempered state. The start of residual stress relaxation, the rates of residual stress relaxation, as well as the position of zero residual stresses correspond well with the experimental results for compressive loading. Regarding the normalized state, only tensile loading and small degrees of compressive loading yield comparably good descriptions of the experimental results. However, high levels of compressive loading reduce residual stresses more rapidly than the model predicts. [5.11,5.20] presume the following cause: As the specimens become noticeably barrel-shaped when subjected to the greatest compressive strains, additional tensile residual stress amounts develop which are not described by the FEM model.

In summary, it may be stated that the level of precision increases from the surface-core model to the FEM model, up to the multiple-layer model. The surface-core model lends itself primarily to a description of the begin of residual stress relaxation and facilitates an understanding of the basic results, such as residual stresses changing mathematical signs. The multiple-layer model and the FEM model, on the other hand, include the multiaxiality of the stress state and allow for examining the initial gradients of residual stresses and workhardening at a higher degree of local resolution. This makes it possible to describe the loading stress-dependency of the longitudinal and transversal residual stresses more accurately. The multiple-layer model even permits a highly precise description of the deformation behavior of the compound. The FEM model should show the same potential if a similarly extensive effort were made in selecting the local workhardening curves of the surface layers, which are required as initial parameters. [5.18] intentionally described this more abstractly, thereby accepting a lower level of precision than [5.6,5.7]. Generally stated, model calculations using the multiple-layer model are the most informative in respect to the surface layer workhardening curves, which are obtained by an iterative adaption to the deformation behavior of the compound and the loading stress-dependency of the residual stresses.

c) **Initiation of Residual Stress Relaxation**

Numerous studies have focused on the initial stage of the relaxation of residual stresses close to the surface due to quasi-static loading. This subject does not require the kind of extensive modeling outlined in the previous sections, as the resistance to the onset of plastic deformation close to the surface can be derived from just the loading stress-dependency of the residual stresses. The initial stage of residual stress relaxation during quasi-static loading is particularly relevant to estimating and assessing the extent of cyclic relaxation of residual stresses. Therefore, examinations of residual stress stability during cyclic loading often focus on residual stress stability for relatively small quasi-static loadings, as these are essential for assessing the influences residual stresses are expected to have on fatigue strength (comp. Sect. 6). This kind of information is crucial for an assessment of modified peening methods, and therefore, the following shall summarize particularly those results which influence the initial stage of residual stress relaxation during single bending loading and during bending loading with inverse loading.

The course shown in Fig. 5.45 indicates surface residual stresses versus applied fictitious loading stress at the surface, as found in conventionally shot peened AISI4140

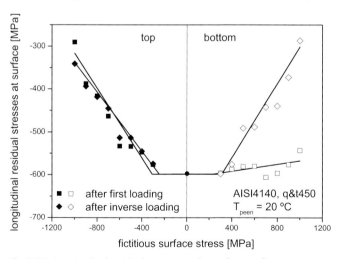

Fig. 5.45: Longitudinal residual stresses at the surface vs. fictitious surface stress after single bending loading or bending loading with subsequent inverse loading for conventionally shot peened AISI4140 in a state quenched and tempered at 450 °C [5.22]

steel quenched and tempered at 450 °C [5.22]. Above a loading stress at the surface of about 310 MPa, there is a strong relaxation of the residual stresses on the compressively loaded top side of the specimen. By contrast, only a small decrease of residual stresses is found on the bottom side undergoing tensile loading. This occurs as a reaction to the residual stress relaxation taking place on the top side and is required for achieving a new stress equilibrium. After loading is inversed, however, pronounced changes of the residual stresses are observed on the side which initially received tensile loading, now experiencing compressive loading. The top side, now receiving tensile loading, shows no significant changes of the residual stress values. This loading is equivalent to one complete cycle of cyclic bending loading. The result is that the loading stress-dependencies of the remaining residual stresses are approximately equal on both sides. Thus, the critical loading stress at the surface required for initiation of residual stress relaxation can be determined as 310 MPa.

Fig. 5.46 shows results for AISI4140 of the same heat treatment state which was warm peened at 290 °C. Subsequent quasi-static loading causes residual stress relaxation which is qualitatively equivalent, except that critical loading stress at the surface is raised to 500 MPa [5.22]. The influence of peening temperature in the context of bending loading and inverse loading is represented in Fig. 5.47 [5.23]. Surface residual stresses are slightly higher for peening temperatures of 310 and 330 °C. They remain stable even when loading stresses at the surface increase noticeably from 500 to 560 and 670 MPa when the peening temperature is raised. Apparently the dislocation structures of the surface layer become pinned and thus, yield strengths at the surface are increased, resulting in residual stresses remaining stable up to great loading stresses. This is caused by the dynamic and static strain aging effects outlined in Sect. 3.3.1a.

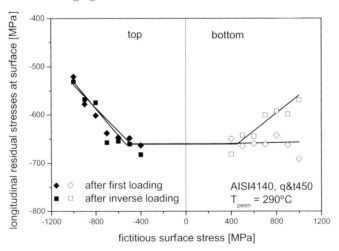

Fig. 5.46: Longitudinal residual stresses at the surface vs. fictitious surface stress after single bending loading or bending loading with subsequent inverse loading for warm peened (T_{peen} = 290 °C) AISI4140 in a state quenched and tempered at 450 °C [5.22]

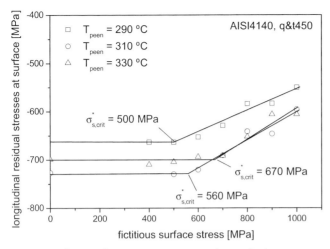

Fig. 5.47: Influence of peening temperature on longitudinal residual stresses at the surface vs. fictitious surface stress during bending loading with subsequent inverse loading for warm peened AISI4140 in a state quenched and tempered at 450 °C, acc.[5.23]

Contrasting results are shown in Fig. 5.48 for a state annealed for 20 minutes at 300 °C, which reduces initial residual stresses. Their stability is increased substantially during bending loading. The result is that critical loading stress at the surface is raised to 690 MPa, while there is no qualitative change of behavior regarding initial loading and inverse loading [5.22]. The influence of annealing conditions on the residual stress stability found after bending loading with inverse loading with a maximum stress at the surface of 1000 MPa is depicted in Fig. 5.49 [5.24]. Increasing annealing temperature decreases the initial residual stresses due to thermal residual stress relaxation (comp. Sect. 4.2.2a). However, there is also a decrease in the residual stress changes which occur during bending loading and inverse loading. The result is that the remaining residual stresses show an initial increase, only to decrease again when the temperature of optimal residual stress stabilization is exceeded. This temperature decreases from about 310, over 260 to 240 °C as annealing time extends from 1, over 5 to 60 minutes. Fig. 5.50 shows these values as a $1/kT_a$ vs. t_a-plot which appears to be degressive. Using the difference quotient in a weighted plot and basing it on a sectionally-defined Arrhenius relation allows for determining activation enthalpies. They are 0.86 and 2.93 eV respectively, for short and long annealing times. This indicates that carbon diffusion within the ferrite contained in AISI4140 is the rate-controlling process during the shorter periods. This shows that conventionally peened and subsequently annealed states also experience the stabilizing effect static strain aging has on residual stress relaxation, as mentioned in Sect. 3.3.1a in the context of warm peening. In respect to longer annealing times, however, residual stress stability appears to be determined rather by the activation enthalpy for self-diffusion, indicating

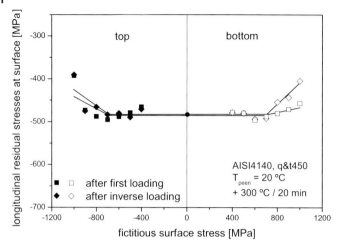

Fig. 5.48: Longitudinal residual stresses at the surface vs. fictitious surface stress after single bending loading or bending loading with subsequent inverse loading for conventionally shot peened and subsequently annealed ($T_a = 300\,°C$, $t_a = 20$ min) AISI4140 in a state quenched and tempered at 450 °C [5.22]

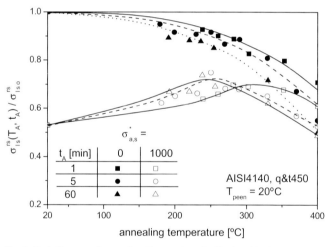

Fig. 5.49: Influence of annealing time on longitudinal residual stresses at the surface vs. fictitious surface stress during bending loading with subsequent inverse loading for shot peened AISI4140 in a state quenched and tempered at 450 °C [5.24]

that creep processes constitute the mechanism which determines the residual stress value. It may be stated that the residual stresses which remain after bending loading are initially determined by static strain aging, while being increasingly affected by thermal residual stress relaxation as annealing times are extended. At

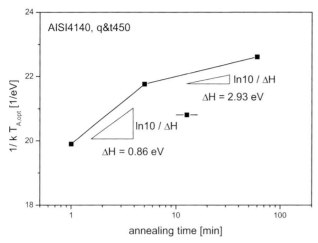

Fig. 5.50: Optimal annealing temperature determined from residual stress stability during quasi-static loading vs. annealing time as an Arrhenius plot for shot peened AISI4140 in a state quenched and tempered at 450 °C [5.28]

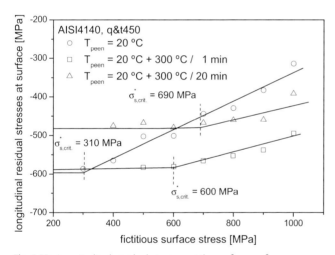

Fig. 5.51: Longitudinal residual stresses at the surface vs. fictitious surface stress after bending loading with subsequent inverse loading for AISI4140 in a state quenched and tempered at 450 °C after conventional shot peening and conventional shot peening plus subsequent annealing for 1 or 20 min at 300 °C, acc. [5.16,5.23]

300 °C, an annealing time of 1 minute should noticeably increase residual stress stability in comparison to the 20-minute time selected for Fig. 5.48. This is confirmed by Fig. 5.51, comparing the dependencies of the longitudinal residual stresses on loading stress at the surface, depicted for conventional peening and

for conventional peening and subsequent annealing at 300 °C for 1 minute or 20 minutes [5.16,5.23]. While the 20-minute annealing treatment already reduces residual stresses significantly, these still remain at their original values after 1 minute, critical loading stress being raised from 310 MPa to 600 MPa. By contrast, a critical loading stress of 690 MPa is found after 20 minutes of annealing, yet this increase is less important, since residual stresses are substantially reduced.

Fig. 5.52 summarizes the dependencies of the longitudinal surface residual stresses which remain after bending loading and inverse loading in respect to the three different shot peening methods [5.17]. It shows that the initial magnitude of the residual stresses is slightly increased and critical loading stress is increased noticeably by warm peening compared to conventional shot peening. However, subsequent annealing causes a significant relaxation of the residual stresses and an increase in critical loading stress, arriving at a value which is substantially higher than the one found for the warm peened state.

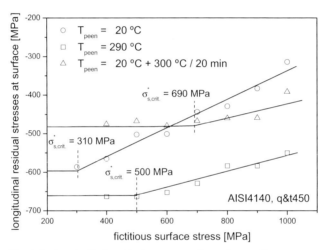

Fig. 5.52: Longitudinal residual stresses at the surface vs. fictitious surface stress after bending loading with subsequent inverse loading for AISI4140 in a state quenched and tempered at 450 °C after conventional shot peening, conventional shot peening plus subsequent annealing for 20 min at 300 °C and warm peening at $T_{peen} = 290$ °C, acc. [5.17]

Figs. 5.53 and 5.54 show how the stability of residual stresses induced by conventional peening and warm peening is influenced by the heat treatment state when bending loading and inverse loading takes place [5.25]. As bulk material hardness increases from the normalized state over the quenched and tempered states to the quenched state, this results in rising levels of surface residual stresses and of the critical loading stress at which residual stress relaxation sets in. The initial amounts of the residual stresses are slightly greater in the warm peened states than in the conventionally peened states.

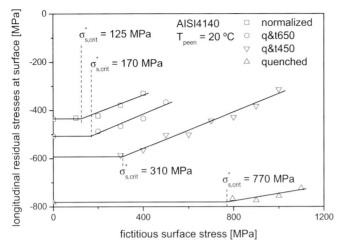

Fig. 5.53: Influence of heat treatment state on longitudinal residual stresses at the surface vs. fictitious surface stress during bending loading with subsequent inverse loading for conventionally shot peened AISI4140, acc. [5.25]

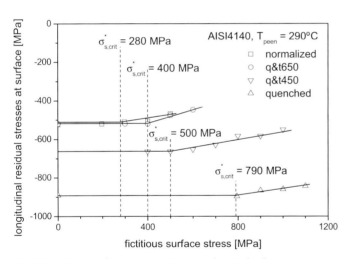

Fig. 5.54: Influence of heat treatment state on longitudinal residual stresses at the surface vs. fictitious surface stress during bending loading with subsequent inverse loading for AISI4140 warm peened at 290 °C, acc. [5.25]

d) **Evaluation of Surface Layer Properties**

Apart from deformation behavior, used in Sects. 5.1.2 and 5.2.2b as an input parameter for modeling residual stress relaxation, the compressive yield strengths at the surface are a focal point in assessing mechanical surface treatments. As the compressive yield strengths at the surface can be derived directly from the relaxation of residual stresses in homogeneous compressive loadings, they are determined easily by means of the surface-core model. Table 5.1 summarizes available data for a number of materials and material states. It shows that the values of the compressive yield strengths at the surface are partly increased and partly decreased, compared to the respective un-peened state. However, the relations of the half widths of the surface layer and the core indicate a shot peening-induced, microstructural workhardening of the surface layer in all cases except for quenched AISI4140. These seeming contradictions are resolved by a closer examination of the surface layer deformations which occur during shot peening. While shot peening treatments primarily cause biaxial plastic stretching in the surface layer, homogeneous compressive loading causes longitudinal plastic compression which runs contrary to the direction of deformation in shot peening. The Bauschinger effect may occur, affecting primarily material states containing semi-coherent and non-conherent precipitations, such as the warm aged aluminum alloy state and quenched and tempered steels (comp. Sect. 3.1.1a). Therefore, it is hardly surprising to find ratios smaller than 1 of the compressive yield strengths of the surface layer and the core in the material states listed. Quenched AISI4140 steel represents a special case in this context. There is evidence of pronounced surface layer workhardening due to the relation of yield strengths, even though the ratio of the half widths of the surface layer and the core indicates significant microstructural worksoftening, as discussed in Sect. 3.1.2d. The ample supply of dissolved carbon facilitates static strain aging effects, creating Cottrell clouds and sometimes also carbon clusters. Apparently, these effects serve to pin the dislocation structure in its post-shot peening position, thereby stopping the Bauschinger effect, as well. As strain aging effects reduce lattice distortions in the surface layer, the half widths close to the surface should show further relaxations when examined in due time after the shot peening treatment. Thus, the compressive yield strengths of the surface are determined by the Bauschinger effect, as well as by the peening-induced microstructural workhardening of the surface layer, which in turn is characterized by the ratio of the half widths of the surface layer and the core, as well as by any static strain aging effects that may occur.

In the case of inhomogeneous bending loading, too, the compressive yield strengths of the surface layers affected by shot peening can be derived from the first stage of residual stress relaxation. This is shown in Table 5.2 for various heat treatment states of AISI4140 steel after conventional shot peening treatments [5.25]. The findings for the normalized state and both of the quenched and tempered states coincide with the values derived and discussed in the case of homogeneous compressive loading. Only hardened state behavior deviates noticeably, showing a pronounced worksoftening of the surface layer. Presumably, the specimens used here did not experience static strain aging effects to the degree

Tab. 5.1: Maximum residual stresses in the surface layer, compound yield strength or critical loading stresses during homogeneous compressive loading, surface compressive yield strength calculated using the surface-core model, ratio of the latter to the compressive yield strength of the undeformed core material and ratio of half widths for surface and core after conventional shot peening for different materials and material states

Material	Material state	$\sigma^{rs}_{s,max}$ [MPa]	$\overline{R}_{cy} = \sigma^{ls}_{crit}$ [MPa]	$R_{cy,s}$ [MPa]	$R_{cy,c}$ [MPa]	$R_{cy,s}/R_{cy,c}$	HWB$_s$ / HWB$_c$	Reference
TiAl6V4	as delivered	−830	−250	−980	−860	1.14	1.87	[5.15]
AlCu5Mg$_2$	as delivered	−340	−175	−454	−375	1.21	1.53	[5.5]
	cold aged	−360	−75	−403	−275	1.47	1.90	[5.5]
	warm aged	−185	−75	−232	−350	0.66	1.34	[5.5]
AISI4140	normalized	−285	−275	−480	−345	1.39	1.70	[5.12]
	quenched and tempered 650 °C	−420	−400	−685	−805	0.85	1.59	[5.6]
	quenched and tempered 450 °C	−585	−600	−953	−1300	0.73	1.11	[5.21]
	quenched	−930	−1420	−2050	−1390	1.47	0.71	[5.12]

discussed above. It seems possible that the period of time that lies between the shot peening treatment and the point at which compressive yield strength is determined has a significant influence on the extent of strain aging-induced workhardening [5.12]. In the case of the specimens which received compressive loading this duration was greater than 9 months, while lasting only a few weeks for the specimens which received bending loading. The possible effects of a few weeks of annealing at room temperature are limited to the formation of Cottrell clouds, reducing lattice distortions and largely immobilizing the dislocations. On the other hand, it is plausible that such annealing treatments lasting several months may see the formation of carbon clusters, effecting a further reduction of lattice distortions and causing significant additional workhardening due to an extended pinning of dislocations. Fig. 5.55 is a graphical representation of the discussed values of the compressive yield strengths of the surface layer and the core. Once again, this confirms that surface layer workhardening occurs in the normalized state, contrary to the surface layer worksoftening found in the quenched and the quenched and tempered states. As shown in Table 5.3, surface layer worksoftening can also be observed after warm peening treatments – again excepting the normalized state. However, due to the dynamic and static strain aging effects occurring during and after the actual shot peening, surface layer worksoftening is significantly lower than after conventional peening treatments [5.25]. Likewise, the

Tab. 5.2: Maximum residual stresses in the surface layer, compound yield strength or critical loading stresses during inhomogeneous bending loading with subsequent inverse loading, surface compressive yield strength calculated using the surface-core-model and ratio of the latter to the compressive yield strength of the undeformed core material after conventional shot peening for different heat treatment states of AISI4140 [5.25]

material	material state	$\sigma^{rs}_{s,max}$ [MPa]	$\overline{R}_{cy} = \sigma^*_{s,crit}$ [MPa]	$R_{cy,s}$ [MPa]	$R_{cy,c}$ [MPa]	$R_{cy,s}/R_{cy,c}$
conventionally peened	normalisized	−436	−125	−510	−345	1.48
	quenched and tempered 650 °C	−507	−170	−610	−885	0.69
	quenched and tempered 450 °C	−600	−310	−801	−1300	0.62
	quenched	−780	−770	−1342	−2065	0.65

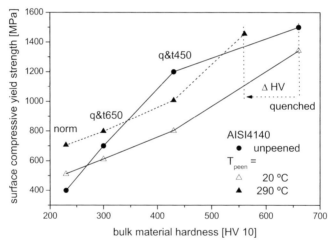

Fig. 5.55: Influence of bulk material hardness on magnitude of surface compressive yield strength before shot peening, after conventional shot peening and after warm peening at 290 °C for AISI4140 [5.25]

result for the state quenched and tempered at 450 °C is a compressive yield strength at the surface which rises as peening temperature is increased. The quenched state must be evaluated separately, as warm peening treatments constitute an annealing treatment of the bulk material, and thus reduce hardness. Therefore, comparing it to the compressive yield strength of the quenched state before shot peening will not yield much useful information. If the annealing effect is resolved, the actual worksoftening effect in the surface layer should, at least, be lower than suggested by the ratio of compressive yield strengths. The values of the compressive yield strengths at the surface as discussed for a peening temperature of 290 °C are also shown in Fig. 5.55, confirming again that the compressive yield strengths at the surface are higher than in the conventionally peened states [5.25]. Taking into account the reduction of hardness which occurs in the quenched state of the bulk material during warm peening, a comparison of the conventionally peened and the warm peened states shows that the strain aging-induced fraction of surface layer workhardening is approximately equal for all the states examined. There is even a significant increase found for the quenched state.

Tab. 5.3: Maximum residual stresses in the surface layer, compound yield strength or critical loading stresses during inhomogeneous bending loading with subsequent inverse loading, surface compressive yield strength calculated using the surface-core-model and ratio of the latter to the compressive yield strength of the undeformed core material after warm peening at 290 °C for different heat treatment states of AISI4140 [5.25]

shot peening modification	material state	peening temperature [°C]	$\sigma^{rs}_{s,\,max}$ [MPa]	$\overline{R}_{cy} = \sigma^*_{s,crit}$ [MPa]	$R_{cy,s}$ [MPa]	$R_{cy,c}$ [MPa]	$R_{cy,s}/R_{cy,c}$
warm peened	normalized	290	−523	−280	−706	−345	2.05
	quenched and tempered 650 °C	290	−520	−400	−799	−885	0.91
	quenched and tempered 450 °C	290	−660	−500	−1008	−1300	0.78
	quenched and tempered 450 °C	310	−725	−560	−1116	−1300	0.86
	quenched and tempered 450 °C	330	−700	−670	−1186	−1300	0.91
	quenched	290	−890	−790	−1456	−2065	0.71

The values for the state quenched and tempered at 450 °C, summarized in Table 5.4, show that the compressive yield strength at the surface can be increased by annealing treatments following conventional peening [5.26]. It should be noted that the increases of the compressive yield strengths at the surface are determined by static strain aging effects, i.e. their kinetics are determined by carbon diffusion. Accordingly, they develop rapidly but may experience a reduction due to effects of over-aging. By contrast, the residual stress relaxation which comes with annealing treatments is always dominated by creep effects. These follow the kinetics of self-diffusion, occurring in volume or in the dislocation core. Accordingly, these creep effects occur more slowly (comp. Fig. 5.50). Consequently, and given the absence of noticeable over-aging effects, an annealing treatment which follows conventional shot peening is always expected to also increase the compressive yield strength of the compound or critical loading stress. In such treatments, however, the compressive yield strengths fail to reach the values achieved at the elevated peening temperatures, 310 and 330 °C, as listed in Table 5.3. Apparently the dynamic strain aging effects which are active during warm peening also have a significant effect on the resulting compressive yield strengths at the surface.

Tab. 5.4: Maximum residual stresses in the surface layer, compound yield strength or critical loading stresses during inhomogeneous bending loading with subsequent inverse loading, surface compressive yield strength calculated using the surface-core-model and ratio of the latter to the compressive yield strength of the undeformed core material after conventional shot peening and subsequent annealing for AISI4140 in a state quenched and tempered at 450 °C [5.26]

shot peening modification	material state	annealing [°C / min]	$\sigma^{rs}_{s,\,max}$ [MPa]	$\overline{R}_{cy} = \sigma^{*}_{s,crit}$ [MPa]	$R_{cy,s}$ [MPa]	$R_{cy,c}$ [MPa]	$R_{cy,s}/R_{cy,c}$
conventionally peened and annealed	quenched and tempered 450 °C	300 / 1	−584	−600	−1025	−1300	0.79
		300 / 20	−480	−690	−1019	−1300	0.78

Table 5.5 compares the compressive yield strengths at the surface determined during loading at 25, 250 and 400 °C in shot peened AISI4140 in a normalized state and in a state quenched and tempered at 450 °C. In the quenched and tempered state, the untreated material and the surface layer show similar decreases of the compressive yield strengths as the temperature rises. Accordingly, their ratio is hardly dependent on temperature. As quenched and tempered steels typically experience a pronounced Bauschinger effect occurring in the pre-deformed surface layer, the ratio is smaller than 1. The normalized state, on the other hand, always shows an increase of the compressive yield strength which is caused by surface layer workhardening. Dynamic strain aging effects being more pronounced than in the quenched and tempered state, rising temperatures hardly decrease flow stress in the surface layer.

The discussion of surface layer flow stresses after mechanical surface treatments was supplemented by [5.27] – as mentioned in Sect. 4.2.3c in connection with Fig. 4.41. These examinations focused on the deformation behavior of the shot peened surface layers of normalized SAE1045. They involved isolating surface layers with a thickness of 0.3 mm by trepanning carried out after the shot peening treatment. The deformation behavior of these cylindrical trepanned

Tab. 5.5: Influence of deformation temperature on maximum residual stresses in the surface layer, compound yield strength or critical loading stresses during inhomogeneous bending loading with subsequent inverse loading, surface compressive yield strength calculated using the surface-core-model and ratio of the latter to the compressive yield strength of the undeformed core material after conventional shot peening for AISI4140 in a normalized state and in a state quenched and tempered at 450 °C [5.11,5.20]

material state	loading temperature [°C]	$\sigma^{rs}_{s,\max}$ [MPa]	$\overline{R}_{cy} = \sigma^*_{s,crit}$ [MPa]	$R_{cy,s}$ [MPa]	$R_{cy,c}$ [MPa]	$R_{cy,s}/R_{cy,c}$
normalized	25	−280	−275	−480	−345	1.39
	250	−275	−300	−496	−300	1.65
	400	−255	−340	−516	−275	1.87
quenched (450 °C)	25	−500	−600	−953	−1300	0.73
	250	−475	−400	−758	−950	0.79
	400	−340	−360	−606	−830	0.73

specimens represents an integral mean value of the deformation behavior of the surface layer. According to Fig. 4.41, workhardening is substantially greater than in the untreated bulk material, and it remains greater up to the highest deformations. The influence of subsequent annealing treatments on the deformation behavior of such specimens was also examined. It was shown that there is a new yield point phenomenon and that significant additional increases of flow stress occur particularly in the range of small strains. This proves that the increases of compressive yield strength at the surface which occur in the case of conventional shot peening and subsequent annealing are effects of static strain aging.

5.2.3
Influences on Workhardening State

The changes of the workhardening state close to the surface are impressively represented by the distributions of characteristics such as the half widths, or from other x-ray interference line characteristics derived from in-depth analyses. However, there are only few studies which focused on this. A systematic study examining tensile loading of three heat treatment states of shot peened AISI4140 steel was carried out by [5.7,5.12]. Fig. 5.56 depicts the distributions of the half widths of the {211}-interference lines for the normalized state, which show that the elevated values close to the surface are maintained, and the half widths of the core have not yet been increased significantly by raising loading stress to 400 MPa. This is hardly surprising, as the loading stresses were selected on the basis of changes of the residual stresses at the surface, and the deformations involved are

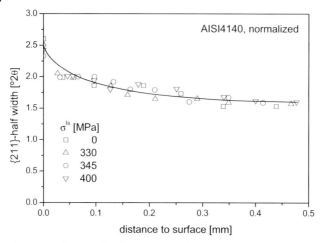

Fig. 5.56: Influence of loading stress on distribution of half widths of {211}-interference lines after tensile loading of shot peened AISI4140 in a normalized state [5.12]

at a very low 2 % for a loading stress of 400 MPa. Fig. 5.57 shows the influence of the loading stresses on the distributions of the half widths of the {211}-interference lines of the material state quenched and tempered at 650 °C. All of the loading stresses selected here again cause deformations which are below 2 %, and the initial workhardening states of the surface region are approximated. This is achieved by an increase of the half widths at distances to surface greater than 0.1 mm, while at the same time the half widths close to the surface do not change significantly. This is shown in a similar manner in Fig. 5.58, which depicts the

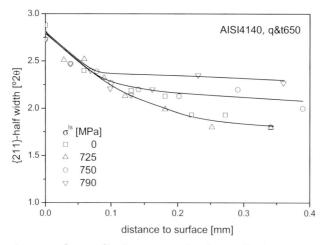

Fig. 5.57: Influence of loading stress on distribution of half widths of {211}-interference lines after tensile loading of shot peened AISI4140 in a state quenched and tempered at 650 °C [5.12]

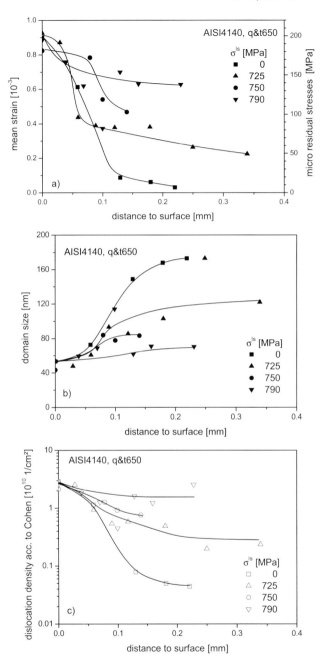

Fig. 5.58: Influence of loading stress on distribution of mean strains or micro residual stresses (a), domain sizes (b) and dislocation densities according to Cohen (c) after tensile loading of shot peened AISI4140 in a state quenched and tempered at 650 °C [5.12]

mean strains, micro residual stresses and domain sizes, as well as the dislocation densities derived according to Cohen (comp. Sect. 3.1.2d, 720). Mean strains and micro residual stresses show noticeable increases close to the surface. Their values remain almost unchanged here, while at greater distances to surface they increase noticeably as loading stress rises. The opposite is true for the domain sizes, which are significantly reduced close to the surface due to shot peening. Increasing loading stress to 790 MPa hardly changes the values at the surface, while the values for distances to surface greater than 0.1 mm largely approximate those of the surface. According to Cohen, and as shown in Fig. 5.58c, dislocation densities therefore exhibit an increase close to the surface. As in the case of mean strains, this increase is largely removed, due to loading using up to 750 MPa and due to the increase of the values in regions with a distance to surface greater than 0.1 mm. As presented in Sect. 3.1.2d, shot peening of a quenched state results in microstructural worksoftening, as evidenced by a reduction of the half widths of the {211}-interference lines close to the surface, shown in Fig. 5.59. As the loading stresses applied during tensile loading involve only very small amounts of plastic deformation, no significant changes of the depth distributions are found. This was confirmed by more precise measurements of the depth distributions of mean strains and micro residual stresses, domain sizes and dislocation densities according to Cohen, as shown in Fig. 5.60. Except for the values of mean strains and micro residual stresses, which are slightly increased after loading but show no defined dependency on loading stress, tensile loading does not effect significant changes of worksoftening close to the surface. Therefore, it can be stated in summary that in virtually all cases, the depth distributions of the workhardening state's characteristics show changes only when greater amounts of plastic deformation are induced. In such cases, the gradients of the workhardening state close

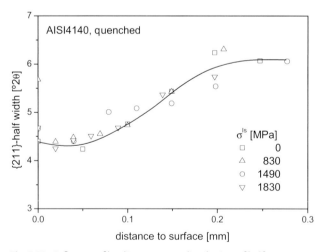

Fig. 5.59: Influence of loading stress on distribution of half widths of {211}-interference lines after tensile loading of shot peened AISI4140 in a quenched state [5.12]

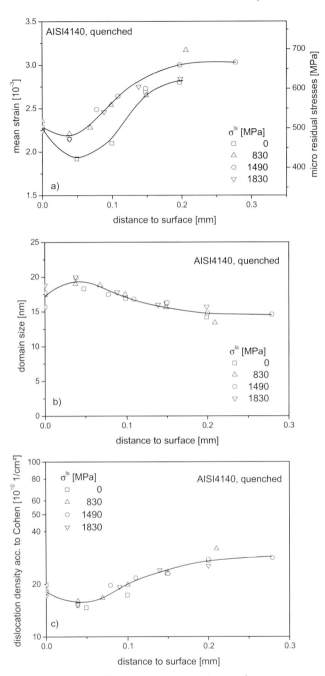

Fig. 5.60: Influence of loading stress on distribution of mean strains or micro residual stresses (a), domain sizes (b) and dislocation densities according to Cohen (c) after tensile loading of shot peened AISI4140 in a quenched state [5.12]

to the surface show decreases. These are more likely to arise from changes of the values for greater distances to surface rather than being caused by changes of the surface values. The examinations presented above employed loading stresses which were selected primarily on the basis of the changes of macro residual stresses at the surface. Therefore, such small plastic deformations hardly change the characteristics of the workhardening state for normalized material, due to the pronounced residual stress relaxation these strains cause. The changes are very pronounced in the quenched and tempered state, whereas the quenched state shows small loading-induced changes concerning mean strains and micro residual stresses only. This is due to elongations to fracture being very small.

The loading stress-dependency of the surface values of the half widths of the {211}-interference lines after compressive or tensile loading is shown in Fig. 5.61 for shot peened AISI4140 in a normalized state [5.12]. It confirms that the shot peening-induced values of the half widths close to the surface do not change, even in cases of great loading stresses and thus, great deformations. However, a closer examination using mean strains and micro residual stresses, domain sizes and dislocation densities according to Cohen – shown in Fig. 5.62 – reveals that the workhardening state close to the surface can be changed by great strains or compressions. Tensile loading stresses of more than 400 MPa and compressive loading stresses of more than roughly 300 MPa cause workhardening effects, almost all of which increase as the loading stress amount rises. Remarkable and fairly abrupt initial changes occur during compressive loading. Fig. 5.63 shows that the surface values of the half widths of the {211}-interference lines for the state quenched and tempered at 650 °C are hardly changed by tensile or compressive loading [5.7,5.12]. Individual half width values show decreases only after applying great tensile stresses. However, increasing loading stress does not establish a specific tendency regarding these half width changes.

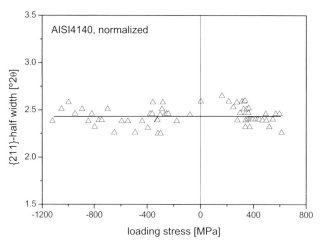

Fig. 5.61: Half widths of {211}-interference lines at the surface vs. tensile or compressive loading stress for shot peened AISI4140 in a normalized state [5.12]

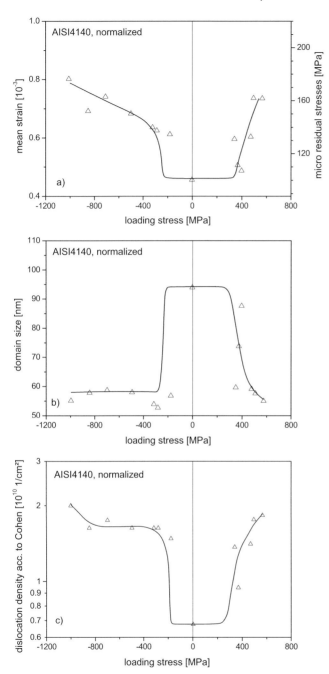

Fig. 5.62: Mean strains or micro residual stresses (a), domain sizes (b) and dislocation densities according to Cohen (c) at the surface vs. tensile or compressive loading stress for shot peened AISI4140 in a normalized state [5.12]

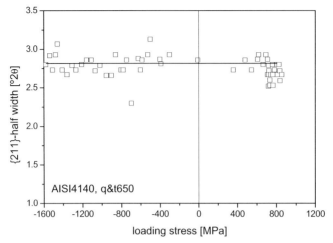

Fig. 5.63: Half widths of {211}-interference lines at the surface vs. tensile or compressive loading stress for shot peened
AISI4140 in a state quenched and tempered at 650 °C [5.12]

Figs. 5.64 and 5.65 show results obtained by a closer examination using mean strains and micro residual stresses, domain sizes and dislocation densities according to Cohen. Starting from the initial value of the respective characteristic of the workhardening state, tensile loading causes small amounts of worksoftening initially, and continuous worksoftening occurs subsequently. For the highest loading stresses examined, worksoftening indicates surface layer regions which are slightly workhardened compared to the initial state. This is particularly evident in Fig. 5.64b, showing mean strains and micro residual stresses as a function of total deformation. This initial worksoftening is presumably caused by a re-arrangement of the dislocation structure in the region close to the surface. Due to the high strain rate deformation during shot peening, dislocation structure shows a tendency to be planar in those examined materials containing medium or high stacking fault energy. The dislocations are initially brought into a tangled arrangement which contains fewer distortions and characterizes quasi-static deformation processes for these materials. Increasing loading stresses further, however, results in a noticeable workhardening throughout the remainder of the region of total deformation. Compressive loading, on the other hand, creates noticeable workhardening of the surface layer as loading stress or compression increases, at the same time causing strong fluctuations of individual values, compared to the initial value of the respective workhardening characteristic.

The dependency of the half widths of the {211}-interference lines on total deformation, as introduced by [5.21], is shown in Fig. 5.66 for AISI4140 in a state quenched and tempered at 450 °C. It confirms the initial worksoftening found for the state quenched and tempered at 650 °C during tensile deformation. Here, great total deformations compensate the initial worksoftening in such a way that the original values are reached again. However, great compressions during com-

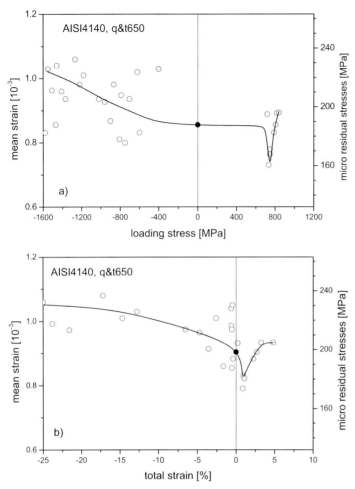

Fig. 5.64: Mean strains or micro residual stresses at the surface vs. tensile or compressive loading stress (a) or total
strain (b) for shot peened AISI4140 in a normalized state [5.7]

pressive loading cause the individual values to fan out, while there is a slight tendency of worksoftening. [5.12] examined quenched AISI4140 in a shot peened
state. During tensile or compressive loading, the characteristics of its workhardening state are dependent on loading stress, as represented by Figs. 5.67 and 5.68.
None of the characteristics show significant changes compared to the initial values during tensile loading, as the tolerable plastic deformations are very small. By
contrast, great compressive loading stresses and associated plastic deformations
of up to 5 % cause a pronounced workhardening of the surface layer. However,
this does not fully compensate the microstructural worksoftening created by shot
peening, as is evident when comparing the data to the information presented by
Fig. 3.36.

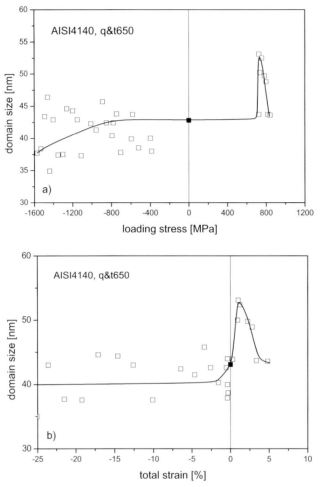

Fig. 5.65: Domain sizes at the surface vs. tensile or compressive loading stress (a) or total strain (b) for shot peened AISI4140 in a normalized state [5.7]

As shown in Fig. 5.69, [5.3] found that both tensile and compressive loading of shot peened TiAl6V4 also causes initial relaxations of the half widths of the {12$\bar{3}$3}-interference lines, minima occurring at amounts of total deformation of about 2 %. At higher loading stresses, the values are constant during tensile loading while showing slight increases during compressive loading. This increase, however, fails to arrive at the value found originally. Fig. 5.70 summarizes the results regarding the dependency of the half widths of the {511/333}-interference lines on total deformation, as found by [5.3] for AlCu5Mg2 aluminum alloy in the as-delivered state, the cold aged and the warm aged state after shot peening. All states show slight initial decreases of the half widths close to the surface during

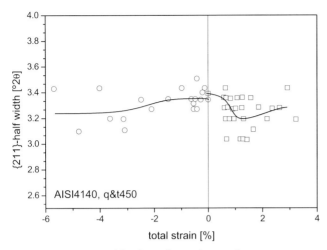

Fig. 5.66: Half widths of {211}-interference lines at the surface vs. total strain during tensile or compressive loading for shot peened AISI4140 in a state quenched and tempered at 450 °C [5.21]

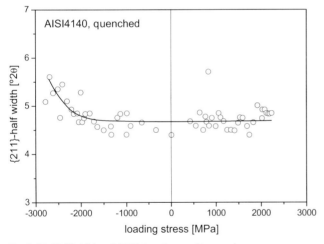

Fig. 5.67: Half widths of {211}-interference lines at the surface vs. tensile or compressive loading stress for shot peened AISI4140 in a quenched state [5.12]

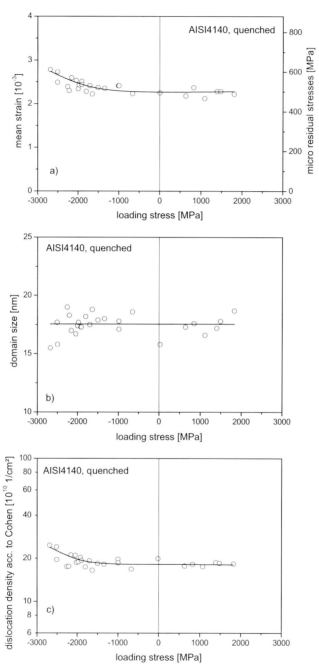

Fig. 5.68: Mean strains or micro residual stresses (a), domain sizes (b) and dislocation densities according to Cohen (c) at the surface vs. tensile or compressive loading stress for shot peened AISI4140 in a quenched state [5.12]

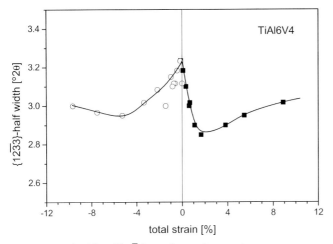

Fig. 5.69: Half widths of {12̄33}-interference lines at the surface vs. total strain during tensile or compressive loading for shot peened TiAl6V4 [5.3]

tensile loading. Half widths pass through a minimum at an amount of total deformation of about 2 %, after which they increase continuously. This increase is least pronounced for the warm aged state. Compressive deformation of the as delivered state and the cold aged state results in an initial relaxation of the half widths, minimum levels occurring at about 0.3 % compression, while greater amounts of compression cause an increase of half widths which is most pronounced for the as delivered state. The warm aged state, on the other hand, shows a slight increase of the half widths, beginning with the start of compressive deformation. Thus, it can be assumed that initial re-arrangements of the shot peening-induced dislocation structure – which involve reductions of distortions and occur more rapidly during compressive deformation than during tensile deformation due to the earlier appearance of plastic deformation on the examined surface – is succeeded by deformation-induced increases of distortions which occur at greater amounts of deformation.

The loading stress-dependency of the half widths was examined by [5.22] for AISI4140 quenched an tempered at 450 °C and shot peened using various methods. The examination also incorporated bending loading which included the range of the onset of macro residual stress relaxation. After conventional shot peening treatments, the specimens' top sides, which were initially compressively loaded, show changes of the half widths close to the surface only after inverse loading, as shown in Fig. 5.71. These changes indicate slight worksoftening. The same is true for the bottom sides of the specimens which initially experienced tensile loading. The half widths of the state warm peened at 290 °C are summarized in Fig. 5.72. During the initial compressive loading of the top side, the half widths already exhibit a small increase, while showing no significant further changes after inverse loading. Corresponding half width increases are found on the bottom

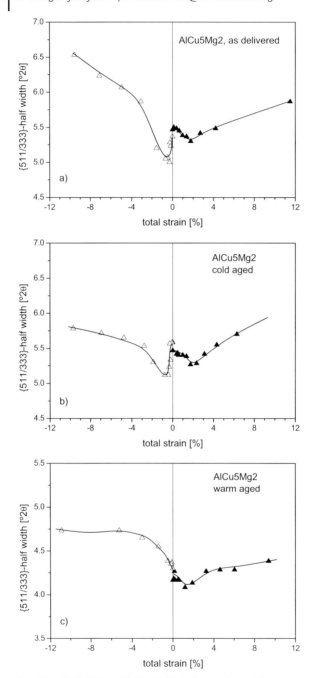

Fig. 5.70: Half widths of {511/333}-interference lines at the surface vs. total strain during tensile or compressive loading for shot peened AlCu5Mg2 in the as-delivered (a), cold aged (b) and warm aged state (c) [5.3]

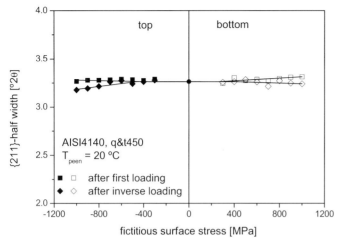

Fig. 5.71: Half widths of {211}-interference lines at the surface vs. fictitious surface stress during bending loading with subsequent inverse loading for conventionally shot peened AISI4140 in a state quenched and tempered at 450 °C [5.22]

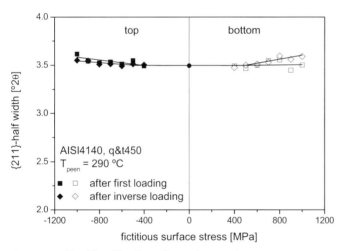

Fig. 5.72: Half widths of {211}-interference lines at the surface vs. fictitious surface stress during bending loading with subsequent inverse loading for warm peened (T_{peen} = 290 °C) AISI4140 in a state quenched and tempered at 450 °C [5.22]

side. These are observed only after inverse loading, caused by the associated compressive loading. The half width increases are attributed to the fact that the strain aging effects caused by warm peening reduce the density of mobile dislocations significantly. As a result, the onset of plastic deformation in the surface layer regions during compressive loading takes place at the same time as a multiplica-

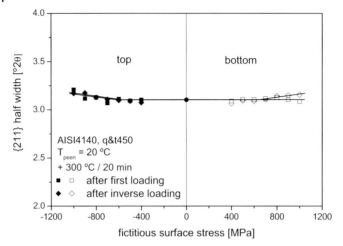

Fig. 5.73: Half widths of {211}-interference lines at the surface vs. fictitious surface stress during bending loading with subsequent inverse loading for conventionally shot peened plus subsequently annealed ($t_a = 20$ min, $T_a = 300\,°C$) AISI4140 in a state quenched and tempered at 450 °C [5.22]

tion of dislocations, along with an increase of the half widths. As shown in Fig. 5.73, this is confirmed by the findings for conventionally shot peened specimens which were subsequently annealed at 300 °C for 20 minutes. Qualitatively, they show the same behavior. This half width increase must be caused by static strain aging effects, as these reduce the density of mobile dislocations and thereby create the need for a formation of new dislocations.

Post-shot peening isothermal tensile or compressive loading, carried out at 250 and 400 °C on AISI4140 normalized or quenched and tempered at 450 °C, show no siginificant influences of loading stress on the remaining half widths. Only the initial values experience slight decreases due to thermal relaxation of micro residual stresses.

In summary, it can be stated that in soft, i.e. ductile, material states such as normalized steels, the workhardening states of the core and of the originally workhardened surface layer region approximate each other initially. Subsequently, microstructural workhardening occurs in the surface layer and in the core, which is clearly identifiable using a sufficiently sensitive method of measurement, such as X-ray interference line profile analysis. In medium-hard states – such as quenched and tempered steels and underaged aluminum alloy states – small relaxations of the micro residual stress state, characterized by the examined properties, are found at the surface during the first stage of plastic deformation. Evidently the microstructure – originally created by shot peening and thus, by high strain rate deformation – is initially re-arranged from its high-distortion state into slightly less distorted states. These are attributed to quasi-static deformation processes. Subsequently, there is the appearance of the kind of microstructural workhardening typical of quasi-static deformation, creating a minimum of the characteristics listed. Compared to tensile loading, compressive loading shows a tendency of shifting this minimum toward smaller amounts of deformation, since deformation begins directly at the surface and is not restricted to the core area initially. During the further course of deformation, these material states experience small increases of the characteristics' values close to the surface, both during tensile and compressive loading. On the other hand, high strength states, such as quenched steels and warm aged aluminum alloys, are so brittle that early failure prevents tensile loading from changing the workhardening state significantly. However, small increases occur during compressive loading which cannot compensate the shot peening-induced worksoftening.

5.2.4
Influences on Microstructure

There are no studies which focus on the influence of quasi-static loading on the microstructure of surface layer states after mechanical surface treatments. Sect. 5.2.3 examines the load-dependency of the distribution of half widths, mean strains or micro residual stresses, domain sizes and dislocation densities according to Cohen as indirect measures of the microstructure. It is shown that increasing loading stress causes a more or less pronounced approximation of the microstructure in the initially non-deformed core region and in the surface layer, the latter having been pre-deformed – sometimes at high strain rates – and usually pre-workhardened. Therefore, at least for ductile material states and large degrees of deformation, the microstructure should become increasingly similar to that of a material state which has not undergone mechanical surface treatment.

References

5.1 O. Vöhringer: Abbau von Eigenspannungen, In: V. Hauk, E. Macherauch (eds.), Eigenspannungen, DGM-Informationsgesellschaft, Oberursel, 1983, pp. 49–83.

5.2 O. Vöhringer: Relaxation of residual stresses by annealing or mechanical treatment, In: A. Niku-Lari (ed.), Advances in surface treatment; International Guidebook on Residual Stresses, Pergamon, New York, 1987, pp. 367–395.

5.3 T. Hirsch: Zum Einfluss des Kugelstrahlens auf die Biegeschwingfestigkeit von Titan- und Aluminiumbasislegierungen, Dissertation, Universität Karlsruhe (TH), 1983.

5.4 B. Scholtes: Eigenspannungen in mechanisch randschichtverformten Werkstoffzuständen, Ursachen-Ermittlung-Bewertung, DGM-Informationsgesellschaft, Oberursel, 1990.

5.5 T. Hirsch, O. Vöhringer, E. Macherauch: Der Einfluss von Zug- bzw. Druckverformungen auf die Oberflächeneigenspannungen von gestrahltem AlCu5Mg2, Härterei Technische Mitteilungen 43 (1988), pp. 16–20.

5.6 V. Schulze, O. Vöhringer, E. Macherauch: Modellierung des Verformungverhaltens und der Makroeigenspannungsänderungen bei quasi-statischer Beanspruchung von vergütetem und kugelgestrahltem 42 CrMo 4,

Zeitschrift für Metallkunde 89 (1998), pp. 719–728.

5.7 V. Schulze: Die Auswirkungen kugelgestrahlter Randschichten auf das quasistatische sowie ein- und zweistufige zyklische Verformungsverhalten von vergütetem 42 CrMo 4, Dissertation, Universität Karlsruhe (TH), 1993.

5.8 R. v. Mises: Mechanik der plastischen Formänderung von Kristallen, Z. angewandte Mathematik und Mechanik 8 (1928) 3, pp. 161–185.

5.9 H. Ismar, O. Mahrenholz: Technische Plastomechanik, Vieweg Verlag, Braunschweig, 1979.

5.10 O. C. Zienkiewicz, S. Vallippan, I. P. King: Elasto-plastic solution of engineering problems "Initial Stress", Finite Element approach, International Journal of Numerical Methods in Engineering 1 (1969), pp. 75–100.

5.11 H. Holzapfel: Das Abbauverhalten kugelgestrahlter Eigenspannungen bei 42 CrMo 4 in verschiedenen Wärmebehandlungszuständen, Dissertation, Universität Karlsruhe (TH), 1994.

5.12 F. Theobald, V. Schulze: unveröffentlicht, Universität Karlsruhe (TH), 1992.

5.13 D. Kirk: Effects of plastic straining on residual stresses induced by shot-peening, In: H. Wohlfahrt, R. Koop, O. Vöhringer (eds.), Shot Peening, Proc. Int. Conf. on Shot Peening 3, DGM-Informationsgesellschaft, Oberursel, 1987, pp. 213–220.

5.14 H. Holzapfel, V. Schulze, O. Vöhringer, E. Macherauch: Residual stress relaxation an an AISI 4140 steel due to quasistatic and cyclic loading at higher temperatures, Materials Science and Engineering A248 (1998), pp. 9–18.

5.15 O. Vöhringer, T. Hirsch, E. Macherauch: Relaxation on shot peening induced residual stresses of TiAl6V4 by annealing of mechanical treatment, In: G. Lütjering, U. Zwicker, W. Bunk (eds.), Titanium – Science and Technology, Proc. Int. Conf. Titanium 5, DGM-Informationsgesellschaft, Oberursel, 1984, pp. 2203–2210.

5.16 A. Wick, V. Schulze, O. Vöhringer: Effects of stress- and/or warm peening of AISI 4140 on fatigue life, Steel Research 71 (2000) 8, pp. 316–321.

5.17 A. Wick, V. Schulze, O. Vöhringer: Effects of warm peening on fatigue life and relaxation behaviour of residual stresses of AISI 4140, Materials Science and Engineering A293 (2000), pp. 191–197.

5.18 H. Holzapfel: unveröffentlicht, Universität Karlsruhe (TH), 1994.

5.19 R. Menig, V. Schulze, O. Vöhringer: Residual stress relaxation and fatigue strength of AISI 4140 under torsional loading after conventional shot peening. stress peening and warm peening, In: L. Wagner (ed.), Shot Peening, Wiley-VCH, Weinheim, 2003, pp. 316.

5.20 H. Holzapfel, V. Schulze, O. Vöhringer, E. Macherauch: Relaxation behaviour of shot peening induced residual stresses in AISI 4140 due quasistatic uniaxial loading at elevated temperatures, In: J. Champaigne (ed.), Proc. Int. Conf. Shot Peening 6, San Francisco, 1996, pp. 385–396.

5.21 H. Hanagarth, O. Vöhringer, E. Macherauch: Relaxation of shot peening residual stresses of the steel 42 CrMo 4 by tensile of compressive deformation, In: K. Iida (ed.), Proc. Int. Conf. Shot Peening 4, The Japan Society of Precision Engineering, Tokyo, 1990, pp. 337–345.

5.22 A. Wick: Randschichtzustand und Schwingfestigkeit von 42CrMo4 nach Kugelstrahlen unter Vorspannung und bei erhöhter Temperatur, Dissertation, Universität Karlsruhe (TH), 1999.

5.23 R. Menig, V. Schulze, O. Vöhringer: Optimized warm peening of the quenched and tempered steel AISI 4140, Materials Science and Engineering A335 (2002), pp. 198–206.

5.24 R. Menig, V. Schulze, O. Vöhringer: Kugelstrahlen und anschließendes Auslagern – Steigerung der Eigenspannungsstabilität und der Wechselfestigkeit am Beispiel von 42CrMo4, Härterei Technische Mitteilungen 58 (2003), pp. 127–132.

5.25 R. Menig, V. Schulze, O. Vöhringer: Shot peening at elevated temperatures – Influence of the material state of AISI 4140 on the effect on fatigue strength, In: A. F.Blom (ed.), Proc. Fatigue 2002, Engineering Materials Advisory Services

Ltd., West Midlands, 2002, pp. 1257–1266.

5.26 R. Menig: Randschichtzustand, Eigenspannungsstabilität und Schwingfestigkeit von unterschiedlich wärmebehandeltem 42 CrMo 4 nach modifizierten Kugelstrahlbehandlungen, Dissertation, Universität Karlsruhe (TH), 2002.

5.27 I. Altenberger: Mikrostrukturelle Untersuchungen mechanisch randschichtverfestigter Bereiche schwingend beanspruchter metallischer Werkstoffe, Dissertation, Universität Gesamthochschule Kassel, 2000.

5.28 R. Menig, V. Schulze: unveröffentlicht, Universität Karlsruhe (TH), 2002.

6
Changes of Surface States during Cyclic Loading

6.1
Process Models

6.1.1
Elementary Processes

The changes of the surface states of mechanically surface treated material states due to cyclic loading can be divided into three phases [6.1–3]. The first phase includes initial loading and subsequent inverse loading. These constitute quasi-static loading. The effects they have on the surface layer state are essentially determined by the compressive yield strength at the surface and the Bauschinger effect, as described in detail in Sect. 5. This is followed by the actual cyclic loading in the crack-free phase. Here, cyclic deformation behavior determines the stability of the surface layer state, and accordingly, cyclic tensile yield strength at the surface is an important parameter. This is exemplified in Fig. 6.1 for shot peened quenched and tempered AISI4140, showing the correlation between residual stress changes and plastic strain amplitudes derived from the stress-strain hysteresis [6.4]. Fig. 6.1a shows cyclic deformation curves for different stress amplitudes, representing the correlation between plastic strain amplitudes and the number of cycles. Following a phase of quasi-elastic behavior, numbers of cycles to incubation occur which decrease as the stress amplitude increases. This causes cyclic worksoftening, resulting in plastic strain amplitudes increasing. As the stress amplitude grows, cyclic worksoftening becomes more pronounced, continuing until the specimen fails. Fig. 6.1b shows the changes of the longitudinal residual stresses which occur at the surface of the specimen during push-pull loading. During the first cycle, quasi-static residual stress relaxation – which increases as stress amplitude rises – occurs as described in Sect. 5. This is followed by the second phase of residual stress relaxation. Here, the changes in residual stress show a linear correlation with the logarithm of the number of cycles. The changes also become more pronounced as the stress amplitude increases. Toward the end of the life span, yet before the phase of crack growth, the two greatest loading amplitudes increase residual stress changes still further. The start of these increases clearly correlates with the onset of the cyclic worksoftening shown in

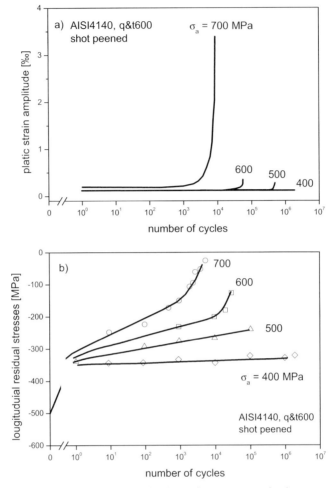

Fig. 6.1: Influence of stress amplitude on plastic strain amplitudes (a) and longitudinal residual stresses at the surface (b) vs. number of cycles in stress controlled push-pull experiments studying the cyclic deformation behavior of shot peened AISI4140 in a state quenched and tempered at 600 °C [6.4]

Fig. 6.1a. Evidently, the cyclic relaxation of residual stress is determined fundamentally by the plastic strain amplitudes that occur. This second phase of surface layer changes during cyclic loading is followed by the phase of crack growth. Here, the changes of the surface layer state most often become even more pronounced, and they are determined by cyclic crack growth behavior.

Focusing on residual stress stability, [6.1] distinguished four different cases of the correlation between residual stresses and number of cycles in the experimental results. They are shown in Fig. 6.2. This division rests not only on a differentiation between the first loading cycle and the actual cyclic loading, as well as the

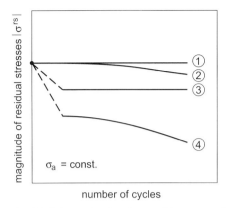

Fig. 6.2: Scheme showing the possible courses of residual
stresses vs. number of cycles during cyclic loading [6.1]

strong influence of loading amplitude, but also on quasi-static yield strength or
cyclic yield strength as the defining factors for stability. The changes of mechanical properties close to the surface, caused by mechanical surface treatment, are
neglected, as the stability of the surface layer state is determined more correctly by
compressive yield strength at the surface and cyclic tensile yield strength at the
surface, due to the presence of great compressive residual stresses. In case 1, residual stresses remain stable throughout cyclic loading, as neither quasi-static
compressive yield strength nor cyclic tensile yield strength are exceeded. Case 2 is
expected to occur in those instances when the examined loading amplitude
exceeds the cyclic, but not the quasi-static tensile yield strength in the course of
cyclic loading. Here, only great numbers of cycles effect a gradual relaxation of
residual stresses. The reverse is true for case 3, in which a significant relaxation of
residual stresses occurs during the first cycle as the quasi-static compressive yield
strength is exceeded, while residual stresses remain unchanged during further cyclic loading. In case 4, however, both quasi-static and cyclic yield strengths are
exceeded. Therefore, the pronounced relaxation of residual stresses which occurs
during the first loading cycle is followed by a continuous, additional relaxation of
residual stresses. [6.2] and [6.5] contain a lot of evidence regarding all four cases.

According to [6.2], the changes of the surface layer state during cyclic loading
are mainly determined by loading type, by mean loading, by the initial amount of
residual stresses and by the material state and its deformation behavior. In accordance with Sect. 5, loading types can be grouped into homogeneous uniaxial
push-pull loading, homogeneous uniaxial bending loading and inhomogeneous
multiaxial loading such as torsional-, notch- or complex component loading,
which will serve as a criterion for structuring the following text. Mean loading
influences surface layer stability considerably, because it causes shifts of the overall stress state – which in turn is composed of loading- and residual stresses –
within the region of vanishing plastic deformations and thus, stable surface layer
states, given by quasi-static and cyclic yield strength. The same is true for the

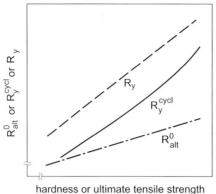

Fig. 6.3: Scheme showing the correlation between hardness or ultimate tensile strength and quasi-static yield strength R_y, cyclic yield strength R_y^{cycl} and fatigue strength R_{alt}^0 of a state free of residual stresses, acc. [6.1]

magnitude of initial residual stresses. The material state has a crucial effect on surface layer- and residual stress stability. In lower strength material states the initial residual stresses are largely removed, regarding both low cycle fatigue and high cycle fatigue. During loading, medium strength material states maintain stable residual stresses in the area of high cycle fatigue, and high strength material states maintain stable residual stresses also in the area of low cycle fatigue [6.35]. The reason for this is the dependency of the difference between fatigue limit and the quasi-static or cyclic yield strength on hardness or tensile strength [6.1], shown schematically in Fig. 6.3. Accordingly, cyclic loadings in the area of the fatigue limit approximate yield strengths. The smaller ultimate tensile strength or hardness is, the closer the approximation becomes. By contrast, cyclic loadings below the fatigue limit are far enough below cyclic yield strength as to preclude plastification and thus, changes of residual stress, as well. The influence of cyclic deformation behavior was already pointed out in the discussion of Fig. 6.1.

6.1.2
Quantitative Description of Processes

a) Description of Residual Stress Relaxation

The division of residual stress relaxation during cyclic loading into three phases – as mentioned in Sect. 6.1.1 – is reflected in the approaches for a quantitative description. While Sect. 5.1.2 is able to provide a detailed description of the behavior during initial loading and inverse loading, no such approaches are available for the third phase of cyclic crack propagation. Therefore, this representation will focus on the second phase of cyclic loading in the crack-free state. Following

[6.1,6.6,6.7], it is reasonable to compare the relaxation of residual stresses to the relaxation of mean stresses. The latter occurs during cyclic loadings with mean strains [6.8,6.9], and it is based on a mechanism which is equivalent to that of the cyclic creep found in cyclic loadings with mean stresses [6.10]. When flow stress is exceeded, the course of cyclic loading sees the centers of locally present stress-strain hystereses shift from the initial position – which runs parallel to the stress axis and is determined by mean strains – to a stable, final position. It may be located at zero mean stresses [6.11,6.12] or at remaining, stabilized mean stresses [6.13–6.15]. According to [6.13], these remaining mean stresses may correlate with the strain amplitudes at half the number of cycles to fracture. Empirical equations describing relaxation of mean stresses were used in multiple ways by numerous authors to assess the relaxation of residual stresses. However, the differing work-hardeninng states of the surface layer and the core, as well as the inhomogeneity of the loading state – which, unlike in experiments with mean stresses, inevitably occurs – can result in residual stresses changing mathematical signs or cause a relaxation toward stabilized final values other than zero [6.16]. The overview studies of [6.16,6.17] contain information on this which essentially represents the relations between residual stresses and the number of cycles as exponential functions, power law functions and logarithmic functions, as shown schematically in Fig. 6.4. Exponential relations plotted versus the logarithm of the number of cycles show relaxation rates which are small initially and then increase continually. The logarithmic relations in the plot selected show constant changes of residual stresses. By contrast, the power law relation shows the relaxation of residual stresses to be particularly steep at first and degressive thereafter. Exponential relations according to

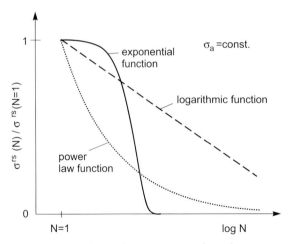

Fig. 6.4: Scheme showing literature-proposed correlations between residual stresses related to their value after the first cycle and number of cycles

$$\sigma^{rs} = \sigma^{rs}(N = 1) \cdot \exp(-k \cdot N) \tag{6.1}$$

were first introduced by [6.18], who pointed out the similarity to creep processes. These relations are based on studies by [6.8] – who applied them in the context of the relaxation of mean stresses – and were later taken up by [6.19–6.22]. [6.21] introduced the concept of relaxation arriving at a final value. Power law relations according to

$$\sigma^{rs} = \sigma^{rs}(N = 1) \cdot N^{-b} \tag{6.2}$$

were suggested by [6.15] as a description of relaxation of mean stresses. However, logarithmic relations in the form of

$$\sigma^{rs} = \sigma^{rs}(N = 1) - \mu \cdot \log N \tag{6.3}$$

have gained general acceptance. [6.15] first used them to describe the relaxation of mean stresses, [6.24,6.25] later utilized them to analyze residual stress relaxation and concurrently present mean stresses in notched specimens. As logarithmic relations are simple to use, [6.17] was able to establish them as a standard. In many instances, the results for slope μ according to

$$\mu = a \cdot \sigma_a - b \tag{6.4}$$

show a linear dependency on loading stress amplitude.

Apart from these entirely empirical models, other approaches started from the aspect of the physical properties of metals. These models describe residual stress relaxation on the basis of microstructural explanations, and thereby include the increase of residual stress relaxation which occurs as the stress gradient increases [6.26,6.27]. [6.28] examined effective shear stresses in the individual slip systems. This is done by averaging the shear components of the possible grain orientations close to the surface versus the loading axis. Workhardening was assumed to increase exponentially as the number of cycles grows. Thus, it became possible to understand the findings for machined states of 2219 aluminum alloy. Here, a growing number of loading cycles causes surface residual stresses to pass through a minimum, and to build up again subsequently. Calculations were carried out by [6.29,6.30] in order to extend the description to also include the changes of work-hardening state caused by induced residual stresses. Thus, it is possible to predict how the residual stress depth profile will be influenced by number of cycles, stress amplitude and stress ratio. The calculations were based on a material model by [6.31] which describes cyclic workhardening or -softening by means of a kinematic workhardening term. However, this method presents a problem in that the parameters required for describing the cyclic deformation behavior of the pre-deformed surface layer are usually not accessible, but must be estimated.

b) Evaluation of Cyclic Yield Strength at the Surface

Sect. 5.1 described a method for determining cyclic yield strength by examining the stability of residual stresses during quasi-static loading. Similarly, the behavior of residual stresses during cyclic loading permits conclusions to be drawn regarding the resistance to cyclic plastic deformation which is present in the surface layer. This resistance is achieved when loading stresses are superposed with the residual stresses that remain after the first loading cycle, resulting in an equivalent stress which is similar to cyclic yield strength. Common fatigue strength theory defines cyclic yield strength as the stress amplitude which causes non-zero plastic strain amplitudes at half the number of cycles to failure. In the context at hand, cyclic yield strength is determined by way of the onset of residual stress relaxation during cyclic loading [6.3]. Applying the von Mises hypothesis makes it possible to define the cyclic yield strength at the surface, which may be calculated according to

$$R_{y,surf}^{cycl} = \sqrt{\left(\bar{R}_y^{cycl} + \sigma_{ax,surf}^{rs}(N=1)\right)^2 + \left(\sigma_{tan,surf}^{rs}(N=1)\right)^2 - \left(\bar{R}_y^{cycl} + \sigma_{ax,surf}^{rs}(N=1)\right)\sigma_{tan,surf}^{rs}(N=1)}$$

(6.5)

from the axial or tangential surface residual stresses $\sigma_{ax,surf}^{rs}(N=1)$ or $\sigma_{ax,surf}^{rs}(N=1)$ – which remain after the first loading cycle – and the cyclic yield strength of the compound $\bar{R}_y^{cycl}$ required for initiation of residual stress relaxation. The difference between axial and tangential residual stresses is usually neglected, and Eq. 6.5 achieves the simplified form

$$R_{y,surf}^{cycl} = \sqrt{\left(\bar{R}_y^{cycl}\right)^2 + \bar{R}_y^{cycl}\left|\sigma_{ax,surf}^{rs}(N=1)\right| + \left(\sigma_{ax,surf}^{rs}(N=1)\right)^2}.$$

(6.6)

The cyclic yield strength of the compound $\bar{R}_y^{cycl}$ was described as the critical loading stress amplitude by [6.32,6.33] and it is essential for determining the cyclic yield strength at the surface. It is calculated by extrapolating the μ-values found at different stress amplitudes in Eq. 6.4 to $\mu=0$. This was done by [6.3], as shown schematically in Fig. 6.5a. An alternative was suggested by [6.32,6.33], which is shown in Fig. 6.5b. It is carried out by comparing the residual stresses' dependence on loading stress amplitude after one loading cycle and, for instance, after 10^4 loading cycles. This allows for determining the cyclic yield strength of the compound as the amplitude at which, after 10^4 loading cycles, the dependency deviates for the first time from the dependency present after the first loading cycle.

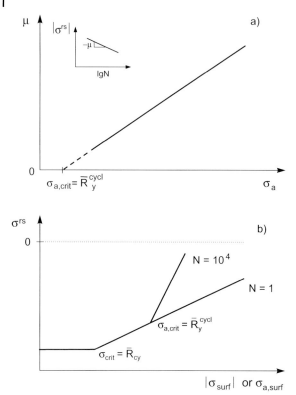

Fig. 6.5: Scheme showing the determination of cyclic yield strength R_y^{cycl} from residual stress measurements and of the critical loading stress amplitude by using extrapolation to $\mu = 0$ (a) and by using the changes of residual stresses between the first and the 10th cycle (b)

c)   Description of Effects on Fatigue Behavior

A comprehensive description of the effects surface layer states have on fatigue strength would be beyond the scope of this study, due to the multitude of influencing factors and the number of individual examinations required in that context. Therefore, this study refers to overview studies, [6.1,6.2,6.34,6.35], and otherwise provides only a short introduction as an approach to the subject and its essential concepts.

Mechanical surface treatments effect changes of the surface layer state. The extent to which these changes affect fatigue limit is determined by the material and the material state, the type of cyclic loading and the loading stress gradient. The expectable overall changes of fatigue limit ΔR_{alt} can be attributed to the influencing factors roughness, residual stress state, workhardening state and, possibly, phase transformations. Neglecting interactions, they can be described by

$$\Delta R_{alt,tot} = \Delta R_{alt}(\Delta R_t) + \Delta R_{alt}(\Delta HV) + R_{alt}(\Delta\sigma^{rs}) + \Delta R_{alt}(\Delta Ph) \tag{6.7}$$

according to [6.34]. Here, ΔR_t is the change of roughness as a measure of surface micro geometry , ΔHV is a measure of the change of workhardening state, $\Delta\sigma^{rs}$ is the change of residual stress state, and ΔPh is a measure of the possible changes of the phase fractions compared to a reference state as homogeneous as possible. Sect. 6.3.5 provides examples which typify relations of this kind. However, hardly any quantitative descriptions which apply universally are available.

The influence on fatigue limit exerted by the residual stresses induced by mechanical surface treatments can be stated formally. This is done by estimating the fatigue limit of a state showing residual stresses R_{alt}^{rs} according to

$$R_{alt}^{rs} = R_{alt}^{o} - m \cdot \sigma^{rs}, \tag{6.8}$$

using the fatigue limit R_{alt}^{o}, which occurs in a residual stress-free state, and the residual stresses σ^{rs}, which are assumed to be uniaxial. In this equation, m is residual stress sensitivity, which increases as ultimate tensile strength grows. Residual stress sensitivity was defined by [6.1] with reference to mean stress sensitivity [6.36]. This relation is based on the assumption that residual stresses have effects similar to those of mean loading stresses, or that the effects of residual stresses can, for the most part, be compensated by loading stresses of the inverse mathematical sign [6.37]. This should be valid for states of higher strength showing residual stresses which are mostly stable during loading in the range of the transition from low to high cycle fatigue or in the high cycle fatigue area itself. In this case the fatigue limit of a state containing residual stresses can be estimated on the basis of the Goodman relation [6.38] according to

$$R_{alt}^{rs} = R_{alt}^{o} \cdot \left(1 - \frac{\sigma^{rs}}{R_m}\right) \tag{6.9}$$

in which R_m is ultimate tensile strength. Accordingly, residual stress sensitivity m is expected to have the value

$$m = \frac{R_{alt}^{o}}{R_m}. \tag{6.10}$$

The differences between mean stresses and residual stresses, as listed in Fig. 6.6, are neglected. In comparison with mean stresses, the particularly important factors are inhomogeneity, the changes of the material state which inevitably occur when residual stresses are induced, and the possible changes of residual stresses during fatigue loading. Nevertheless, residual stresses can be treated as loading stresses in many cases. Therefore, their effects can be assessed by means of fatigue limit diagrams, analogous to the way in which the effects of mean stresses are evaluated. Representing this by using residual stress Haigh-diagrams – a method introduced by [6.1] – has proven to be effective. As shown schematically in Fig. 6.7, these diagrams represent the correlation between the fatigue limit of a state containing residual stresses and the initial value of the residual stresses.

Feature	Mean stresses	Residual stresses
Distribution over cross section	Homogeneous or inhomogeneous with known distribution	Inevitably inhomogeneous with usually unknown distribution
Multi-axiality at unnotched geometries	Mostly uniaxial	Mostly biaxial at the surface and triaxial in the component
Constancy during fatigue loading	controllable	At most controllable in surface layers
Adjustability	Adjustable and variable without changes in the material state	Basically not adjustable without changes in the material state and only exceptionally variable without such changes
Removability	Due to load removal	Due to thermal, quasi-static or cyclic loading and plastic deformations combined with them
Notch effect	Increase by elastic notch effect	No increase by notch effect

Fig. 6.6: Different characteristics of mean stresses and residual stresses during fatigue loading, acc. [6.1]

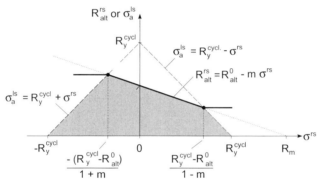

Fig. 6.7: Scheme showing a residual stress-Haigh-diagram for estimating the influence of residual stresses on the fatigue strength of smooth material states [6.1]

Starting from the fatigue limit R_{alt}^o of a state without residual stresses, the fatigue limit R_{alt}^{rs} of a state containing residual stresses is reduced initially as tensile residual stresses increase. This follows Eq. 6.9 and is analogous to the Goodman relation. To the same extent, compressive residual stresses should have a positive effect on fatigue limit. The regions of a residual stress-induced decrease and increase of fatigue limit are represented in a limited way by straight lines of the form

$$\sigma_a^{ls} = R_y^{cycl} \pm \sigma^{rs} \tag{6.11}$$

representing the boundary of the cyclic yield strength exceeded by the sum of loading- and residual stresses and thus, describing the onset of residual stress relaxation. This permits deriving the residual stress interval, in which stability of the residual stresses and thus, applicability of the Goodman relation, can be assumed, as $[(R_y^{cycl} - R_{alt}^o)/1 + m), (R_y^{cycl} - R_{alt}^o)/(1 - m)]$. It is also assumed that for initial residual stress values outside this interval, residual stress values are decreased to the values found at the respective interval boundaries, and that these residual stresses are present in a stabilized form. Therefore, fatigue limits outside the interval of residual stresses should not show further changes compared to the values reached at the interval boundaries. Accordingly, the influence of residual stresses on fatigue limit can be described by the doubly bent solid curve shown in Fig. 6.7. It is obvious that alternating bending strength cannot be influenced in a positive or negative way solely by the residual stresses outside the residual stress interval mentioned above. In the meantime, corresponding diagrams have been analyzed and comparatively assessed for numerous material states, e.g. for steels in [6.35]. The essential influencing parameters and their effects are described in detail in Sect. 6.3.5.

The formal description of the influence of mechanical surface treatments on fatigue limit by means of residual stress Haigh-diagrams is inevitably limited in two ways. Firstly, its direct form describes the influence of residual stresses exclusively, thereby neglecting the additional parameters listed in Eq. 6.7 which influence the fatigue limit found after mechanical surface treatments. Secondly, residual stress Haigh-diagrams do not offer a means of describing or predicting crack initiation which may occur beneath the surface in higher strength materials (comp. Sect. 6.3.3). Consequently, residual stress Haigh-diagrams run the risk of overestimating the effect of mechanical surface treatment on fatigue limit. Therefore, the available approaches were extended by [6.7] on the basis of [6.39] and transferred to the so-called concept of the local fatigue limit, which in the meantime has been refined e.g. by [6.1,6.40,6.41]. Falling back on Eq. 6.9, which is extended by distance to surface as

$$R_{alt}(z) = \omega \cdot n(z) \cdot R_{alt}^o(z) \cdot \left(1 - \frac{\sigma^{rs}(z)}{R_m(z)} \right), \tag{6.12}$$

the effects of surface micro- and macro geometry, stress gradient, as well as distributions of residual stresses and workhardening on the distribution of fatigue limit $R_{alt}(z)$ can be assessed. Here, ω characterizes the surface state dependent on roughness, notch factor and surface hardness, while ω is set to 1 for $z > 0$ and therefore is effective only directly at the surface. $n(z)$ is the support figure which also describes the notch effect present in deeper regions and is dependent on the local related stress gradient and on local hardness. According to Fig. 6.8, $R_{alt}(z)$ is determined by the notch factor, which is governed by notch geometry, and by the stress gradient, which is determined by loading type and notch geometry. Additional factors are roughness and the distributions of residual stresses and hardness found after mechanical surface treatments. Due to the known distortion of the hardness values by the residual stress values present – as mentioned in

Sect. 3.1.2d – the measured hardness distributions must be corrected, using the residual stress distributions, in order to obtain the changes of hardness $HV_{corr}(z)$ which are caused solely by the microstructure. According to e.g. [6.41], these changes of hardness can be used for determining local ultimate tensile strength by employing various correlations. In an analogous way, the local fatigue limits of the state without residual stresses can be estimated from local hardness, according to the correlations listed e.g. in [6.42], depending on loading type. Thus, all parameters required for applying Eq. 6.12 are available and it is possible to calculate the distribution of fatigue limit, $R_{alt}(z)$. By subsequently comparing it with local loading stress states, the critical loading stress state and the depth at which cracks may initiate can ultimately be determined. The fact that this kind of estimate applies to smooth and to notched states has been proven numerous times (see e.g. [6.5,6.29]). The concept's limitations, however, are evident in Fig. 6.9, which shows the local fatigue limit and the distribution of loading stresses in the region of the specimen's fatigue limit, as well as experimentally determined positions of cracks, for smooth and notched specimens made of quenched SAE1045 after shot peening with shot of varying hardness [6.42]. Independent of shot hardness, crack initiation at the surface is predicted for notched specimens (Fig. 6.9a and b), and confirmed experimentally. Smooth specimens, on the other hand, are expected to show crack initiation beneath the surface, according to the concept of local fatigue limit. Again, this agrees with experimental results. In the experiment, however, the estimated local fatigue limit at the surface was clearly exceeded in the two notched states which showed crack initiation at the surface. Therefore, it can be assumed that small loading stress amplitudes cause surface crack initiation which, however, are unable to spread [6.42].

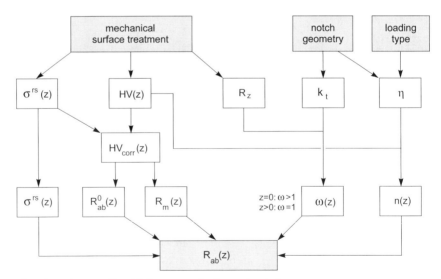

Fig. 6.8: Evaluation of local fatigue strength after mechanical surface treatment, acc. [6.1, 6.40]

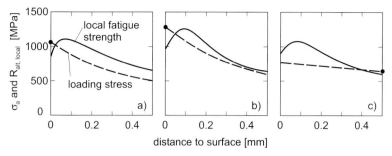

Fig. 6.9: Local fatigue strength and distribution of loading stresses in the range of the specimen's alternating bending strength for notched (a,b) and smooth (c) specimens made of quenched SAE1045 after shot peening using 46–50 HRC (a) and 54–58 HRC (b,c) shot [6.42]

Thus, highest strength and notched states additionally require an assessment of the ability of propagation of initiating cracks. [6.35] introduced a concept which incorporates this objective. It assumes that a crack is stopped when its propagating force, which is determined by the effective width of stress intensity ΔK_{eff}, drops below the threshold value $\Delta K_{th,eff}$, i.e.

$$\Delta K_{eff} = K_{max} - \max\{K_{min}, K_{op}\} < \Delta K_{th,eff} \qquad (6.13)$$

is valid. Here K_{max} is maximum- and K_{min} is minimum stress intensity, while $K_{op} = 0$ is assumed as a rough approximation of the stress intensity required for crack opening. In notched components or specimens which contain residual stresses, ΔK_{eff} is assumed as

$$\Delta K_{eff} = \left(k_t \cdot \sigma_{n,max} + \sigma^{rs}\right) \cdot \sqrt{\pi \cdot a} \cdot Y \qquad (6.14)$$

for $K_{min} < K_{op} = 0$ and as

$$\Delta K_{eff} = k_t \cdot \Delta\sigma_n \cdot \sqrt{\pi \cdot a} \cdot Y \quad (6.15)$$

for $K_{min} \geq K_{op} = 0$. In these equations, $\sigma_{n,max}$ is maximum nominal stress, $\Delta\sigma_n$ is the width of nominal stresses, k_t is the notch factor, $\sigma_{n,a}$ is nominal stress amplitude, a is crack length and Y is a geometry factor which will be dependent on crack length. Assuming a crack initiating at the surface and propagating toward the interior of the specimen, correlations between ΔK_{eff} and distance to surface can be calculated. These can be compared to threshold values and thus, crack propagation ability can be estimated. The concept was successfully employed in [6.35,6.42] for an assessment of ability of crack propagation. It showed that instead of using the residual stress state as found before loading, if possible, the residual stress state after loading with an amplitude similar to the fatigue limit is to be used. This is due to the fact that, although using the initial residual stress state allows for conservative estimates in the case of tensile residual stresses, the residual stresses' effect on fatigue limit may be over-estimated in the case of compressive residual stresses [6.35]. The concept has not yet been extended to incorporate the method of weight functions, which would permit, in particular, an improved description of the residual stresses' effect and of the concurrent changes of the residual stresses.

An additional possibility for a quantitative assessment the influence of mechanical surface treatments on fatigue strength is basically provided by the so-called local concept, an approach described in [6.44]. It can be subdivided into a calculation of number of cycles to crack initiation and a calculation of crack propagation, while pertaining finite-element calculations allow for an inclusion of varying material behavior in the regions close to the surface, given the availability of reliable data. [6.45] applied the concept to nitrided and shot peened steels, assuming total strains in the surface layer to be equal to total strains in the core, despite changed material behavior.

6.2
Experimental Results and their Descriptions

6.2.1
Influences on Residual Stress State

a) **Basic Results**

Analogous to the description in Sect. 5, the following discussion of the basic results regarding the changes of residual stress distribution and of surface residual stresses distinguishes between homogeneous uniaxial tensile or compressive loading, inhomogeneous uniaxial bending loading, biaxial torsional loading and loading of notched components.

There is only little information available on the distribution of residual stresses in mechanically surface-treated material states after **push-pull loading**. According to [6.46], the distributions of longitudinal residual stresses which remain in normalized SAE1045 after half the number of cycles to failure, using different stress amplitudes, show clear differences in magnitude after shot peening and after deep rolling. Fig. 6.10 shows that the residual stress values – as long as they remain sufficiently high – are similarly distributed after loading as they are immediately after surface treatment. However, deep rolling-induced residual stresses are reduced to a noticeably greater extent by the same stress amplitude than shot peening-induced residual stresses. The probable reason for this is the greater microstructural workhardening found in shot peened specimens, effecting a higher resistance to quasi-static and cyclic dislocation movements. Examining austenitic AISI304 in a shot peened state, [6.46] found a zone close to the surface, about 80 μm thick and partially martensitically transformed, in which the relaxation of residual stresses was clearly greater than it was at greater distances to surface.

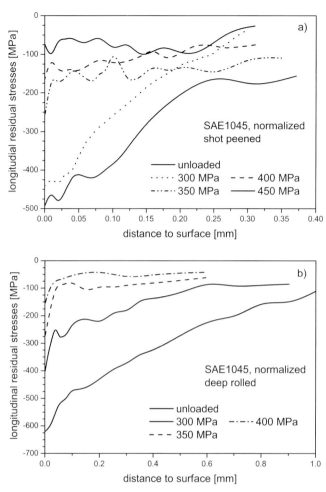

Fig. 6.10: Distribution of longitudinal residual stresses in shot peened (a) or deep rolled (b) SAE1045 in a normalized state after push-pull loading with different stress amplitudes up to half the number of cycles to failure [6.46]

Austenite shows a relaxation by up to two thirds, while relaxation in the martensitic phase does not exceed 50 %.

Regarding the dependency of residual stresses on the number of cycles during push-pull loading, first, a piece of evidence shall be provided which shows that the differences between residual stresses and mean stresses listed in Fig 6.6 in Sect. 6.1.2c actually are important. Fig. 6.11 shows results obtained by [6.47] for shot peened smooth specimens made of normalized AISI4140 subjected to homogeneous uniaxial tensile-compressive loading without mean stresses. The interruption of the experiment, required when measuring residual stresses, was varied by type. Measurements carried out after unloading from compressive loads show

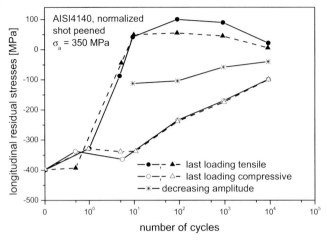

Fig. 6.11: Influence of unloading condition on longitudinal residual stresses at the surface after push-pull loading with a stress amplitude of 350 MPa for shot peened AISI4140 in a normalized state [6.47]

a gradual relaxation of longitudinal residual stresses as the number of cycles increases, longitudinal residual stresses staying within the compressive range. On the other hand, the residual stresses present after unloading from tensile loads disappeared completely between the 10th and the 100th loading cycle, becoming tensile thereafter and increasing again slightly during the further course of the experiment. After an interruption with a decreasing amplitude, however, small compressive residual stresses were measured which were decreased only slightly by a growing number of loading cycles. These results are due to differences in the deformation behavior of the surface layer and the core of the specimen. These differences are inextricably linked with the induction of residual stresses and result in lower plastic strain amplitudes in the workhardened surface layer compared to the core [6.47]. This constitutes indirect evidence of the different effects which mean stresses and residual stresses have. The experimental procedure used influences the residual stresses' dependency on the number of cycles for different loading amplitudes. This is exemplified by a comparison of Figs. 6.1 and 6.12, both of which examine AISI4140 quenched and tempered at 600 °C and subsequently shot peened [6.4]. As Fig. 6.1 shows for nominal stress-controlled tests, residual stresses show initial decreases within the first loading cycle, further linear decreases with increasing number of cycles and rising stress amplitudes, while the residual stress relaxation rate shows renewed increases only when significant cyclic worksoftening occurs. Fig. 6.12 shows similar results for total strain-controlled tests. After the first loading cycle, initially linear correlations are found only for amplitudes up to 3 ‰. However, at 4.5 ‰, which is the greatest amplitude examined, cyclic worksoftening already sets in after the first loading cycle, and therefore no linear region is found. Nevertheless, both experimental proce-

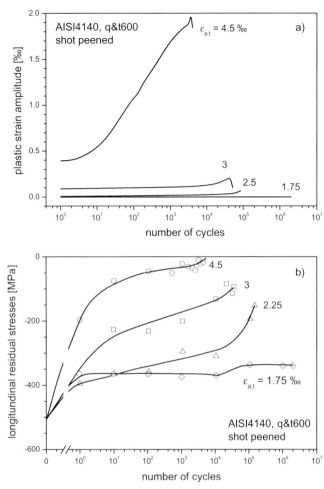

Fig. 6.12: Influence of total strain amplitude on plastic strain amplitudes (a) and longitudinal residual stresses at the surface (b) vs. number of cycles in total strain-controlled push-pull experiments studying the cyclic deformation behavior of shot peened AISI4140 in a state quenched and tempered at 600 °C [6.4]

dures show a similar residual stress relaxation behavior, as evidenced by Fig. 6.13, showing the residual stress- and plastic strain amplitude values present at half the number of cycles to crack initiation. As these two experimental procedures do not show any discernable differences, it is evident that the cyclic relaxation of residual stresses is determined by plastic strain amplitudes [6.4]. Residual stress relaxation can be described by an equation of the form

$$\sigma_{ax}^{rs} = \sigma_{ax,0}^{rs} + \sigma_{ax}^{rs} \cdot \lg \frac{\varepsilon_{a,p}}{\varepsilon_{0}^{rs}} \qquad (6.16)$$

in which $\sigma_{ax,o}^{rs} = -500$ MPa is the original residual stress value, $\sigma_{ax}^{rs} = 137$ MPa is the change of longitudinal residual stresses dependent on the respective decade of plastic strain amplitude, and $\varepsilon_o^{rs} = 5.5 \cdot 10^{-4}$ ‰ is a characteristic strain value. According to [6.46], the approach also applies to shot peened normalized SAE1045, at least in the case of loading without mean stresses. Using the same states and a stress amplitude of 450 MPa, [6.46] additionally confirmed that the dependency of plastic strain amplitudes on number of cycles is closely coupled with residual stress relaxation. This is shown in Fig. 6.14. Following the quasi-static relaxation of residual stresses during the first loading cycle, residual stresses initially stay approximately constant and then swiftly drop to values below 100 MPa when worksoftening sets in.

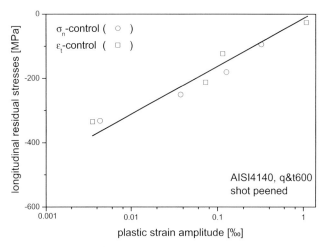

Fig. 6.13: Influence of the controlled variable in push-pull experiments on longitudinal residual stresses at the surface vs. plastic strain amplitude at half the number of cycles to failure for shot peened AISI4140 in a state quenched and tempered at 600 °C [6.4]

The distribution of residual stresses after **alternating bending loading** shall be described primarily by example of quenched and tempered AISI4140 in various shot peened states. As shown in Fig. 6.15 for a state annealed at 450 °C, loading using a fictitious loading stress amplitude at the surface, $\sigma_{a,surf}^* = 1000$ MPa, effects a noticeable relaxation of the longitudinal residual stresses induced by conventional shot peening during the first loading cycle – as discussed in Sect. 5.2.2a for Fig. 5.18 – and reduces them continuously thereafter [6.48]. Corresponding to bending loading, during which loading decreases as the distance to surface grows, residual stress relaxation is always most pronounced in the region within 0.1 mm distance to surface, and is even more intense for $N \geq 100$ at a distance to surface of 0.025 mm. Thus, a minimum of residual stress values develops close to the surface which [6.49] found also in shot peened quenched and tempered SAE1045.

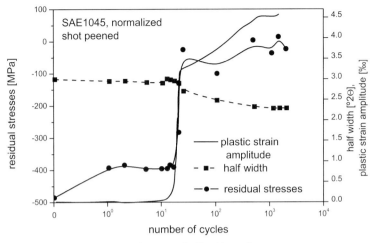

Fig. 6.14: Longitudinal residual stresses, half widths at the surface and plastic strain amplitudes vs. number of cycles for shot peened SAE1045 in a normalized state [6.46]

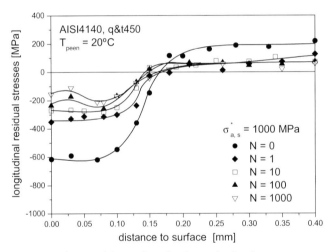

Fig. 6.15: Influence of the number of cycles on the distribution of longitudinal residual stresses during alternating bending with a fictitious stress amplitude at the surface of 1000 MPa after conventional shot peening, for AISI4140 in a state quenched and tempered at 450 °C [6.48]

Fig. 6.16 shows a corresponding plot for the same material state after stress peening using a tensile prestress of 500 MPa [6.48]. After the first loading cycle, which – as discussed in Sect. 5.2.2a for Fig. 5.19 – effects a particularly pronounced residual stress relaxation due to very high initial stresses, significant residual stress changes at the surface occur only up to the 10th loading cycle. Subsequently,

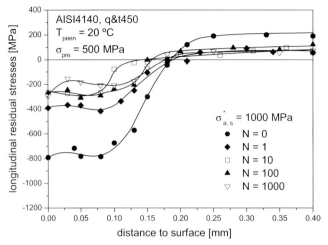

Fig. 6.16: Influence of the number of cycles on the distribution of longitudinal residual stresses during alternating bending with a fictitious stress amplitude at the surface of 1000 MPa after stress peening with a tensile prestress of 500 MPa, for AISI4140 in a state quenched and tempered at 450 °C [6.48]

significant additional changes of residual stresses occur only beneath the surface. Here, too, a minimum of residual stress values develops at medium distances to surface after 1000 loading cycles. Fig. 6.17 shows corresponding distributions in specimens warm peened at 290 °C [6.48]. While residual stress relaxation is reduced during the first loading cycle – as discussed in Sect. 5.2.2a for Fig. 5.20 – the residual stresses decrease continuously during the further course of cyclic loading, in turn causing the minimum of the residual stress values which develops beneath the surface. The distribution of residual stresses after a combined stress/warm peening treatment at 290 °C using a tensile prestress of 500 MPa – as discussed in Sect. 5.2.2a for Fig. 5.21 – shows a noticeable relaxation of residual stresses during the first loading cycle. Subsequently, the residual stresses remain approximately constant up to the 100th loading cycle. After 1000 loading cycles, a renewed and marked relaxation of residual stresses is observed. Examining conventionally shot peened TiAl6V4 specimens after alternating bending loading, [6.50] found the most pronounced relaxation of residual stresses in the compressive residual stress maximum. According to Fig. 5.19, the greatest stress amplitude examined, 850 MPa, roughly creates a plateau of residual stresses in a region close to the surface which is about 0.1 mm thick. [6.50] observed similar behavior during alternating bending loading of shot peened AlCu5Mg2 when using a cold aged, as-delivered state of the material.

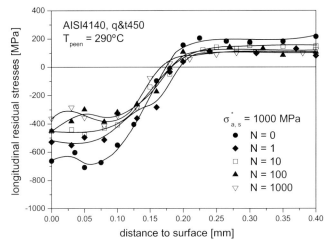

Fig. 6.17: Influence of the number of cycles on the distribution of longitudinal residual stresses during alternating bending with a fictitious stress amplitude at the surface of 1000 MPa after warm peening at 290 °C, for AISI4140 in a state quenched and tempered at 450 °C [6.48]

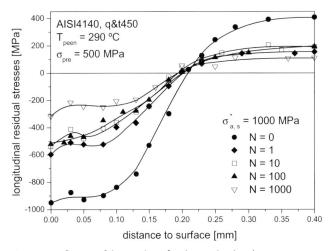

Fig. 6.18: Influence of the number of cycles on the distribution of longitudinal residual stresses during alternating bending with a fictitious stress amplitude at the surface of 1000 MPa after stress peening with a tensile prestress of 500 MPa at 290 °C, for AISI4140 in a state quenched and tempered at 450 °C [6.48]

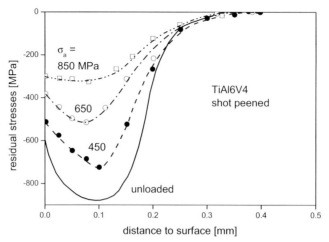

Fig. 6.19: Influence of fictitious stress amplitude at the surface on the distribution of longitudinal residual stresses during alternating bending for shot peened TiAl6V4 [6.50]

The influence which loading amplitude and the number of cycles have on residual stresses during alternating bending loading has been examined extensively. Therefore, the following contains only a selection of results, chosen primarily from the aspect of highest possible residual stress stability. First, however, Fig. 6.20 uses the example of shot peened quenched and tempered AISI4140 to show that during cyclic bending loading there are hardly any differences between residual stresses after unloading from tensile loads or after unloading from compressive loads – apart from the first loading cycle, which should be assessed from the viewpoint of quasi-static loading. However, the differences between longitudinal and transversal residual stresses which were introduced in the first loading cycle remain up to failure [6.3]. Depicting the influence of amplitude on the loading stress-dependence of residual stress relaxation for the same material state, Fig. 6.21 shows a pronounced residual stress relaxation during the first loading cycle, followed by a relaxation which is linear to the logarithm of the number of cycles and shows a relaxation rate which increases with growing amplitudes [6.3]. Contrastingly, an increased relaxation of residual stresses is observed during the phase of cyclic crack propagation. Similar behavior was reported for shot peened AISI4140 in a normalized state [6.3] and in a state quenched and tempered at 650 °C [6.51]. [6.52–54], for instance, examined the stability of shot peening-induced residual stresses during loading with amplitudes in the range of the fatigue limit for TiAl6V4 and various aluminum-base alloys. Most of the material states and loading stress amplitudes examined showed residual stress changes which occurred in the range of cyclic crack propagation, or merely during the first loading cycle, i.e. due to quasi-static yield strength being exceeded – which corresponds to case 3 in Fig. 6.2. By contrast, AlCu5Mg2 in a warm aged state shows an additional residual stress relaxation during the course of cyclic loading – which corresponds to case 4 in Fig. 6.2. Evidently, this state is workhardened to an extent at which shot peening will not effect further workhardening of the surface layer, and thus, the cyclic yield strength is exceeded locally during the course of cyclic loading. By contrast, the other materials and material states examined seem to experience sufficient additional workhardening, and therefore the cyclic yield strength is not exceeded during cyclic loading.

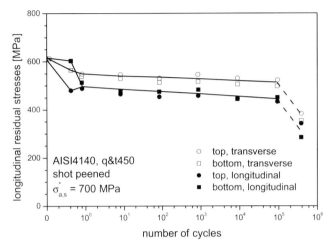

Fig. 6.20: Magnitudes of longitudinal and transversal residual stresses on specimens' top and bottom sides vs. number of cycles during alternating bending loading, for conventionally shot peened AISI4140 in a state quenched and tempered at 450 °C [6.3]

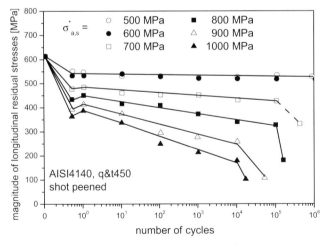

Fig. 6.21: Influence of fictitious stress amplitude at the surface on the magnitudes of longitudinal residual stresses vs. number of cycles during alternating bending loading, for conventionally shot peened AISI4140 in a state quenched and tempered at 450 °C [6.3]

To date, there have been only few studies which examined the effects of **isothermal cyclic loading at elevated temperatures** on the residual stress state. Shown in Fig. 6.22 for AISI4140 quenched and tempered at 450 °C is the residual stresses' dependency on the number of cycles during loading, using a fictitious loading stress amplitude of 800 MPa at the surface, at room temperature, 250 °C and 400 °C [6.3]. Residual stresses at the various temperatures are differentiated, in essence, by the thermal relaxation of residual stresses which occurs before loading (comp. Sect. 4) and to a lesser degree by their behavior during the first loading cycle (comp. Sect. 5). At an increasing number of cycles, however, residual stress behavior is essentially similar – disregarding the outlier value found after the first loading cycle at 400 °C. [6.55] determined the residual stresses' dependency on loading temperature for AISI304 at half the number of cycles to failure during cyclic push-pull loading using a stress amplitude of 280 MPa. According to Fig. 6.23, the remaining residual stresses decrease almost linearly with rising temperature, and there are significant residual stresses still present after loading at 650 °C.

The influence of various **modifications of the peening process** on the longitudinal residual stresses' dependency on the number of cycles is shown in Fig. 6.24 for a fictitious loading stress at the surface of 1000 MPa. It shows that the residual stress relaxation which occurs during the first loading cycle sets the general level of residual stresses which is the starting point for additional residual stress changes at a similar relaxation rate. Only the state stress peened at 290 °C exhibits contrasting behavior, its relaxation rate increasing after the 1000th loading cycle. Apart from this, the differences of the residual stress values are essentially caused by the superpositions of increased residual stresses and loading stresses in the case of the stress peened states – as mentioned in Sect. 5.2.2a –, and by the effects of static and dynamic strain aging in the case of the states peened at elevated temperatures and the conventionally shot peened and annealed state. [6.56] determined the optimal conditions for warm peening treatments of quenched and tempered AISI4140 by varying peening temperature and analyzing cyclic residual stress relaxation for numbers of cycles up to $N = 1000$. The results plotted in Fig. 6.25 show that the maximum of the residual stress values, found immediately after warm peening using a temperature of about 310 °C, is only faintly present after one loading cycle, while it builds up again clearly during further loading cycles up to $N = 1000$. Therefore, a peening temperature of about 310 °C can be viewed as optimal in respect to residual stress stability during alternating bending loading, as it achieves maximum pinning of the dislocation structure due to static strain aging effects. Analogously, [6.57] varied annealing conditions for conventional peening and subsequent annealing, and examined the relaxation of residual stresses using a fictitious loading stress amplitude at the surface of 1000 MPa. Fig. 6.26 use the example of the influence of annealing temperature for annealing times of 1 min (a), 5 min (b) and 60 min (c), to show that the stabilization of residual stresses discussed in Sect. 5.2.2c causes maximum residual stress values at temperatures which decrease as annealing time increases. The dependency of the optimal annealing temperature on annealing time is represented as an Arrhenius

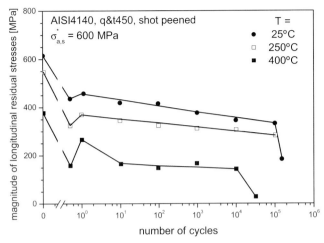

Fig. 6.22: Influence of loading temperature on the magnitudes of longitudinal residual stresses vs. number of cycles during alternating bending loading with a fictitious stress amplitude at the surface of 800 MPa, for conventionally shot peened AISI4140 in a state quenched and tempered at 450 °C [6.3]

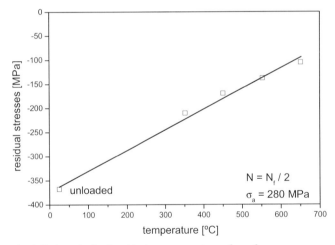

Fig. 6.23: Longitudinal residual stresses at the surface after push-pull loading with a stress amplitude of 280 MPa up to half the number of cycles to failure vs. loading temperature, for deep rolled AISI304 [6.55]

plot in Fig. 6.26d. Analyzing it in respect to the effective apparent activation enthalpy results in $\Delta H = 1.06$ eV, which is an indication of carbon diffusion in a-iron being the rate-controlling process [6.58]. In contrast to residual stresses after the first loading cycle, the optimal stabilization of residual stresses during cyclic loading is uniform for all annealing times examined, occurring due to static strain aging effects. It is no longer determined significantly by thermal residual stress relaxation, which is controlled by diffusion effects.

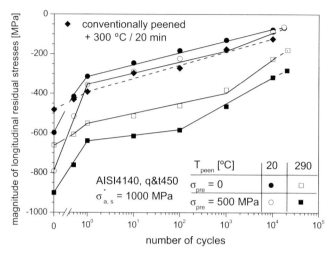

Fig. 6.24: Longitudinal residual stresses vs. number of cycles at alternating bending loading with a fictitious stress amplitude at the surface of 1000 MPa after conventional and various modified shot peening treatments of AISI4140 in a state quenched and tempered at 450 °C [6.32]

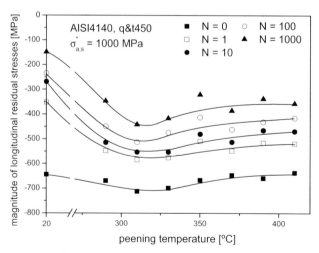

Fig. 6.25: Influence of the number of cycles on longitudinal residual stresses at the surface vs. peening temperature during alternating bending loading with a fictitious stress amplitude at the surface of 1000 MPa after warm peening of AISI4140 in a state quenched and tempered at 450 °C [6.66]

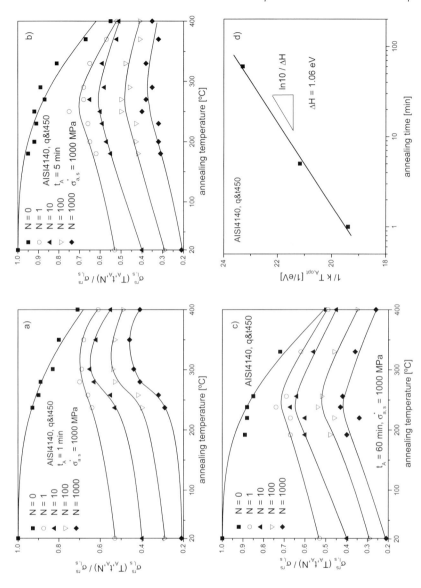

Fig. 6.26: Influence of the number of cycles on longitudinal residual stresses at the surface vs. annealing temperature during alternating bending loading with a fictitious stress amplitude at the surface of 1000 MPa after conventional shot peening plus subsequent annealing for 1 min (a), 5 min (b) and 60 min (c) of AISI4140 in a state quenched and tempered at 450 °C, and optimal annealing temperature vs. annealing time for these material states (d) [6.58]

Additional information is provided in [6.33,6.59] regarding the influence of the heat treatment state of AISI4140 during alternating bending loading with amplitudes in the high cycle fatigue range. According to Fig. 6.27, the shot peening-induced residual stress magnitudes are reduced significantly during the first loading cycle in the case of soft material states, while remaining approximately constant in the case of the quenched state. After 10^7 loading cycles all states show a similar extent of additional residual stress relaxation. Thus, the quenched state – starting out with particularly high residual stress magnitudes – shows a relaxation of residual stresses which is far smaller than the relaxation found in other states. Fig. 6.28 shows a corresponding plot for a state stress peened using a prestress of $\sigma_{pre} = 0.4\ R_y$. For soft states, residual stress relaxation during the first loading cycle is approximately twice as high as the relaxation observed after conventional shot peening. For the quenched state, residual stress relaxation during the first loading cycle is many times higher than the relaxation found for conventionally shot peened states. While the residual stresses in soft states, particularly in the state quenched and tempered at 650 °C, are reduced further by loading up to $N = 10^7$, they remain approximately constant in the quenched state, following a pronounced initial relaxation. In all cases, however, the relaxation of residual stresses is clearly more pronounced in the stress peened states compared to conventionally shot peened states. Fig. 6.29 shows a corresponding plot for a state warm peened at 290 °C, indicating that after warm peening, only the normalized state and the state quenched and tempered at 650 °C experience a relaxation of residual stresses during the first loading cycle. By contrast, the state quenched and tempered at 450 °C shows constant residual stresses, as does the state which was originally

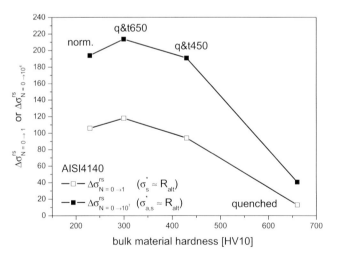

Fig. 6.27: Changes of longitudinal residual stresses in the first and up to the 10th cycle vs. bulk material hardness for conventionally shot peened AISI4140 during alternating bending loading with a fictitious stress amplitude at the surface in the range of the fatigue limit [6.33]

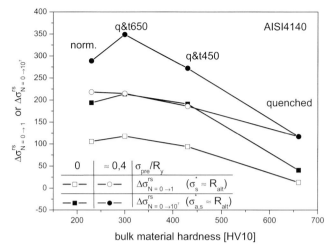

Fig. 6.28: Changes of longitudinal residual stresses in the first and up to the 10^{7th} cycle vs. bulk material hardness for conventionally shot peened and stress peened AISI4140 during alternating bending loading with a fictitious stress amplitude at the surface in the range of the fatigue limit [6.33]

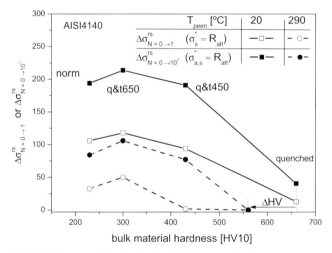

Fig. 6.29: Changes of longitudinal residual stresses in the first and up to the 10^{7th} cycle vs. bulk material hardness for conventionally shot peened and warm peened AISI4140 during alternating bending loading with a fictitious stress amplitude at the surface in the range of the fatigue limit [6.33]

quenched. This state was annealed during the peening treatment and therefore reduced in hardness. In the originally quenched state residual stresses remain constant even after 10^7 loading cycles, while the other states show a moderate residual stress relaxation which is smaller than the relaxation observed during the first loading cycle in conventionally peened states. This again underscores the fact that the stability of residual stresses is significantly increased in the warm peened states. In summary it can be stated that increasing material hardness results in a higher stability of residual stresses, due to the increase of the ratio of cyclic yield strength and high cycle fatigue during loading in the range of the fatigue limit, as seen in Fig. 6.3. However, at the same values of cyclic yield strength or fatigue limit, great residual stresses, as found e.g. in stress peened states, are more prone to relaxation. However, if the cyclic yield strength is increased for similar residual stress states, e.g. due to dynamic or static strain aging effects caused by warm peening or annealing after conventional shot peening, the residual stresses remain much more stable.

Fig. 6.30 shows AISI4140 in a state quenched and tempered at 450 °C as an example for examinations of the stability of residual stresses during **torsional swelling loading** and how it is influenced by stress- or warm peening [6.60,6.61]. After conventional shot peening treatments residual stresses show decreases in the compressively loaded 135°-direction which are pronounced during the first loading cycle and continuous thereafter. On the other hand, the residual stress component in the tensilely loaded 45°-direction remains almost unchanged, as residual stresses decrease effective stresses in this direction. The warm peened state shows a similar behavior, albeit that the relaxation of residual stresses in the compressively loaded direction is significantly reduced and therefore, much greater residual stresses remain in this direction throughout the course of cyclic loading. For a specimen peened using a torsional prestress, however, the residual stresses measured in the direction of future compressive loading show smaller residual stresses even before loading, which are not reduced during the first loading cycle. Residual stresses in the tensilely loaded direction are increased, and they, too, show no relaxation during the first loading cycle. During the further course of loading, a continuous residual stress relaxation occurs in both directions. As a result, the residual stresses which remain in the compressively loaded direction are increased only in comparison with the conventionally peened state, while the residual stress magnitude in the tensilely loaded direction is even smaller than it is for the other two states. [6.62] confirmed these results for SAE5155 spring steel of varying strength. Thus, the findings discussed for alternating bending loading are essentially confirmed for modified shot peening treatments, as well. However, the results also show that the assessment of the loading situation is more complex due to the biaxial and swelling loading stress state.

The relaxation of residual stresses was examined for **notched states**, as well. Fig. 6.31 shows the distribution residual stresses in quenched and shot peened SAE1045 after alternating bending loading using 10^7 loading cycles and an amplitude in the range of the fatigue limit [6.42,6.63]. While the smooth specimen and the notched specimen peened using soft shot show a residual stress state which is stable – within the scope of measured value scattering –, the notched specimen peened using hard shot showed a noticeable relaxation of residual stresses which is particularly strong close to the surface, resulting in the compressive residual

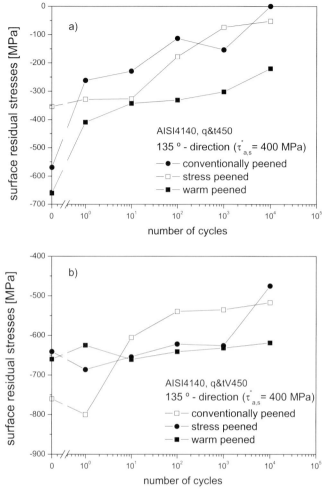

Fig. 6.30: Influence of shot peening treatment modifications on surface residual stresses in the direction of compressive (a) and tensile principal stress (b) vs. number of cycles during torsional swelling loading with a surface shear stress amplitude of 400 MPa for AISI4140 in a state quenched and tempered at 450 °C [6.60,6.61]

stress maximum being shifted from 100 μm to 170 μm. This was caused by the greater residual stresses which had been induced by shot peening. Both notched specimens show a relaxation of residual stresses to values which result in maximum compressive stresses of -2030 MPa when superposed with the loading stresses. Evidently, alternating bending loading also creates residual stress profiles that effect similar maximum compressive stress amounts during loading. For quenched and tempered SAE1045, however, [6.35,6.63] found a significant relaxa-

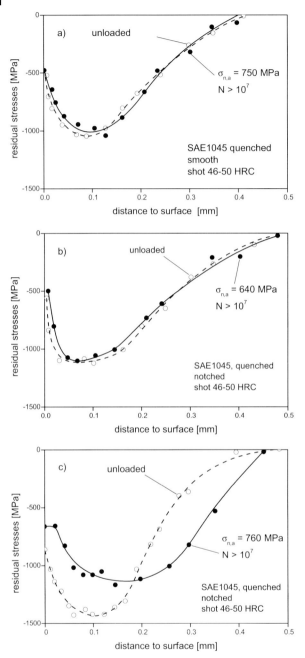

Fig. 6.31: Distributions of longitudinal residual stresses before and after alternating bending loading up to 10^7 cycles with the amplitudes in the region of the fatigue limit indicated for shot peened smooth specimens (a), for notched specimens shot peened using soft shot (46–50 HRC) (b) and hard shot (54–58 HRC) (c) made of SAE1045 in a quenched state [6.42, 6.63]

tion of residual stresses in both smooth and notched specimens, as shown in Fig. 6.32. For the notched specimens, this relaxation is particularly pronounced in the surface layer as it was loaded the most. [6.64] reported on the changes of the notch residual stresses found in shot peened notched specimens made of quenched and tempered AISI4140 after tensile, compressive, alternating and swelling loading. Alternating loading causes a relaxation of residual stresses which is noticeably stronger in the longitudinal direction than in the transversal direction, which is a typical course of residual stress relaxation. Swelling loading, on the other hand, causes a noticeable relaxation of the transversal residual stress-

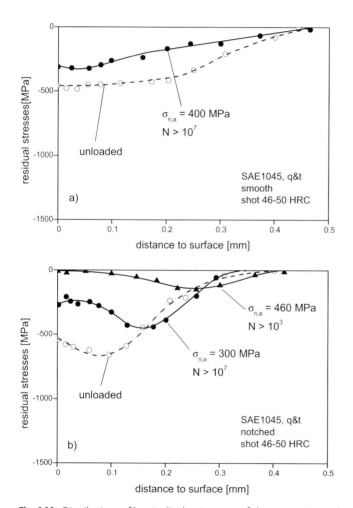

Fig. 6.32: Distributions of longitudinal residual stresses before and after alternating bending loading up to 10^7 cycles with the indicated amplitudes in the region of the fatigue limit and up to the number of cycles to failure, occurring at about 10^3 cycles, for shot peened smooth specimens (a) and for notched specimens shot peened using soft shot (46–50 HRC) (b) made of SAE1045 in a quenched and tempered state [6.35, 6.63]

es, while the longitudinal residual stresses are hardly changed. This behavior initially comes as a surprise, yet it is explained by Fig. 6.33, plotting transversal residual stresses versus longitudinal residual stresses. It shows the effective mean strains (symbols) created by residual stresses and mean loading stresses, as well as the cyclic changes of the stress state (arrows) which occur as a result. Due to the biaxial stress state found in the notch ground, the cyclic stress changes show a slight incline against the longitudinal stress axis. Also indicated are the flow surfaces according to von Mises' flow criterion for quasi-static and cyclic yield strength. This, together with the assumption of the normal rule – according to which the direction of the plastic strain vector is perpendicular to the flow surface –, allows for deriving the residual stress changes expected for the direction contrary to that of the plastic strains. This agrees with the data depicted for the loaded states and leads to effective mean stresses showing a tendency to relax toward the origin. Additionally, the orientation of the plastic strain vectors provides an understanding of the relaxation of longitudinal residual stresses during alternating loading (Fig. 6.33a) and their stability (Fig. 6.33b) during swelling loading. Aforementioned method for assessing of the relaxation of residual stresses which uses the effective mean strains and the flow surface was also verified by examining the residual stress changes induced by single instances of overloading.

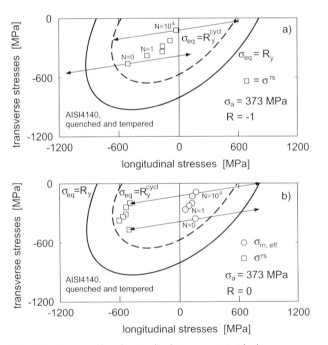

Fig. 6.33: Transversal vs. longitudinal stresses or residual stresses during alternating push-pull loading (a) or tensile swelling loading (b) after loading with a nominal stress amplitude of 373 MPa up to different numbers of cycles for shot peened AISI4140 in a quenched and tempered state, and comparison with quasi-static and cyclic yield surface according to the v. Mises criterion, acc. [6.64]

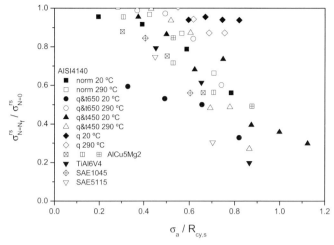

Fig. 6.34: Residual stresses after failure related to their initial values vs. loading stress amplitude related to compressive yield strength at the surface for different materials and material states, completed acc. [6.52]

In summary, Fig. 6.34 depicts the residual stresses which remain after loading, using various stress amplitudes related to cyclic yield strength (comp. Sect. 6.2.1c), for various materials and material states. Related to their initial values, the courses of the residual stresses are similar, and they decrease as loading increases [acc. 6.52]. It is striking that even the curves of the warm peened states of AISI4140 do not exhibit any systematic increases. Relating stress amplitude to cyclic yield strength at the surface apparently suffices to describe the stabilization of the residual stresses. This provides a method of assessing the residual stresses which remain after cyclic loading. Compared to Eq. 6.16, this method is crude but easier to implement.

b) Modeling of Residual Stress Relaxation

The relaxation of shot peening-induced residual stresses during **push-pull loading** of AISI4140 was examined by [6.47] for a normalized state and for two quenched and tempered states, and described using the concept introduced in Eqs. 6.3 and 6.4. Fig. 6.35 shows the resulting correlation of slope μ, related to the residual stress value found after the first loading cycle, and the stress amplitude, which is related to the cyclic yield strength of the unpeened material. For loading amplitudes below the cyclic yield strength, residual stresses decrease only marginally, while the extent of residual stress relaxation increases as stress amplitudes rise. Growing material hardness results in increasing related relaxation rates and a similar influence of related stress amplitudes for medium loadings and all degrees of material hardness. The reason for this initially surprising influence of material hardness is probably due to the fact that relating the unpeened material to cyclic yield strength does not account for the workhardening found in the surface layer after shot peening. Apart from this, the

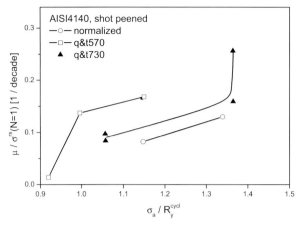

Fig. 6.35: Related relaxation rates of residual stresses vs. loading stress amplitude in push-pull loading related to compressive yield strength at the surface for shot peened AISI4140 in three different heat treatment states [6.47]

quenched and tempered states are expected to show a much more pronounced, typical cyclic worksoftening for half the number of cycles to fracture – which is the basis for cyclic yield strength – than for the small numbers of cycles which are the primary focus when assessing residual stress stability. Therefore, when both effects are taken into account, the curves for the individual material states are, at least, closer to each other. According to Fig 6.36, those points which in Fig. 6.35 show similar dependencies on stress amplitude, but on different levels, will largely be located on a straight line when plotted versus amplitude related to high

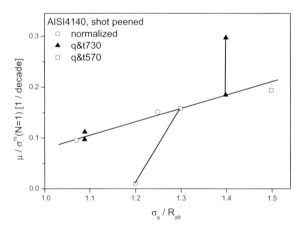

Fig. 6.36: Related relaxation rates of residual stresses vs. loading stress amplitude in push-pull loading related to alternating fatigue strength for shot peened AISI4140 in three different heat treatment states [6.47]

cycle fatigue in the shot peened state. Thus, fatigue limit seems to be a more suitable parameter for describing cyclic residual stress relaxation for these loading states and the relatively soft material states which show a pronounced relaxation of residual stresses during the first loading cycle.

[6.47] carried out analogous studies for **alternating bending loading**, arriving at the correlations represented in Figs. 6.37 and 6.38. Here, too, slope μ increases

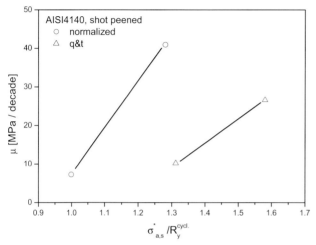

Fig. 6.37: Related relaxation rates of residual stresses vs. fictitious stress amplitude at the surface related to alternating bending fatigue strength for shot peened AISI4140 in different heat treatment states [6.47]

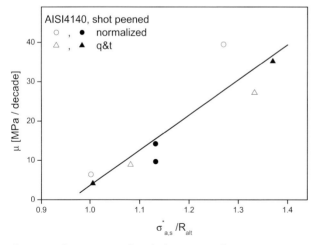

Fig. 6.38: Relaxation rates of residual stresses vs. fictitious stress amplitude at the surface related to alternating bending fatigue strength for shot peened AISI4140 and SAE1045 in different heat treatment states [6.5, 6.47, 6.65]

with rising amplitudes, but does not decrease again as material hardness rises, despite being plotted in absolute terms and not as related to the residual stress value after the first cycle. Plotting the slope μ versus amplitude related to the fatigue limit yields a uniform description of all the amplitudes examined and some additional values – obtained on SAE1045 by [6.5,6.65] and presented as a supplement. As expected, alternating bending loading results in greater residual stress stability for harder states, which again is described more uniformly using the fatigue limit rather than cyclic yield strength.

Fig. 6.39 shows slopes μ, as determined by [6.3,6.32], for **various heat treatment states of quenched and tempered AISI4140 steel**, dependent on fictitious stress amplitude at the surface. All heat treatment states show linear relations with the same slopes. Differences occur only in respect to levels, reflecting the decrease of the residual stress relaxation rate which occurs as hardness increases. Apart from a description of residual stress relaxation using Eq. 6.3, analyses of the amplitude-dependence of the residual stresses after 1 and 10^4 loading cycles were carried out for some of the material states shown in Fig. 3.9, which are depicted in Fig. 6.40 [6.59]. All heat treatment states show sectionally linear correlations. As fictitious loading stress at the surface rises, these correlations show initially constant residual stresses, and residual stress amounts which decrease linearly at $N = 1$ and when the critical loading stress or compressive yield strength of the compound is exceeded. At $N = 10^4$, a renewed and intensified residual stress relaxation – showing a linear dependency on loading stress amplitude – occurs above the critical loading stress amplitude or the cyclic yield strength of the compound, which is identical to the amount of the quasi-static compressive yield strength of the compound for all material states except the state quenched and tempered at 450 °C. Increasing material hardness results in rising initial residual stress values, along with increasing critical loading stresses and loading stress amplitudes. The values for critical loading stress amplitudes are always smaller than the values derived by extrapolating to stress amplitudes, resulting in slopes $\mu = 0$, as shown in Fig. 6.39. The reason may be found in the predominant influence of great stress amplitudes when slope μ is examined. For the examined material states which show cyclic worksoftening, the influence of great stress amplitudes results in cyclic worksoftening being more pronounced and therefore, cyclic yield strength may be underestimated. By contrast, the loading stress increases required in the first cycle for achieving similar degrees of residual stress relaxation remain at 270 MPa / 100 MPa for almost all material states, with the exception of the 650 MPa / 100 MPa of the quenched state, indicating noticeably greater residual stress stability. At $N = 10^4$, increases of the loading amplitude occur, which show a drop from about 200 MPa / 100 MPa for the normalized state and the state quenched and tempered at 650 °C to roughly 120 MPa / 100 MPa for the state quenched and tempered at 450 °C, and which again are considerably greater for the quenched state, at 650 MPa / 100 MPa. Regarding the quenched state, this shows that the stability of residual stresses is very high for low cycle fatigue loading, as well.

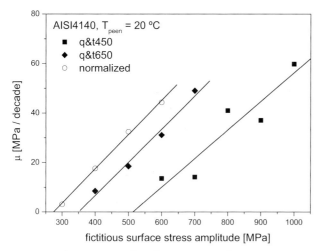

Fig. 6.39: Relaxation rates of residual stresses vs. fictitious stress amplitude at the surface for conventionally shot peened AISI4140 in different heat treatment states [6.32]

The influence of **modified peening treatments** on the stability of residual stresses is depicted in Fig. 6.41 for AISI4140 quenched and tempered at 450 °C, showing slopes μ dependent on fictitious stress amplitude at the surface. The slopes increase above a critical loading stress amplitude, which again is increased for the warm peened and annealed states, compared to the conventionally peened state. Starting from values around zero, slope increase is approximately linear to the growth of fictitious loading stress at the surface. The increase is noticeably more pronounced than for the conventional state, probably due to the fact that the effects of static and dynamic strain aging – which raise the cyclic yield strength of the compound – are reduced quickly and drastically when residual stress relaxation and thus, dislocation movement, set in. The same values for cyclic yield strength of the compound can be derived from Fig. 6.42, which shows the amplitude-dependency of residual stress relaxation at $N = 1$ and $N = 10^4$ for the warm peened state and the annealed state. Also, it is evident – as is the case when examining the slopes – that the influence of amplitude above the cyclic yield strength of the compound is clearly more pronounced than it is for the conventionally peened state. These statements were confirmed by [6.33] for different warm peening temperatures and annealing conditions.

For shot peened AISI4140 in a normalized state and in a state quenched and tempered at 450 °C, [6.3] used slope μ for describing cyclic residual stress relaxation **at elevated temperatures**. As shown in Fig. 6.43, linear relations of μ-values and stress amplitudes are found. At the same stress amplitudes, μ-values are noticeably greater for the normalized state than for the quenched and tempered state. However, the influence of stress is similar for both states. At the same time, temperature has practically no influence, up to the point at 400 °C and a stress amplitude of 500 MPa for the normalized state at which determining μ becomes

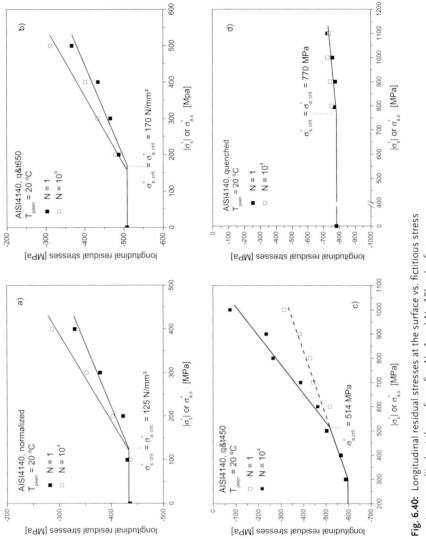

Fig. 6.40: Longitudinal residual stresses at the surface vs. fictitious stress or stress amplitude at the surface after $N = 1$ and $N = 10^4$ cycles for conventionally shot peened AISI41410 in a normalized state (a), in states quenched and tempered at 650 °C (b) or 450 °C (c) and in a quenched state (d) [6.59]

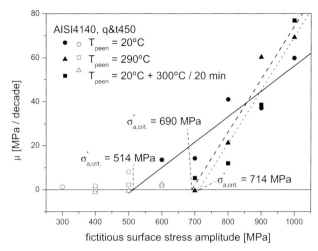

Fig. 6.41: Relaxation rates of residual stresses vs. fictitious stress amplitude at the surface after conventional shot peening, warm peening and conventional shot peening plus subsequent annealing for AISI4140 in a state quenched and tempered at 450 °C [6.32]

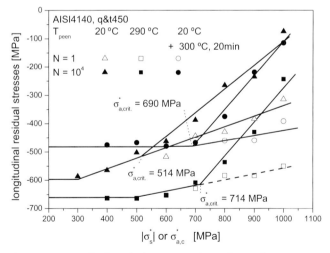

Fig. 6.42: Longitudinal residual stresses at the surface vs. fictitious stress or stress amplitude at the surface after $N=1$ and $N=10^4$ cycles after conventional shot peening, warm peening and conventional shot peening plus subsequent annealing for AISI41410 in a state quenched and tempered at 450 °C [6.33]

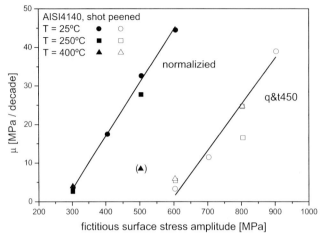

Fig. 6.43: Relaxation rates of residual stresses vs. fictitious stress amplitude at the surface at 25, 250 and 400 °C for shot peened AISI4140 in a normalized state and in a state quenched and tempered at 450 °C [6.3]

rather error-prone, due to the low level of remaining residual stresses. Cyclic residual stress relaxation at elevated temperatures therefore is distinguished from relaxation at room temperature only by thermal residual stress relaxation, and by effects which may appear during the first loading cycle. Overall, the results presented in this section show that both the concept according to Eqs. 6.3 and 6.4 – by means of which the residual stresses' dependency on the number of loading cycles can be described by a log-N relation with the amplitude-dependent slope μ – and the concept of the residual stresses' dependency on loading stress amplitude for one loading cycle and an exemplary number of cycles – set here at 10^4 - can be employed for a description of residual stress relaxation. Some deviations are found in the resulting values of cyclic yield strength of the compound, which represents the loading stress amplitude required for cyclic residual stress relaxation to set in. These deviations are attributed to the varying extent of cyclic work-softening.

c) **Evaluation of Surface Layer Properties**

Statements regarding the cyclic properties of the surface layer have been made only for alternating bending, determining cyclic yield strength at the surface -as introduced in Sect. 6.1.2b – according to Eq. 6.6. [6.3] was the first to determine cyclic yield strength at the surface for shot peened AISI4140 in a normalized and a quenched and tempered state. He applied slopes μ, as obtained by means of Eqs. 6.3 and 6.4, to different fictitious loading stress amplitudes at the surface and extrapolated to the amplitude present at $\mu = 0$. In the process, the critical loading stress amplitude, in conjunction with the residual stresses present at this amplitude after one cycle, was inserted into Eq. 6.6 as the cyclic yield strength of the compound. These values of cyclic yield strength at the surface were always compared with cyclic yield strength in a conventional sense, i.e. with the stress amplitude determined for the bulk material at $N_i/2$, at which plastic strain amplitudes other than zero start to occur. This made it possible to assess the changes of the surface layer state. More detailed examinations are found in [e.g. 6.32,6.33] for the example of variously shot peened AISI4140 in different heat treatment states. Here, cyclic yield strength of the compound was determined by analyzing the amplitude-dependence of residual stress relaxation. These examinations were summarized by [6.33], incorporating the influence of bulk material hardness on cyclic yield strength at the surface – as plotted in Fig. 6.44 – for conventionally peened states. The residual stresses found after the first loading cycle – which increase with growing hardness – and the cyclic yield strengths of the compound – increasing even more strongly – result in cyclic yield strengths at the surface increasing strongly. However, the increase exhibited by the latter is noticeably smaller than that of the cyclic yield strengths of the bulk material, also shown in Fig. 6.44 as far as they are available. The result is that cyclically effective workhardening occurs close to the surface of both of the materials of lower hardness, while the harder state quenched and tempered at 450 °C shows a decrease of cyclic yield strength at the surface, as compared to the bulk material. The stabilization of dislocation structures found after warm peening is evidenced not only by stabilized residual stresses, but also by cyclic yield strength at the surface. [6.66] was able to show this when he determined the optimal warm peening temperature for AISI4140 quenched and tempered at 450 °C (comp. Fig. 6.45). While specimens peened at room temperature exhibited cyclically effective worksoftening in the re-

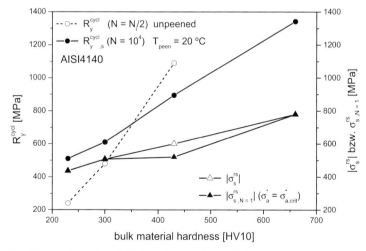

Fig. 6.44: Cyclic surface yield strength vs. bulk material hardness and comparison with the cyclic yield strength of the bulk material for conventionally shot peened AISI4140 [6.59]

gion close to the surface, all warm peened states show cyclically effective workhardening – due to static and dynamic strain aging effects – which is slightly greater than at 310 °C than at the other two temperatures. Cyclic yield strengths at the surface, found after warm peening at 290 °C, are plotted versus bulk material

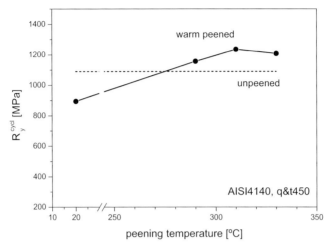

Fig. 6.45: Cyclic surface yield strength vs. peening temperature and comparison with the cyclic yield strength of the bulk material for warm peened AISI4140 in a state quenched and tempered at 450 °C [6.59]

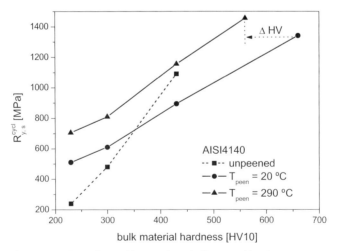

Fig. 6.46: Cyclic surface yield strength vs. bulk material hardness and comparison with the cyclic yield strength of the bulk material for AISI4140 warm peened at 290 °C [6.59]

hardness in Fig. 6.46 and compared to the values found after conventional peening and, in part, to the cyclic yield strengths of the bulk material. Nearly independent of bulk material hardness, the values for 290 °C peening temperature are increased by about 200 MPa – for the quenched state, the warm peening-induced decrease of hardness must be taken into account. Due to the increase of cyclical yield strength at the surface, cyclically effective workhardening occurs in all cases. According to [6.57,6.58], annealing allows for residual stress stability to be increased, compared to conventionally shot peened states. This is expressed also by an increase of cyclic yield strength at the surface. For AISI4140 quenched and tempered at 450 °C, however, cyclic yield strength at the surface stays lower than it does after warm peening, therefore resulting in cyclically effective worksoftening of the surface layer. Accordingly, it can be stated that the workhardening effect of static and dynamic strain aging combined is more pronounced than that of static strain aging alone.

6.2.2
Influences on Worhardening State

The distribution of workhardening states induced in normalized SAE1045 by shot peening or deep rolling changes during **push-pull loading** up to half the number of cycles to fracture, as shown in Fig. 6.47 [6.46]. The increases of half widths close to the surface are reduced to smaller surface values as stress amplitude increases, while the depth of the affected surface layer regions decreases. This reduction of the depth effect is more pronounced in the shot peened state, related to the initial depth effect of workhardening, than it is in the deep rolled state. By contrast, hardly any changes of the shot peening- or deep rolling-induced distribution of

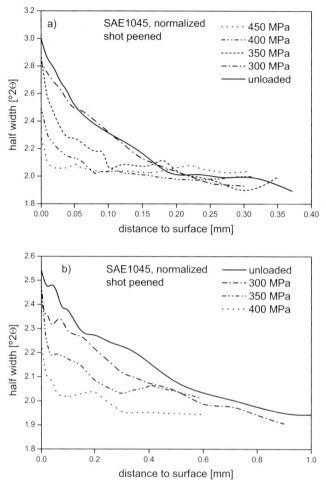

Fig. 6.47: Distributions of half widths after push-pull loading up to half the number of cycles to failure with different stress amplitudes for shot peened (a) and deep rolled (b) SAE1045 in a normalized state [6.46]

half widths are found for austenitic AISI304 steel [6.46]. Only the shot peened state shows slight decreases of half widths directly at the surface. The dependency of the macro residual stress state on the type of unloading, as depicted in Fig. 6.11, cannot be verified by measuring half widths in normalized AISI4140, according to Fig. 6.48. Evidently, the changes of the micro residual stress state that occur during a single loading cycle are too small to be detected by measuring the half widths of x-ray interference lines. As depicted in Fig. 6.49 for shot peened AISI4140 quenched and tempered at 600 °C, nominal strain- or total strain-controlled tensile-compressive loading causes systematic changes of half widths with the number of cycles only at those amplitudes which, according to Fig. 6.1a or

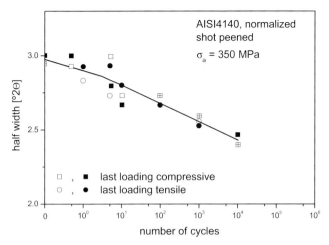

Fig.6.48: Influence of unloading condition on the half widths at the surface after push-pull loading with a stress amplitude of 350 MPa for shot peened AISI4140 in a normalized state [6.47]

6.12a, show significant plastic strain amplitudes before the start of the phase of cyclic crack propagation [6.4]. The changes of half widths at the surface that occur as the number of cycles increases, shown in Fig. 6.14 for shot peened normalized SAE1045, were not discussed yet. These changes, too, correlate strongly with the cyclic deformation curves. The onset of pronounced inhomogeneous cyclic work-softening causes small but important half width decreases. These show further decreases in the phase of homogeneous, small-scale worksoftening [6.46]. In agreement with the results of [6.9,6.47,6.67,6.68], a decrease of the shot peening-induced surface layer workhardening due to cyclic deformation is observed.

As was the case for residual stresses, the changes of workhardening state which occur during **alternating bending loading** shall be discussed primarily by examining the influence exerted by modifications of the conventional peening process. Fig. 6.50 shows the influence of the number of cycles on the distribution of half widths after loading using a fictitious stress amplitude at the surface of 1000 MPa for specimens made of AISI4140 quenched and tempered at 450 °C which were conventionally peened (a), stress peened (b), warm peened (c) or treated using combined warm- and stress peening (d) [6.32]. This compilation impressively demonstrates that the increase of half widths close to the surface of conventionally peened specimens is rapidly reduced during the course of cyclic loading. Thus, after 1000 loading cycles, workhardening states cease to show significant changes with increasing distance to surface. After stress peening, too, the half widths at the selected fictitious stress amplitude at the surface are greatly reduced, leaving behind only small amounts of workhardening in the surface layer. By contrast, half widths show no significant changes in distribution with rising number of cycles after warm peening. Evidently, not only the macro residual stress state, but

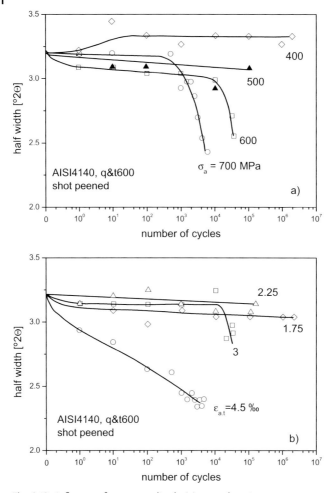

Fig. 6.49: Influence of stress amplitude (a) or total strain amplitude (b) on half widths vs. number of cycles in stress-controlled or total strain-controlled push-pull experiments for shot peened AISI4140 in a state quenched and tempered at 600 °C [6.4]

also the micro residual stress state, remains much more stable due to static and dynamic strain aging effects, thereby preventing the micro-plastic deformations typical of fatigue processes. As the number of cycles increases, the specimens which received a combined warm-/stress peening treatment show initial, unsystematic changes of the distribution of half widths, and slight decreases of the work-hardening states close to the surface.

By example of normalized shot peened AISI4140, Fig. 6.51 shows that during cyclic loading it is not only the residual stresses – as depicted in Fig. 6.21 for AISI4140 in a state quenched and tempered at 450 °C –, but also the half widths

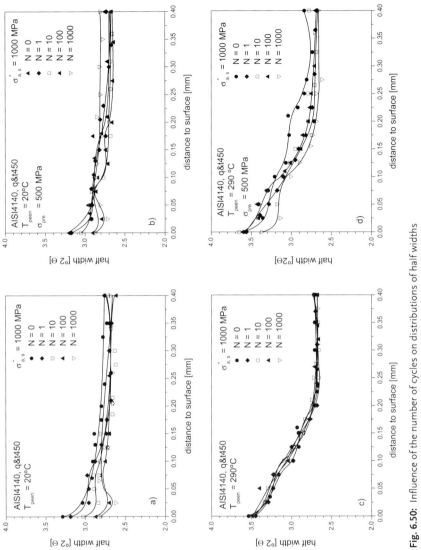

Fig. 6.50: Influence of the number of cycles on distributions of half widths after alternating bending loading with a fictitious stress amplitude at the surface of 1000 MPa after conventional shot peening (a), stress peening (b), warm peening (c) and stress peening at elevated temperature (d) for AISI4140 in a state quenched and tempered at 450 °C [6.32]

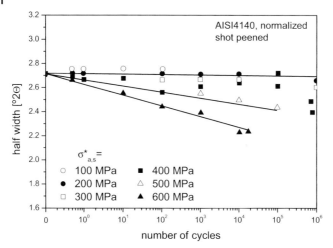

Fig. 6.51: Influence of fictitious stress amplitude at the surface on half widths at the surface vs. number of cycles during alternating bending loading for conventionally shot peened AISI4140 in a normalized state [6.69]

which show decreases linear to the logarithm of the number of cycles. The higher the loading stress amplitude, the more pronounced these decreases become [6.69]. Due to higher residual stress stability, and thus, smaller plastic deformations, AISI4140 in a state quenched and tempered at 450 °C shows significant changes of half widths only right before specimen fracture. For the same material and alternating bending loading using a fictitious stress amplitude at the surface of 1000 MPa, Fig. 6.52 shows the correlation found between half widths at the surface, related to their starting value, and the number of loading cycles, after different peening treatments [6.32]. After conventional shot peening and after stress peening, reduction of the half widths begins as soon as cyclic loading starts, leaving only about 75 % of the initial values at the point of specimen fracture. On the other hand, after warm peening and combined warm-/stress peening half widths are stable and show greater values after fracture than the other states up to 1000 loading cycles. The same is true for the conventionally peened and annealed state up to 100 loading cycles. This additionally confirms the statements made in the context of Fig. 6.50 regarding the stabilization of the workhardening state due to warm peening. Fig. 6.25 shows that 290 °C is the ideal warm peening temperature for AISI4140 in a state quenched and tempered at 450 °C [6.66]. This temperature creates the highest values of half widths after peening, and at the same time, half widths are absolutely stable during loading using a fictitious stress amplitude at the surface of 1000 MPa. As peening temperature is increased, half widths decrease slightly, due to thermally-induced effects. The half widths change unsystematically with the number of cycles, indicating workhardening states that are cyclically stable. Therefore, the temperature of 320 °C, seen in Fig. 6.25 as optimal in respect to residual stresses, appears to be almost equally suitable in regard to

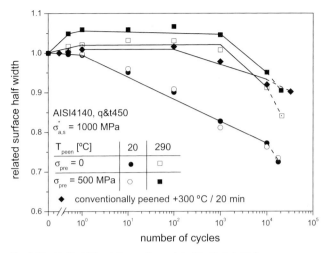

Fig. 6.52: Half widths at the surface related to their initial values vs. number of cycles during alternating bending loading with a fictitious stress amplitude at the surface of 1000 MPa after conventional shot peening, stress peening, warm peening, stress peening at an elevated temperature and conventional shot peening plus subsequent annealing for AISI4140 in a normalized state [6.32]

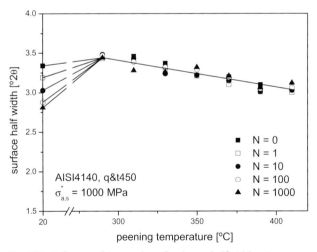

Fig. 6.53: Influence of the number of cycles on half widths at the surface vs. peening temperature during alternating bending loading with a fictitious stress amplitude at the surface of 1000 MPa after warm peening of AISI4140 in a state quenched and tempered at 450 °C [6.66]

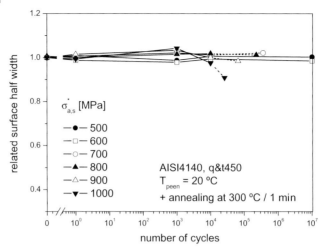

Fig. 6.54: Influence of fictitious stress amplitude at the surface on half widths at the surface vs. number of cycles after conventional shot peening plus subsequent 1-minute annealing at 300 °C for AISI4140 in a state quenched and tempered at 450 °C [6.58]

half widths. As an example for conventional peening and subsequent annealing, Fig. 6.54 summarizes the dependencies of the half widths on the number of cycles, as found after 1-minute annealing at 300 °C and subsequent loading using various fictitious loading stresses at the surface [6.57,6.58]. Except for the highest loading amplitude of 1000 MPa, for which slightly reduced half widths are found after specimen fracture, half widths remain absolutely stable during cyclic loading. [6.33] found warm peening treatments to have similar effects on the stability of half width changes for other heat treatment states of AISI4140, as well. The stabilizing effect on dislocation structure and thus, on workhardening state, which static and dynamic strain aging have during cyclic loading can therefore be proven by measuring half widths.

[6.55] examined shot peened or deep rolled AISI304 subjected to **isothermal push-pull loading**. As depicted in Fig. 6.55, half widths decrease with rising temperature. These decreases are slight at first and grow increasingly pronounced during the course of loading. As is the case for room temperature, all other temperatures show lower values for the deep rolled states than for the shot peened states.

6.2.3
Influences on Microstructure

Systematic studies of the influence of cyclic loading on microstructure in mechanically surface-treated surface layers were presented primarily in [6.46]. Here, the main focus was on normalized SAE1045 and austenitic AISI304 after shot peening and deep rolling. In normalized SAE1045 shot peening or deep rolling

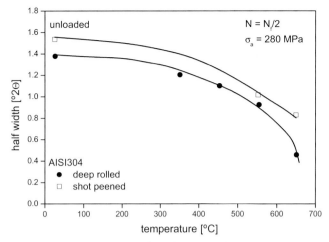

Fig. 6.55: Half widths at the surface after push-pull loading with a stress amplitude of 280 MPa up to half the number of cycles to failure vs. loading temperature for deep rolled AISI304 [6.55]

induces a tangled dislocation structure in the surface layer, which is turned into a cell structure – as in the unpeened state – by push-pull loading using a stress amplitude of 450 MPa and half the number of cycles to fracture, as shown in Fig. 6.56. However, the cell structure is not as well-defined, the interior of the cells still showing elevated dislocation densities in comparison with the unpeened state. At lower stress amplitudes, the surface layer shows cells which are not only larger but also far more diffuse, or even dislocation structures which remain tangled. Accordingly, some remnants of the dislocation structure resulting from the pronounced deformation which occurs during shot peening or deep rolling are preserved after cyclic loading. This agrees with the examinations of homogeneously pre-deformed states in [6.70–74]. Fig. 6.57 again confirms this by example of shot peened α-iron, showing that after 20 cycles or after 400 cycles of push-pull loading using a stress amplitude of 200 MPa, the surface layer's tangled dislocation structure was not removed. [6.46] also found that the cell structure is dependent on distance to surface. According to the schematic depiction in Fig. 6.58, shot peened states and, in particular, deep rolled states show cells which are smaller in the surface layer and larger in the core region than cells in non-workhardened states. As cell size is inversely proportional to applied loading amplitude in homogeneously deformed states – according e.g. to [6.75] –, this can be viewed as an indication that the workhardened surface layer and the non-workhardened core are coupled during alternating loading, the expected result being a locally greater loading of the surface layer and an unloading of the core. This also corresponds to [6.65]'s results for deep rolled SAE1045. Additional examinations of the influence of mean stresses show that when mean stresses occur, larger cells are found than when they are absent, while positive mean stresses show a slightly stronger effect than negative mean stresses.

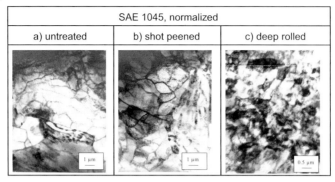

Fig. 6.56: Dislocation structure in untreated (a), shot peened (b) and deep rolled (c) SAE1045 in a normalized state after push-pull loading with a stress amplitude of 450 MPa up to half the number of cycles to failure [6.46]

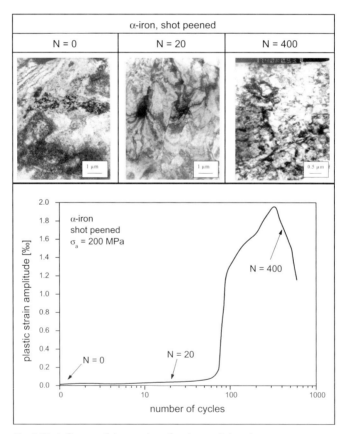

Fig. 6.57: Influence of the number of cycles on dislocation structure and plastic strain amplitudes in shot peened α-iron after push-pull loading with a stress amplitude of 200 MPa [6.46]

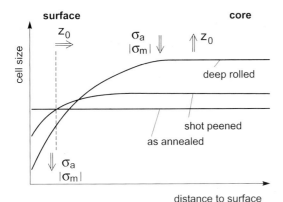

Fig. 6.58: Scheme showing cell size vs. distance to surface in mechanically surface treated SAE1045 after push-pull loading in the low cycle fatigue range and the influence of certain properties in the surface and core region, acc. [6.46]

Examining austenitic AISI304 steel after shot peening and deep rolling, [6.46] found nano-crystalline surface layers, transformations into α'-martensite, and tangled dislocation arrangements with a high dislocation density in the austenite. According to Fig. 6.59, alternating tensile-compressive loading barely changes this microstructure, in contrast to the unpeened state, which clearly shows cell formation. While the nano-crystalline surface layers exhibit no discernable changes after shot peening, the localized cell formation shown in Fig. 6.59b is found in the austenite at distances to surface greater than 2 µm. As shown in Fig. 6.60, increases of the martensite content are found after deep rolling. They appear in the border area between the surface layer, which was partially transformed by the deep rolling

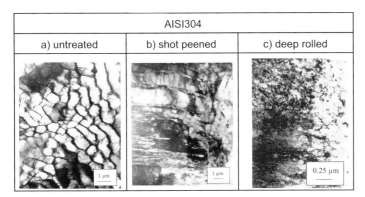

Fig. 6.59: Dislocation structure in untreated (a), shot peened (b) and deep rolled (c) AISI304 after push-pull loading with a stress amplitude of 320 MPa up to half the number of cycles to failure [6.46]

treatment, and the core, which was not transformed initially. According to [6.46], most of the increases occur at the crossing points of slip bands.

Fig. 6.61 shows surface layer microstructures of deep rolled AISI304, observed by [6.55] using a transmission electron microscope, before loading (a) and after isothermal, push-pull loading at 450 °C (b). The nano-crystalline surface layer remains stable up to 450 °C and does not show any signs of re-crystallization yet. In-situ TEM examinations showed that the isolated nano-crystalline surface layer would re-crystallize only at temperatures above 600 or 650 °C. For AZ31 magnesium alloy, on the other hand, the surface layer showed re-crystallization already at 100 °C.

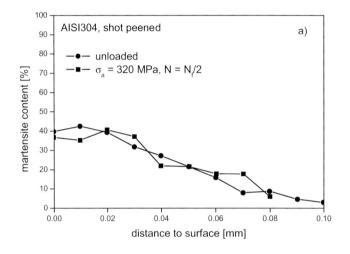

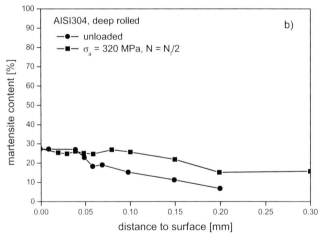

Fig. 6.60: Distribution of α'-martensite content for shot peened (a) and deep rolled (b) AISI304 before and after push-pull loading with a stress amplitude of 320 MPa up to half the number of cycles to failure [6.46]

AISI304, deep rolled	
a) as rolled	b) loaded

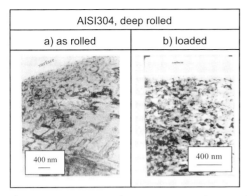

Fig. 6.61: Transmission electron microscope images of the microstructure of deep rolled AISI304 before (a) and after (b) stress-controlled push-pull loading at 450 °C with a stress amplitude of 280 MPa up to half the number of cycles to failure [6.55]

6.3
Effects of Surface Layer Stability on Behavior during Cyclic Loading

6.3.1
Basic Results

The surface layer states created by mechanical surface treatments, and their stability, have a significant effect on fatigue strength, i.e. on the number of cycles to failure during low cycle- and high cycle fatigue or on the fatigue limit. As this subject area is extremely relevant to technological practice, a vast number of original and overview studies have been published, which shall be represented here by [6.2,6.35,6.76] alone, as these provide access to a multitude of other studies. The essential influencing parameters of the surface state – as discussed in the context of Eq. 6.7 in Sect. 6.1.2c – are changes of roughness, workhardening of the surface layer region, residual stresses and phase transformations which might occur. According to [6.35], a more detailed examination should also include the residual stress gradient close to the surface and, for an evaluation of the loading stress state, notch factor and the loading stress gradient close to the surface. The individual parameters have different effects, depending on the material state. Regarding steels, [6.35] therefore differentiated between low-, medium- and high strength states. This subdivision shall be taken up in the following, separate discussion of the individual phases of fatigue, i.e. the phase of cyclic deformation which incorporates cyclic workhardening and worksoftening behavior, the crack initiation phase, the crack propagation phase and, finally, the achievable number of cycles to failure or fatigue limit.

6.3.2
Effects on Cyclic Deformation Behavior

The effects of mechanical surface treatments on cyclic deformation behavior can be quite significant, as the cyclic deformation curves in Fig. 6.62 show for 7 mm-diameter, round specimens of SAE1045 in an untreated, a shot peened, and a deep rolled state [6.46,6.77]. The deep rolled specimens, in particular, show a decrease of the number of cycles to incubation for the onset of the initial cyclic worksoftening that is typical of normalized states. This is probably due to the

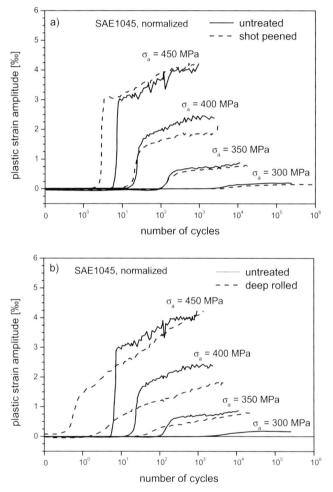

Fig. 6.62: Influence of stress amplitude on cyclic deformation curves determined during push-pull loading of normalized SAE1045 in a state shot peened with an intensity of 0.120 mmA (a) and in a state deep rolled with a pressure of 150 bar [6.46]

deep-reaching residual stresses which are superposed with the loading stresses. Actual cyclic deformation is decreased for the surface-treated states, due to the occurrence of pronounced surface layer workhardening. In a deep rolled state, this workhardening affects a much greater cross-sectional area, and therefore, this state shows the smallest plastic strain amplitudes. Surface layer workhardening is particularly evident in the stress-strain curves at half the number of cycles to failure, obtained on specimens of which a number were trepanned after surface treatment. As shown in Fig. 6.63, the cyclic stress-strain curves for shot peened solid specimens are changed only insignificantly, compared to the unpeened state. As wall thickness is reduced, however, the curves for shot peened and trepanned states are shifted toward much higher stress amplitudes. An important aspect regarding the cyclic deformation of surface treated notched specimens is shown in Fig. 6.64 [6.78,6.79]. Due to the pronounced relaxation of residual stresses in the notch ground during compressive loading half cycles, specimens initially show cyclic contractions as the number of cycles is increased. These contractions are particularly pronounced at high nominal stress amplitudes. Specimen elongations found at higher numbers of cycles are a sign of micro crack initiation and - propagation, observed for those curve sections indicated by broken lines. As shown in Fig. 6.65, a medium strength state of AISI4140, quenched and tempered at 570 °C for 2 hours, shows changes in cyclic deformation behavior which are less pronounced than the changes found for normalized SAE1045 [6.47,6.80,6.81]. At first, shot peened specimens show smaller or even disappearing plastic strain amplitudes, which in the unpeened state are small initially. After the number of cycles for incubation, worksoftening sets in and remains continuous until specimen fracture, reaching an extent which is almost independent of distance to surface. For shot peened states, this worksoftening behavior is shifted to higher numbers of loading cycles, permitting higher numbers of cycles to failure.

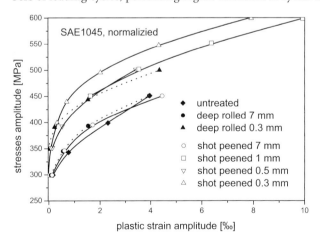

Fig. 6.63: Cyclic stress-stress-curves at half the number of cycles to failure, some of which were determined using trepanned specimens, for normalized SAE1045 in an unpeened state, a state shot peened with an intensity of 0.750 mmA and a state deep rolled with a pressure of 150 bar [6.46]

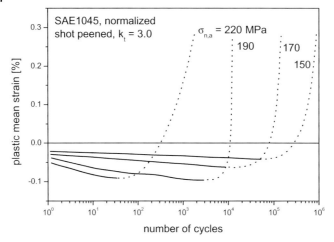

Fig. 6.64: Influence of nominal stress amplitude on mean plastic strains vs. number of cycles for shot peened notched specimens with a notch factor $k_t = 3.0$ made of normalized SAE 1045 [6.78,6.79]

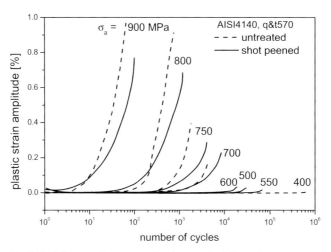

Fig. 6.65: Influence of stress amplitude on cyclic deformation curves determined during push-pull loading for untreated and shot peened AISI4140 in a state quenched and tempered at 570 °C [6.47,6.80]

The influence of surface layer workhardening on cyclic deformation behavior is particularly evident in cyclic stress-strain curves, determined by [6.46,6.77] for deep rolled or shot peened austenitic AISI304 steel. As in the case of normalized SAE1045, the cyclic stress-strain curves for solid specimens are raised only slightly in comparison with the untreated state. For trepanned specimens, on the other hand, decreasing wall thickness causes stress levels to show marked increases, which are particularly pronounced after deep rolling. This behavior shows that cyclic loading of a compound composed of a workhardened surface layer and an uninfluenced core can cause pronounced plastic deformations in the core, while the surface layer can experience loading which is almost elastic despite the additional load transfer in the compound. Accordingly, an averaged behavior is observed for the compound. [6.82] reported that clearly decreased plastic strain amplitudes were also found for AZ31 magnesium alloy after shot peening or deep rolling. These were attributed to the surface layer workhardening induced during surface treatment.

When shot peening is followed by annealing treatments, the cyclic deformation behavior which is observed subsequently for loading at room temperature can be changed significantly, as well. [6.83] showed this by example of cyclic deformation curves for normalized SAE1045 resulting from push-pull loading using a stress amplitude of $\sigma_a = 350$ MPa. According to Fig. 6.66, the curves show that short-term annealing treatments using temperatures up to 550 °C shift cyclic worksoftening to higher numbers of cycles and reduces its extent considerably. This is caused by static strain aging effects – as mentioned in Sect. 4 –, pinning the shot peening-induced dislocation structure of the surface layer and thereby reducing plastic strain amplitudes, as well as increasing fatigue life. By contrast, the dominant effect caused by annealing treatments using longer times and higher temperatures – exemplified by the 1-hour annealing treatment at 650 °C – is the reduction of the surface layer workhardening induced previously. Therefore, cyclic worksoftening becomes significantly more pronounced, compared to the state which was shot peened only, approximating the cyclic worksoftening of the untreated material state.

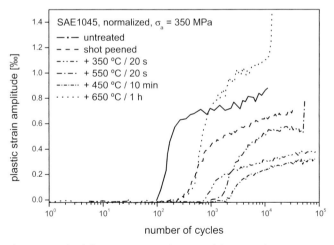

Fig. 6.66: Cyclic deformation curves determined during push-pull loading with a stress amplitude of 350 MPa for normalized SAE1045 in an untreated state, after shot peening and after conventional shot peening plus subsequent annealing under the conditions indicated [6.46]

Surface layer workhardening induced by mechanical surface treatments can also affect **cyclic deformation behavior at elevated temperatures**, if the dislocation structures induced close to the surface show sufficient thermal stability. As an example, the cyclic deformation curves obtained for 450 °C are shown in Fig. 6.67 for untreated, shot peened, and deep rolled states of austenitic AISI304 steel and

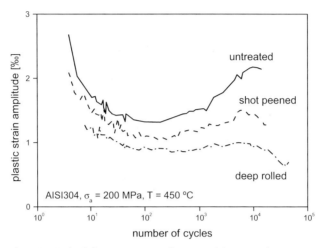

Fig. 6.67: Cyclic deformation curves determined during push-pull loading at 450 °C with a stress amplitude of 200 MPa for AISI304 in an unpeened, a shot peened and a deep rolled state [6.55]

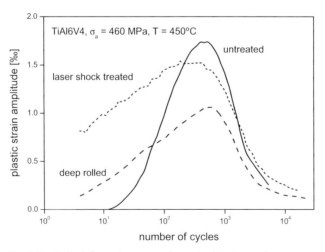

Fig. 6.68: Cyclic deformation curves determined during push-pull loading at 450 °C with a stress amplitude of 460 MPa for TiAl6V4 in an unpeened, a shot peened and a deep rolled state [6.55]

a stress amplitude of 200 MPa, in comparison with similar curves for untreated, shot peened, and laser shock treated states of TiAl6V4 titanium alloy and a stress amplitude of 460 MPa [6.55]. While the kind of cyclic deformation behavior found in the untreated material also occurs in the surface treated states in a qualitatively equivalent way, the magnitudes of the plastic strain amplitudes are clearly reduced in the surface treated states. Only the initial plastic strain amplitudes are increased in the surface treated TiAl6V4 alloy, compared to the untreated state. The small maximum values of the plastic strain amplitudes found for the deep rolled states of both materials are due to surface layer worksoftening, which is more pronounced, compared to laser shock treatment, or affects a greater cross-sectional area, compared to shot peening treatment.

Studies have focused also on the influence of mechanical surface treatments on deformation behavior during **thermo-mechanical fatigue loading** [6.84]. For AISI H11 tool steel in various states of mechanical surface treatment, Fig. 6.69 shows the tensile mean strains' dependency on the number of cycles. Tensile mean strains develop when there are temperature changes in the range of 200 to 625 °C, and they are a measure of the compression of the compound which accumulates during thermo-mechanical cycles. These are the greater, the greater the depth of the surface treatment is. However, they lead to negligible differences in fatigue life, due to a rapid relaxation of the macro- and micro residual stresses close to the surface.

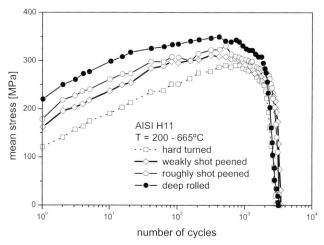

Fig. 6.69: Mean stresses vs. number of cycles determined during thermo-mechanical fatigue of AISIH11 tool steel after different mechanical surface treatments [6.84]

6.3.3
Effects on Crack Initiation Behavior

Due to microstructural changes during the crack-free phase of cyclic deformation, metallic materials subsequently experience crack initiation, which can be caused by different mechanisms [e.g. 6.85,6.86]. These mechanisms are based on dislocation movement, which in turn is caused by shear stresses that are effective in characteristic slip planes and -directions. Quantitative statements regarding the effects of mechanical surface treatment on crack initiation are very few in number, and virtually impossible to compare, as they are highly dependent on crack length, which is used as the criterion for crack initiation, and because shear stresses are a decisive factor only for crack lengths up to several multiples of grain size. For greater crack lengths, on the other hand, normal stresses and occurring residual stresses are crucial. Actual crack initiation takes place at positions of great slip activity, which in untreated states may be given in the form of extrusions or intrusions. This leads to the conclusion that crack initiation sets in at numbers of cycles which are the smaller, the higher plastic strain amplitudes are. As discussed in the previous section, mechanical surface treatments influence cyclic deformation behavior. Accordingly, the number of cycles to crack initiation will either be increased, reduced, or left unchanged by mechanical surface treatments.

Examining normalized SAE1045 in an untreated state, [6.46] found cracks formed at intrusions and extrusions and at grain- or phase boundaries in the vicinity of former austenite grain boundaries. No slip localization in the form of intrusions and extrusions was found for shot peened or deep rolled states, due to the great number of slip obstacles, such as high dislocation density. As shown in Fig. 6.70a, this results in the number of cycles to crack initiation, related to the number of cycles to failure, being clearly higher, compared to the untreated state. [6.47] reported similar results for SAE1080 and normalized AISI4140. [6.46] found a similar setting for austenitic AISI304 steel, in which the formation of in- and extrusions on persistent slip bands in the shot peened state and the deep rolled state is obstructed by heightened dislocation density and, additionally, by twin boundaries and the increased number of grain boundaries and phase boundaries to the α'-martensite phase, due to the nano-crystalline surface layer. Again, the number of cycles to crack initiation, related to the number of cycles to failure, is increased, as shown in Fig. 6.70b. However, this increase of the related number of cycles to crack initiation is not a universal result. For quenched and tempered SAE1045, [6.87] found that crack initiation occurred earlier in a shot peened state than in an untreated state, due to the micro-notch effect of the roughened surface. Crack initiation can be delayed by subsequent polishing. For quenched SAE1045, [6.88] found that inspite of an increase of the number of cycles to failure after shot peening, the number of cycles to crack initiation is reduced for loading in sea water, where corrosion scars act as micro-notches and influence crack initiation. According to [6.89], shot peened states of 7010 AlZnMgCu alloy can also show accelerated crack initiation in shot peening-induced notches and laminations. The same applies to TiAl6V4 titanium alloy in a state with a fine lamellar structure,

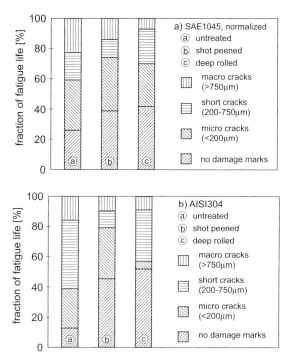

Fig. 6.70: Influence of mechanical surface treatments on development of damage during push-pull loading with a stress amplitude of 450 MPa for normalized SAE1045 (a) and during push-pull loading with a stress amplitude of 320 MPa for AISI304 (b) [6.46]

which after shot peening shows earlier crack initiation due to the micro-notch effect of the rough surface [6.90]. Careful polishing can improve roughness, while a surface layer which is still workhardened is able to delay crack initiation to a point at which the related number of cycles to crack initiation becomes greater than in the unpeened state.

Crack initiation behavior after laser shock treatment was examined by [6.91] for notched, 3-point-bending specimens made of 7075 aluminum alloy. Crack initiation was clearly delayed, owing to smooth surfaces and the great depth effect of the laser shock treatment.

The following general statements can be made: In mechanically surface treated material states, the number of cycles to crack initiation can be increased by surface layer workhardening, compressive residual stresses, and by smoothing – a possibility provided by deep rolling or laser shock treatment. The number of cycles to crack initiation can be decreased by surface roughening caused by shot peening, by the notch effect of shingles and laminations, and by local tensile residual stresses which promote crack initiation at interfaces to hard phases.

Mechanical surface treatments can also change the site of crack initiation. This is the case when sufficiently high compressive residual stresses impede crack initiation at the surface in such a way that it is shifted to positions below the surface [6.92]. This is shown in Fig. 6.71 for two shot peened states of quenched SAE1045, which were created using shot with an average diameter of 0.3 mm and 0.6 mm, respectively, and which show an alternating bending strength of 960 and 1050 MPa. For loading amplitudes that are greater than the fatigue limit, crack initiation is always observed at the surface. By contrast, loading amplitudes which cause numbers of cycles to failure of around 10^7 result in crack initiation occurring about 0.3 mm beneath the surface. Sub-surface crack initiation is found even for loading amplitudes below the fatigue limit. However, the propagation of these cracks is evidently limited. The state peened using shot with an average diameter of 0.3 mm is depicted in Fig. 6.72, which shows depth of crack initiation – visible as the centers of the rosette-like structures. These results spurred the development of the concept of the local fatigue limit, introduced in Sect. 6.1.2c. For great compressive residual stresses, smooth specimens made of high strength steel always show crack initiation beneath the surface, due to high residual stress stability. This results in residual stress sensitivity becoming insignificant for great compressive residual stresses. An increasing notch factor can stop crack initiation beneath the surface. This makes it possible to achieve even greater fatigue limits than for smooth specimens if the compressive residual stress maximum beneath the surface prevents developing cracks from propagating further. For medium and high strength peened states, too, rising loading stress gradients can shift the depth of crack initiation closer to the surface [6.35].

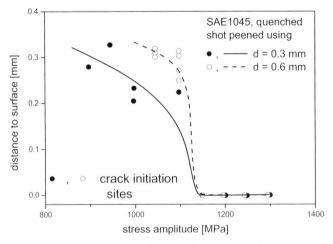

Fig. 6.71: Distances to surface of crack initiation sites vs. fictitious stress amplitude at the surface for quenched SAE1045 shot peened using shot with a diameter of 0.3 and 0.6 mm [6.92]

Fig. 6.72: Scanning electron microscope images of fracture surfaces of quenched SAE1045 in a state shot peened using 0.3-mm diameter shot after alternating bending loading with a fictitious stress amplitude at the surface of 1300 (a), 1250 (b), 1200 (c), 1100 (d), 1000 (e), 950 (f) and 900 MPa (g) [6.92]

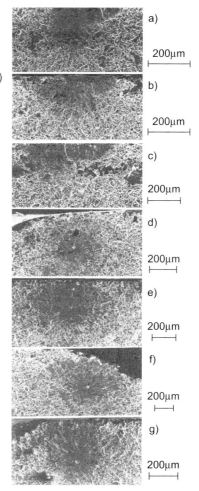

a)
200μm

b)
200μm

c)
200μm

d)
200μm

e)
200μm

f)
200μm

g)
200μm

6.3.4
Effects on Crack Propagation Behavior

When cyclic loading takes place, fatigue life is determined to a great degree by the phase of crack propagation. In particular, this applies to cases in which macro residual stresses are present. A description of crack propagation in states originally lacking residual stresses – these states, however, inevitably incorporate the plastic zones induced by crack propagation – is usually carried out using the Forman equation, which predicts reduced crack propagation rates if compressive mean stresses are present [6.93]. By specifically increasing the compressive residual stress fields and -amounts close to the crack tips by means of interspersed overloading, it was shown that the crack propagation rate can be reduced by compressive residual stresses [6.94,6.95].

Workhardening and residual stresses clearly influence the crack propagation rate in mechanically surface treated material states. [6.87] compared crack length development at an increasing number of cycles for an unpeened and a shot peened state of SAE1045 quenched and tempered at 600 °C. Fig. 6.73 shows that while crack initiation occurs earlier in the shot peened state, due to the micronotch effect of the roughened surface, crack propagation, on the other hand, is clearly decelerated for cracks longer than roughly 0.1 mm, as their tips now are located in the zone of workhardening and compressive residual stresses. Above a crack length of about 0.3 mm, the crack propagation rate shows values similar to those found for the unpeened state. Fig. 6.74 shows results obtained by [6.46] for deep rolled AISI304. Accordingly, for small crack lengths, in particular, the propagation rate shows decreases which become more pronounced as rolling pressure and its depth effect increase. Due to the smooth surface, no acceleration of crack initiation is observed. [6.89] found contrasting results for Al7010 AlZnMgCu alloy, depicted in Fig. 6.75. Here, the shot peened state and the untreated state show approximately equal crack propagation rates for very small and very high crack lengths. Crack propagation rates are clearly reduced for crack lengths between roughly 0.05 and 0.3 mm in the shot peened state, showing a minimum in the region of the compressive residual stress maximum. Thereby, early crack initiation is over-compensated and working life increased. This was confirmed by [6.96] for AZ80 magnesium alloy after shot peening, which showed decelerated crack propagation particularly for small crack lengths or low stress intensity, due to compressive residual stresses and workhardening states being present in the crack tip region.

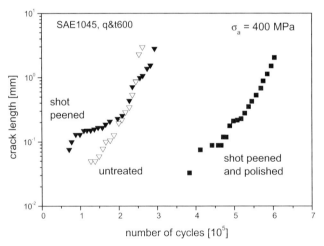

Fig. 6.73: Crack length vs. number of cycles for SAE1045 quenched and tempered at 600 °C in an unpeened state, a shot peened state and a state shot peened and subsequently polished during rotating bending loading with a fictitious stress amplitude at the surface of 100 MPa [6.87]

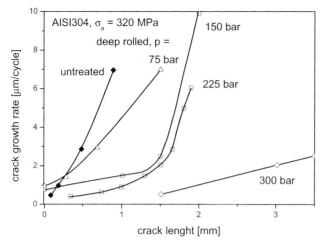

Fig.6.74: Crack propagation rate vs. crack length for AISI304 in an untreated state and states deep rolled at different pressures during push-pull loading with a stress amplitude of 320 MPa [6.46]

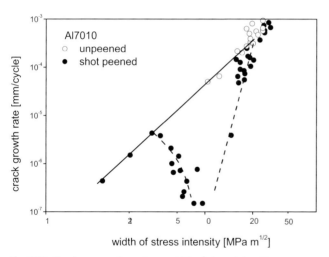

Fig. 6.75: Crack propagation rate vs. width of stress intensity for Al7010 AlZnMgCu-alloy in an unpeened and a shot peened state during alternating bending loading [6.89]

Examining laser shock treated, notched 3-point-bending specimens made of 7075 aluminum alloy, [6.91] found that crack initiation was clearly delayed, and additionally, crack propagation was decelerated. This was attributed to the method's great depth effect. Similar results were reported by [6.97] for laser shock treated 5456 aluminum alloy and by [6.98–100] for laser shock treated TiAl6V4 titanium alloy. For the latter, [6.100] found the $da/dN,\Delta K$ curves for 3-point bending shown in Fig. 6.76.

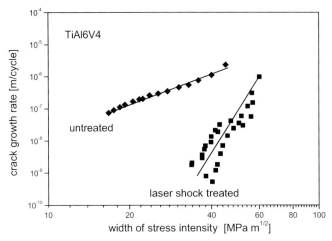

Fig. 6.76: Crack propagation rate vs. width of stress intensity for TiAl6V4 in an unpeened and a laser peened state during swelling 3-point bending loading [6.100]

For a small ΔK, laser shock treated states are expected to show practically no crack propagation, and therefore, cracks may be stopped. For a medium ΔK, crack propagation rates are reduced by almost 3 orders of magnitude, compared to the untreated state. A great ΔK clearly reduces this effect, as the process zone of the crack moves out of the zone which was workhardened by the laser shock treatment and contains compressive residual stresses. [6.98,6.99] reported that laser shock treatments increased the fatigue life or strength of fan vanes made of TiAl8V1Mo1, even when these additionally bore notches as a simulation of foreign-object-damage. Due to its great depth effect, laser shock treatment creates states with a much higher damage-tolerance than achievable by shot peening. Another example showing the pronounced effect of laser shock treatment on crack propagation was introduced in [6.101] using sheet metal specimens with a center hole which consisted of 2024 aluminum alloy in a T351 state. The hole simulated a drilled mounting hole which fails due to crack initiation and -propagation. As shown in Fig. 6.77, irradiation of the annularly shaped area around the hole resulted in a decreased crack propagation rate, allowing for considerable fatigue life increases. It was also reported that additional laser shock treatment, due to its great depth effect, was able to stop crack propagation in states in which crack initiation had already occurred.

[6.102] compared crack propagation behavior in TiAl6V4 with a lamellar microstructure for a shot peened state and a deep rolled state, at room temperature and at 500 °C. Loading at room temperature (comp. Fig. 6.78a) shows that shot peening strongly decelerates crack propagation. As a result, cracks are stopped almost completely in the region of the compressive residual stress maximum, and only great crack lengths show propagation rates that compare to the electropolished state. Stress-relieving after shot peening allows for increasing the crack propagation

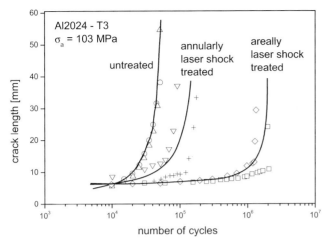

Fig. 6.77: Crack length vs. number of cycles at a nominal stress amplitude of 103 MPa for Al2024-T351 aluminum alloy in an unpeened state, and after areal and annular laser peening [6.101]

rate to values higher than those of the electropolished state, as roughness effects are predominant. Loading at 500 °C, on the other hand, yields a completely different result (comp. Fig. 6.78b), as residual stresses have largely relaxed and thus, the shot peened state shows the same behavior as the state which was additionally stress-relieved. Due to greater roughness, the shot peened state shows accelerated crack propagation in comparison with the electropolished state. Therefore, an electropolishing treatment subsequent to shot peening can extend fatigue life to values higher than those achieved by electropolishing alone, if there is surface layer workhardening still remaining [6.102]. Fig. 6.79 depicts contrasting results obtained by [6.103] for deep rolled TiAl6V4, which shows decelerated crack propagation both at room temperature and at 450 °C, as smooth surfaces and workhardening are in effect in both cases.

In summary, it may be stated for cyclic loading of mechanically surface treated material states that the crack propagation rate can be reduced primarily by compressive residual stresses, but also by the effects of workhardening. Regarding achievable fatigue life, however, the fact must be taken into account that crack initiation may be clearly increased after shot peening treatments, in particular, due to the micro-notch effect of the roughened surface. A proven method for quantitatively assessing the crack propagation rate in large-range residual stress fields is to use effective stress intensities, as crack propagation is expected to occur only during the crack opening phase [6.104]. For negative K_{min} values, the result is that residual stresses clearly influence the effective width of stress intensity. Thus, it is possible to describe the observed deceleration of the crack propagation rate which is caused by compressive residual stresses and which may even result in cracks being stopped.

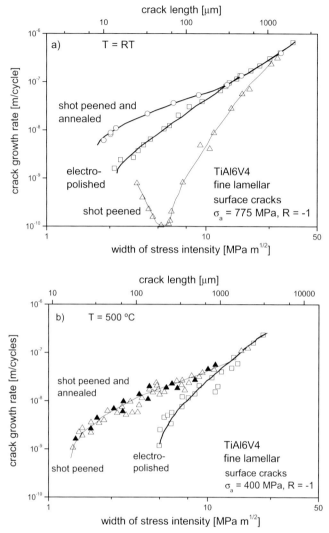

Fig. 6.78: Crack propagation rate vs. width of stress intensity at room temperature (a) and 500 °C (b) for fine lamellar TiAl6V4 in an electropolished state, a shot peened state and a state shot peened and subsequently stress-relieved [6.102]

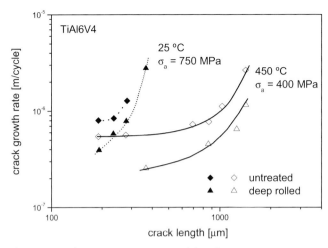

Fig. 6.79: Crack propagation rate vs. crack length at room temperature and 450 °C for TiAl6V4 in an untreated state and a deep rolled state [6.103]

6.3.5
Effects on Fatigue Behavior

Steels have been the primary focus of systematic studies which examined how fatigue strength is affected by surface layer states that were induced, for the most part, by mechanical surface treatments. The studies presented separate assessments for the influence of roughness, workhardening and residual stresses on fatigue limit. Low strength states show a small decrease of fatigue limit related to the change in roughness. This decrease grows more pronounced with increasing hardness – as shown in Fig. 6.80, which compiles results for different heat treatment states of SAE1045 and SAE5115 [6.105]. This is caused by a growing notch sensitivity of the various material states, as well as by ductility decreasing with rising hardness, or a descending threshold of the effective width of stress intensity for crack propagation. It must be taken into account, however, that mechanical surface treatments generally cause smaller changes of roughness in higher strength states than in lower strength states, and thus, the scope of changes of the fatigue limit is restricted. As shown in Fig. 6.81, workhardening effects close to the surface have a particularly strong influence on fatigue limit in the case of lower strength states, while this influence is extant for medium strength states and practically negligible for high strength states. Soft and medium strength states show the greatest increases of hardness achievable by mechanical surface treatments. These increases determine achievable fatigue strength. The influence of phase transformations – for steels, specifically transformations of retained austenite after case-hardening and subsequent mechanical surface treatment – on fatigue limit is difficult to access and can hardly be separated from the associated

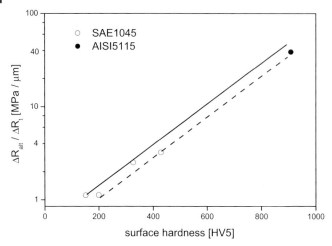

Fig. 6.80: Influence of shot peening-induced changes in roughness on the fatigue limit of steels with different degrees of surface hardness [6.105]

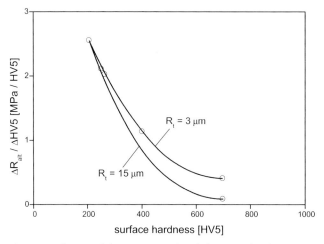

Fig. 6.81: Influence of shot peening-induced changes in hardness on the fatigue limit of steels with different degrees of surface hardness [6.105]

changes of the residual stress state. Therefore, no universally applicable statements can be made in this context. The influence that residual stresses created by mechanical surface treatments have on fatigue limit is strongly dependent on their stability, which is generally noticeably greater in high strength states than it is in low strength states. As shown in Fig. 6.3, the reason for this is that with rising hardness or tensile strength, fatigue limit increases to a lesser extent than quasi-static or cyclic yield strength. Accordingly, fatigue loading in the range of

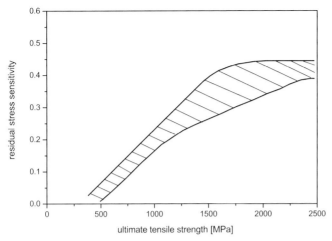

Fig. 6.82: Residual stress sensitivity of alternating bending fatigue strength of steels vs. ultimate tensile strength [6.1]

the fatigue limit is closer to yield strength for low strength states than it is for high strength states. For this reason, plastifications and thus, residual stress changes are likely to be far less pronounced [6.1]. An additional measure of the change in fatigue limit related to the change in residual stresses which can be used is residual stress sensitivity, as defined by [6.1] in a manner analogous to mean stress sensitivity according to [6.36]. When residual stresses are stable, residual stress sensitivity is given by the ratio of fatigue limit in a state without residual stresses to ultimate tensile strength, according to Eq. 6.10. As depicted in Fig. 6.82, residual stress sensitivity initially shows an increase which is linear to the growth of ultimate tensile strength, before reaching its maximum value of about 0.4 at ultimate tensile strength values between 1500 and 2000 MPa [6.1].

Behavior varies, due to the effect of hardness on the individual influencing parameters. For steels, therefore, it has proven useful to differentiate between low strength, medium strength and high strength states. Making use of this division, [6.35] summarized the essential aspects of the effects of residual stresses, workhardening state and surface topography. For **steels in a low strength state**, cyclic loading causes cyclic plastic deformation incorporating processes of workhardening and -softening when loading amplitude reaches the fatigue limit. This results in a pronounced relaxation of residual stresses which occurs the earlier, the higher loading amplitude is. Consequently, the influence of residual stresses and their gradient on fatigue limit is reduced to near insignificance.

This is shown in Fig. 6.83, using a modified Haigh plot to summarize alternating bending strengths (comp. Sect. 6.1.2c) found for normalized SAE1045 after various cutting treatments on smooth and notched specimens with different notch factors k_t and related loading stress gradients at the surface

$$\eta = \frac{1}{\hat{\sigma}} \frac{d\sigma}{dz} \bigg|_{z=0} \qquad (6.17)$$

[6.5,6.39,6.106–109]. The kind of loading which results in a finite service life causes an increased and thorough plastic deformation and thus effects a residual stress relaxation which is even more pronounced. Therefore, the residual stress state shows no significant influence on those branches of the Wöhler lines which describe low cycle fatigue. In low strength states, however, mechanical surface treatments usually cause pronounced surface layer workhardening, thereby also increasing dislocation density. On the one hand, this increases resistance to the onset of residual stress relaxation, and on the other hand, it counteracts cyclic plastic deformation. As a result, the development of typical fatigue microstructures is impeded, the number of cycles to crack initiation or the propagation rate of short cracks is reduced, and thus, local fatigue strength is increased. Thus, given sufficient wokhardening depth – provided, for instance, by deep rolling – the fatigue limit of a soft state, too, can be increased considerably if loading stress gradients are sufficiently high [6.65]. Owing to low hardness and high ductility, micro notches inherent in topography rapidly reduce stress peaks. As resistance to cyclic crack propagation is sufficiently great, as well, the influence of surface topography on the fatigue strength of low strength material states is comparably small. Beyond this, Fig. 6.83 shows that due to adaptive cyclic plastic deformations and stress redistributions, the effective notch factor is much smaller than the notch factor, and therefore, notches affect fatigue limit only marginally. Additionally, notch factor remaining the same while loading stress gradients rise serves to increase fatigue limit, as the volume subjected to the highest loading is decreased.

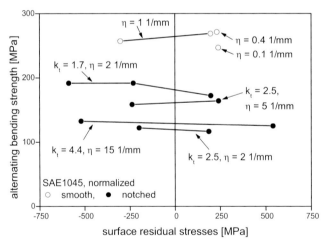

Fig. 6.83: Alternating bending fatigue strength of smooth and notched specimens made of normalized SAE1045 vs. surface residual stresses after different milling processes [6.5,6.39,6.106–109]

Medium strength states of steels show a markedly higher residual stress stability than low strength states, and therefore have a greater influence on the resulting fatigue limit. Smooth specimens show pronounced fatigue limit decreases for tensile residual stresses and slight increases for compressive residual stresses. Notched specimens, on the other hand, show only slight decreases in fatigue limit as residual stress values increase algebraically. These decreases are not dependent on the residual stresses' mathematical sign. This can be attributed to the fact that loading which results in 10^7 cycles to failure already causes residual stress relaxation at the surface. As notched specimens experience greater plastic deformations in the notch ground, creating a greater relaxation of residual stresses, residual stress sensitivity is lower in notched states. For medium strength states, residual stress penetration depth is an important additional factor. This is shown in Fig. 6.84 by example of residual stress distribution (a) and Wöhler lines for alter-

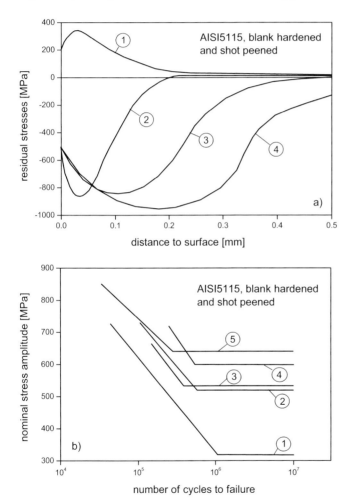

Fig. 6.84: Distributions of residual stresses (a) and S-N-curves caused by alternating bending loading (b) after different shot peening treatments for blank-hardened SAE5115 [6.34, 6.110, 6.111]

nating bending (b) for blank-hardened SAE5115 after blankhardening, after shot peening using different shot intensities and after additional electropolishing [6.34,6.110,6.111]. For the same surface residual stresses in states 2 to 4, marked increases in fatigue limit are observed as the depth effect of the peening treatment increases. The reason for this that at different numbers of cycles compressive residual stresses reduce the propagation rate of surface cracks which probably were initiated at similar numbers of cycles [6.35]. Therefore, the residual stresses' depth effect, or their gradient, must be smaller than the loading stress gradient if pronounced increases in fatigue limit are to be achieved. State 5, which was electropolished after shot peening of typ. 3, shows the highest fatigue limit, as roughness was clearly reduced by electropolishing and surface residual stresses are increased compared to the peened-only states. Evidently, this impedes crack initiation and the propagation of micro cracks to such an extent that state 4's greater residual stresses and their greater depth effect are over-compensated. In the range of low cycle fatigue, however, the electropolished state shows shorter fatigue life than state 4 and a similar fatigue life as state 3. Again, the number of cycles to crack initiation is increased for the electropolished state. This evidently has a similar effect as the reduced crack propagation rate in shot peened state 3, and a lesser effect in comparison with state 4, where the crack propagation rate shows a stronger decrease due to the residual stresses' depth effect and magnitude. It should also be noted for medium strength steels that the influence of residual stresses is almost totally diminished in the range of short-term fatigue. Here, an almost complete residual stress relaxation occurs, and cyclic worksoftening effects generally overcompensate workhardening induced by surface treatment.

The changes in fatigue limit induced in medium strength states by mechanical surface treatments can be represented by residual stress Haigh diagrams (comp. Sect. 6.2.1c). This is shown in Fig. 6.85 for smooth and notched specimens of SAE1045, quenched and tempered at 600 °C, after various grinding, milling and shot peening treatments [6.5,6.39,6.106–109]. Here, cyclic yield strength was derived from residual stress relaxation in the smooth and notched specimens and from fatigue limit in the ground states with tensile residual stresses. In the case of smooth specimens, results show that the diagram overestimates fatigue limit for the ground specimens with compressive residual stresses and that residual stresses show a slight relaxation, although cyclic yield strength should not be reached at any point. [6.35] attributed this to the fact that this state shows less workhardening compared to the other ground states, and therefore, both cyclic yield strength and fatigue limit are lower. The shot peened state shows a noticeably smaller residual stress relaxation than expected, due to the presence of greater and deeper workhardening. Fatigue limit seems to correspond well. However, if the increase in roughness found after shot peening and its effects on fatigue limit are corrected – according to Fig. 6.80, and as indicated by the second arrow –, it becomes evident that the workhardening effect of the shot peened state increases the fatigue limit significantly.

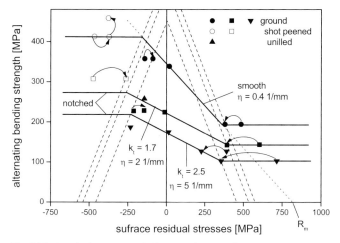

Fig. 6.85: Residual stress-Haigh diagram showing alternating bending fatigue strength for smooth and notched specimens with different surface states made of SAE1045 in a state quenched and tempered at 600 °C for 2 h [6.35]

Despite an observable relaxation of residual stresses, the notched specimens show lower residual stress sensitivity than the smooth specimens. The fatigue limit of the state showing compressive residual stresses after grinding is overestimated, as in the case of the smooth specimens, due to a smaller workhardening effect than that of the ground states with tensile residual stresses. By contrast, the fatigue limit of the milled state can be predicted very well. Experimental results for the shot peened state, on the other hand, show a higher fatigue limit than expected from the Haigh diagram, as the workhardening effect cannot be incorporated in the estimate. The problems associated with estimating fatigue limits by means of Haigh diagrams allow for certain conclusions to be drawn regarding medium strength states. On the one hand, cyclic residual stress relaxation must be taken into account when assessing the influence of residual stresses on fatigue limit, and on the other hand, size and distribution of the workhardening states, as well as the increases in roughness caused by mechanical surface treatments, can have a significant bearing on the resulting fatigue limit. Thus, fatigue limit estimates will only be successful if they are based on a sensitivity of the fatigue limit to relaxed residual stresses, instead of the usual residual stress sensitivity, and if they use a cyclic yield strength which, at least, corresponds with the workhardening state.

Applying the concept of the local fatigue limit (comp. Sect. 6.2.1c) is complicated by residual stress relaxation, as this necessitates obtaining information on the initial distribution of residual stresses, and on residual stress sensitivity or the distribution of relaxed residual stresses. Fig. 6.86 shows estimates for notched (a) and smooth (b) specimens of quenched and tempered SAE1045 (600 °C / 2h) in a shot peened state. While surface crack initiation is predicted and found for the

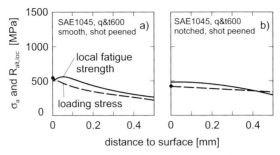

Fig. 6.86: Distribution of local fatigue limit and loading stresses, and crack initiation site for notched (a) and smooth (b) shot peened specimens made of SAE1045 in a state quenched and tempered at 600 °C for 2 h [6.5, 6.63, 6.106, 6.107]

notched specimens, smooth specimens show unexpected crack initiation at the surface. This probably indicates that surface roughness, and its micro notch effect, should not be neglected [6.5,6.63,6.106,6.107].

The concept of assessing crack propagation ability on the basis of linear-elastic fracture mechanics (comp. Sect. 6.2.1c) was successfully applied to medium strength states of steel, as well. First, aforementioned results for quenched and tempered SAE1045 shall serve as an example for an assessment of the loading situation for loading at each state's fatigue limit [6.35]. Resulting distributions of the effective width of stress intensity are shown in Fig. 6.87 for ground notched specimens, with and without regard to grinding-induced tensile residual stresses. Based on the anticipated range of the threshold value for crack propagation, the expected result for both of the depicted distributions of stress intensity is that cracks initiated at the surface are able to propagate when they have reached a length of 15 or 30 µm. Accordingly, fatigue limit is given by the stress amplitude which is just below the value for crack initiation. In the case of notched and shot peened specimens of the same material state, and assuming shot peening-induced residual stresses according to Fig. 6.88, cracks initiated at the surface should be able to propagate only when their length exceeds 170 µm. However, if the distribution of relaxed residual stresses is used, results in the range of 30 to 200 µm show effective widths close to the threshold value. Accordingly, fatigue limit correlates with the propagation ability of cracks initiated at the notch ground which have a length of at least 30 µm. As seen in Fig. 6.89, however, the concept fails in the case of smooth, shot peened specimens. Due to the concept's drastic simplifications, and even when a distribution of relaxed residual stresses is assumed, a crack would have to be at least 100 µm long to be able to propagate. A similar concept was introduced in [6.112–114] aimed at estimating the fatigue limit of deep rolled crankshafts, which is determined by crack arrest phenomena. The method involves carrying out loading calculations utilizing a finite-element simulation, into which the original residual stress states are inserted. This allows for determining the distributions of loading stresses and of the residual stresses that remain after loading. These are then assessed by means of linear-elastic frac-

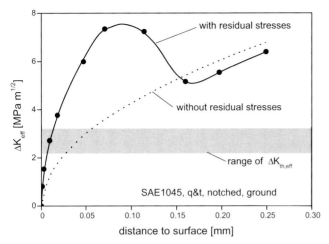

Fig. 6.87: Width of effective stress intensity with and without consideration of residual stresses vs. distance to surface for ground notched specimens made of SAE1045 in a state quenched and tempered at 600 °C for 2 h ($\sigma_a = R_{alt} = 143$ MPa) [6.35]

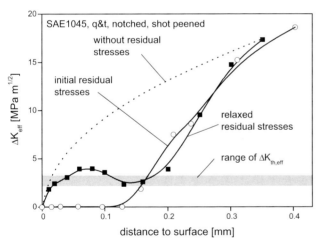

Fig. 6.88: Width of effective stress intensity with and without consideration of initial and relaxed residual stresses vs. distance to surface for shot peened notched specimens made of SAE1045 in a state quenched and tempered at 600 °C for 2 h ($\sigma_a = R_{alt} = 310$ MPa) [6.35]

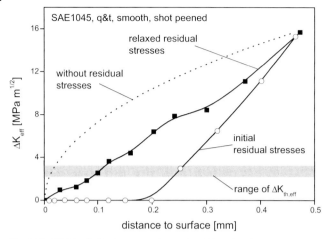

Fig. 6.89: Width of effective stress intensity with and without consideration of initial and relaxed residual stresses vs. distance to surface for shot peened smooth specimens made of SAE1045 in a state quenched and tempered at 600 °C for 2 h ($\sigma_a = R_{alt} = 414$ MPa) [6.35]

ture mechanics, assuming the occurrence of crack propagation close to the surface. As indicated by Fig. 6.90, basing the assessment on residual stress profiles also results in surface cracks which are able to propagate. As loading of the crankshaft increases, the cracks are stopped by an unspecified normalizing momentum at increasing depths, where the effective widths of stress intensity no longer reach the threshold value for further crack propagation. These results agree well with the crack lengths determined experimentally for corresponding amplitudes of momenta, thereby permitting fatigue limit to be determined as the momentum amplitude at which the threshold value is reached.

The residual stresses which occur in **high strength states of steels** are generally very stable. In particular, tensile residual stresses in combination with only a small loading at the fatigue limit, may even remain almost unchanged in the range of a finite but technically relevant fatigue life. Here, the residual stress sensitivity of the fatigue limit is equal to mean stress sensitivity. Compressive residual stresses, on the other hand, are combined with greater loading at the fatigue limit. For small loading stress gradients – as in thick, smooth specimens, for instance –, resulting in crack initiation beneath the surface. This is influenced only slightly by residual stress magnitude if the absolute penetration depth of the residual stresses is not great enough, or if their gradient is sufficiently small. Thus, it is assumed that grinding-induced residual stresses, due to their small penetration depth, usually cannot prevent crack initiation beneath the surface and therefore have only little influence on fatigue limit. Due to greater penetration depth, shot peening-induced compressive residual stresses, on the other hand, allow fatigue limit and fatigue life to be increased significantly. In particular, it is notched specimens that show great loading stress gradients and compressive residual stresses

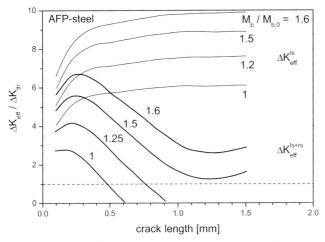

Fig. 6.90: Width of effective stress intensity related to threshold value vs. distance to surface with and without consideration of residual stresses vs. crack length for different multiples M_b of the reference momentum $M_{b,o}$ in the deep rolled chamfer of a crank shaft pin bearing made of 38MnS6BY AFP-steel [6.112–114]

that run deep enough and/or have small gradients. Thus, crack initiation is shifted to the surface, and a decreased crack propagation rate, combined with the effect of the compressive residual stresses, achieves a marked increase in fatigue limit. This is the case e.g. for shot peened notched specimens. Due to notch effect and a high fatigue limit, however, these specimens can also show a relaxation of residual stresses, within certain limits. In summary, it can be stated that residual stresses alone do not constitute a suitable parameter for assessing fatigue limit. The reason is that if surface layer states are optimized, e.g. maximum compressive residual stresses are beneath the surface, fatigue limit is no longer determined by maximum loading, but by loading at the sites where cracks can be stopped. Here, effective notch factors can decrease to values smaller than 1 for great residual stresses, while they approximate the notch factors only when residual stresses are small. In the case of high strength steels, worksoftening processes caused by mechanical surface treatment are an additional factor, which, in combination with a notch effect due to roughness, at least explains the initiation of cracks at the surface. In principle, roughness itself is very important. However, as mechanical surface treatments cause only small changes in roughness, the effects which occur are not significant, even after shot peening treatments.

In the case of high strength steels, residual stress Haigh diagrams are a suitable means of assessing the influence of mechanical surface treatments on fatigue limit. This is exemplified in Fig. 6.91 for smooth and notched specimens of quenched SAE1045 in various surface states [6.5,6.63,6.106,6.107]. After shot peening, smooth specimens show a smaller increase of fatigue limit than in the medium strength state. In addition, the Haigh diagram overestimates this

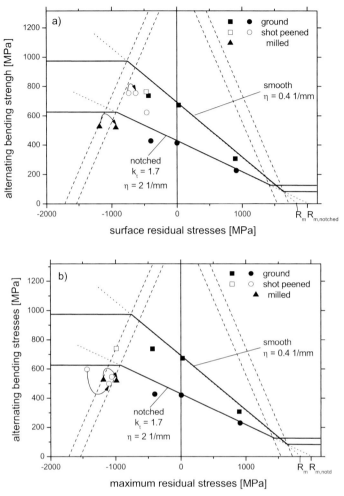

Fig. 6.91: Residual stress-Haigh diagram showing alternating bending fatigue strength for smooth and notched specimens with different surface states made of quenched SAE1045, evaluated at the surface (a) and at the position of greatest residual stresses [6.35]

increase. Residual stresses have only little influence on alternating bending strength, as crack initiation occurs beneath the surface and its influence cannot incorporated in the Haigh diagram. By contrast, notched specimens show significant fatigue limit increases after shot peening and smaller ones after milling, as crack initiation occurs at the surface. In the case of smooth specimens, the residual stress effect is small, and virtually eliminates the notch effect in the range of great compressive residual stresses. For notched shot peened states, the correlation of residual stresses at the surface and nominal loading at the surface represented in the Haigh diagram stands in contradiction to residual stress behavior, as residual stresses are mostly stable or show only small changes. By contrast, the residual stress change is described better if maximum residual stresses and nominal loading at the position of maximum residual stresses are examined. High cycle fatigue loading causes a relaxation of very high residual stress maxima. This relaxation is particularly pronounced beneath the surface. Again, however, fatigue limit is not described optimally. Thus, these results lead to the conclusion that for these states, fatigue limit is determined by the propagation ability of cracks initiated at or beneath the surface, rather than by the magnitude of residual stresses at a given distance to surface.

Therefore, an assessment of the positions of crack initiation calls for utilizing the concept of the local fatigue limit. As shown schematically in Fig. 6.92 for smooth specimens of high strength steel after mechanical surface treatments, during high cycle fatigue loading, crack initiation is expected to occur beneath the surface, and during low cycle fatigue loading it is expected to occur at the surface [6.35]. By contrast, notched specimens are expected to show crack initiation at the surface for both levels of loading. As shown in Fig. 6.93, this was confirmed for experimentally determined crack locations in quenched SAE1045 [6.1,6.92]. However, as crack propagation ability cannot be evaluated within the concept, fatigue limits were assessed only in respect to crack initiation, inevitably resulting in conservative estimates.

Supplementary use of the concept of crack propagation ability based on linear-elastic fracture mechanics should permit an improved description. According to [6.35,6.42], this method can be applied to notched specimens in ground states and in shot peened states if the relaxed residual stresses are used, as in the case of the

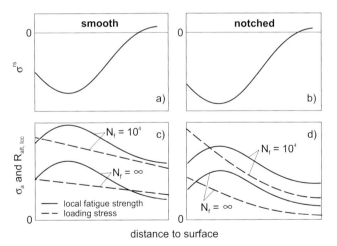

Fig. 6.92: Scheme showing the distribution of local fatigue strength and loading stresses for smooth (a) and notched (b) states after loading in the range of fatigue strength and in the range of low cycle fatigue [6.35]

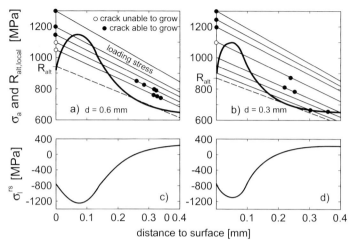

Fig. 6.93: Distribution of local fatigue strength and loading stresses with different fictitious stress amplitudes at the surface and indication of crack initiation sites for quenched SAE1045 shot peened using shot with a diameter of 0.6 mm (a) and 0.3 mm (b) [6.1, 6.92]

medium strength states. Again, however, the applicability of this strongly simplified model has its limits, as evidenced by a shot peened state in which cracks need to reach 170 μm in length to be able to propagate, even if relaxed residual stresses are utilized and loading is in the range of the fatigue limit. The concept was also applied to a crack initiated beneath the surface of a smooth specimen [6.35,6.42]. As shown in Fig. 6.94, at first, the propagation rate toward the surface is decreased, and ultimately, the crack is stopped. At that point, however, crack length will be high in comparison with the compressive residual stresses' penetration depth. Therefore, an occurrence of residual stress redistributions and -relaxation must be assumed. As a result, the crack will propagate all the way to the surface and lead to failure. Accordingly, the stress amplitude just high enough to initiate the crack beneath the surface can be correlated with the fatigue limit. In that case, residual stress magnitude is of secondary importance. [6.115] introduced another example which shows that the concept of crack propagation inability can be used to describe fatigue limit for high strength states of steel after shot peening treatments. After bending swelling loading, notched flat specimens of case-hardened SAE5115 in an unpeened and in a shot peened state show internal oxidation. This is a result of gas carburization, and it creates surface cracks with a maximum length of about 20 μm. Assuming cracks propagating downward from the surface, Fig. 6.95 shows the distribution of stress intensity found after loading in the range of the respective fatigue limit. In agreement with experimental results, distributions in the unpeened specimens show that surface cracks can propagate to the interior, given continuously increasing stress intensity. In the shot peened specimens, on the other hand, stress intensities in the region of the compressive residual stress maximum are reduced to the threshold value of crack propagation. Thus, loading in the range of the fatigue limit will stop initiated cracks at a length of about 100 μm. Experiments confirm this, as shown in Fig. 6.95b.

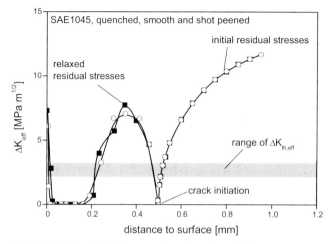

Fig. 6.94: Width of effective stress intensity with and without consideration of initial and relaxed residual stresses vs. distance to surface for shot peened smooth specimens made of quenched SAE1045 [6.35]

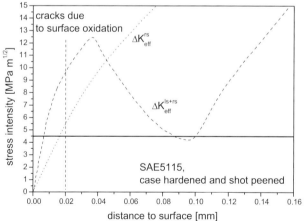

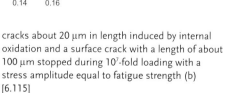

Fig. 6.95: Width of effective stress intensity with and without consideration of residual stresses vs. distance to surface for unpeened and shot peened notched specimens made of case-hardened SAE5115 (a) and microstructure showing surface

cracks about 20 μm in length induced by internal oxidation and a surface crack with a length of about 100 μm stopped during 10^7-fold loading with a stress amplitude equal to fatigue strength (b) [6.115]

Fig. 6.96 is a schematic representation of the influence of roughness R_t, work-hardening state ΔHV, residual stress- or loading stress gradient η^{rs} or η^{ls}, and notch factor k_t, on residual stress Haigh diagrams for low strength, medium strength and high strength states of steel, providing a summary of the previous discussion of the different ways these factors influence the individual strength levels of steel. Low strength material states show a pronounced relaxation of residual stresses during loading using amplitudes in the range of the fatigue limit. Therefore, the influence of macro residual stresses and their gradients on fatigue limit is small. The workhardening state close to the surface, however, strongly influences fatigue limit, as workhardening increases resistance to cyclic plastic deformation, which may indirectly increase residual stress stability, as well. The influence of surface topography is quite small. For notched states, effective notch factor is markedly smaller than notch factor, as cyclic plastic deformations cause stress redistributions in the notch ground. If notch factor remains the same, but loading stress gradients are increased, fatigue limit rises, as the volume subjected to the highest loading is decreased. Medium strength states, on the other hand, always show a clear decrease in fatigue limit as tensile residual stresses increase. Consequently, compressive residual stresses can effect fatigue limit increases if they – despite loading stress amplitudes which are increased in the range of the fatigue limit, compared to tensile residual stresses – are sufficiently stable and their gradient is low compared to the loading stress gradient, or their depth of penetration is great. Workhardening close to the surface can also cause significant fatigue limit increases in medium strength states, and increases of surface roughness have a pronounced effect on fatigue limit. As notch-free states generally show lower loading

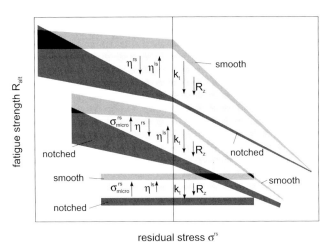

Fig. 6.96: Scheme summarizing the influence of roughness R_z, workhardening state ΔHV, the gradient of residual or loading stresses η^{rs} or η^{ls} and notch factor k_t on the residual stress-Haigh diagrams of low strength, medium strength and high strength steels, acc. [6.35]

stress gradients, crack initiation may occur below the surface layer influenced by mechanical surface treatment. For notched states, on the other hand, significant increases in fatigue limit are possible if the position of crack initiation is shifted to the surface. This can cause residual stress sensitivity to be as high in the compressive range as it is in the tensile range, and may even result in the fatigue limits of notched states being increased in comparison to smooth specimens, i.e. effective notch factors smaller than 1. At low residual stress levels, however, the effective notch factor approximates notch factor, given that crack initiation is not shifted to the surface by particularly high loading stress gradients. In the case of high strength material states, notch-free states show tensile residual stresses which always remain stable during high cycle fatigue loading, causing marked decreases in fatigue limit. Except for the extremely high values, compressive residual stresses usually stay stable, as well. However, they can effect fatigue limit increases only if the occurring stress gradients are high enough, or if residual stress gradients are small enough. Otherwise, crack initiation will occur beneath the surface layer and the state close to the surface will become less important. Surface roughness does effect clear changes in fatigue limit, yet this influence is only partially relevant, as mechanical surface treatments create only small changes of roughness in high strength states. For notched states, great compressive residual stresses with a sufficient penetration depth cause pronounced fatigue limit increases. This can result in a higher fatigue limit, compared to smooth states, as in the case of medium strength material states. Cracks which are stopped in the region of maximum residual stresses are highly relevant in this context.

It is not possible to make similarly well-founded and systematic statements regarding **light metals**, as there are hardly any published studies which include the stability of surface layer states. In some areas, this necessitates drawing indirect conclusions in respect to the processes that occur. [6.116] provided a current and concise overview of the effects of mechanical surface treatments on the fatigue strength of titanium, aluminum and magnesium alloys. The criterion used for the stability of residual stress states is cyclic deformation behavior. Here, cyclic workhardening is equated with stable residual stresses, and cyclic worksoftening is equated with strongly relaxing residual stresses. Accordingly, an examination of titanium alloys must focus individually on α-alloys – which show cyclic worksoftening – or meta-stable β-alloys, $(\alpha+\beta)$ alloys – which show an almost neutral cyclic deformation behavior – and γ-TiAl alloys, which presumably show cyclic workhardening. After shot peening, cyclically worksoftening Ti-2.5Cu α-alloy [6.116], or meta-stable titanium β-alloys TiV10Fe2Al3 [6.117], BetaC and LCB (low cost beta) [6.118], show only small increases – or even a decrease – in fatigue limit, due to the strong residual stress relaxation expected. Therefore, examinations of the stability of residual stresses in these alloys were carried out by means of additional annealing treatments, the results of which will be presented below. By contrast, $(\alpha+\beta)$-alloys TiAl6V4 and TiAl6Nb7 show an almost neutral cyclic behavior, with only minor cyclic worksoftening taking place. Thus, a marked influence of residual stresses on fatigue strength must be assumed. However, an abnormal mean stress sensitivity is found for TiAl6Nb7 [6.119], which has the effect of preventing increases of maximum stresses in the range of the fatigue limit when tensile mean stresses grow. Obviously, cyclic crack initiation is very much determined by maximum stress. Therefore, if the majority of the compressive residual stresses in regions close to the surface are stabilized by mechanical surface treatments, one must assume internal tensile residual stresses to be equally stable and to cause crack initiation. As in the case of the aforementioned group of alloys, this prevents mechanical surface treatments from achieving increases in fatigue strength, given the absence of extreme loading stress gradients. However, a heat treatment method was developed recently which does not show this abnormal mean stress sensitivity, as the material is quenched in water instead of being air-cooled. This kind of cooling from β-transus temperature strongly decreases the width of the α-region, which in turn increases strength and resistance to crack initiation [6.120]. As a result, crack initiation is suppressed in the region of tensile residual stresses, and fatigue limit therefore shows a moderate increase. The cyclic deformation behavior of Ti47Al3.7 (Nb,Cr,Mn,Si)0.5B γ-TiAl alloy has not yet been examined. However, workhardening is expected. Therefore, the residual stresses induced by mechanical surface treatment are assumed to be stable, and noticeable fatigue limit increases are found [6.121]. The cyclic deformation behavior of age-hardening aluminum alloys can be categorized in terms of annealing state. While underaged microstructures, found e.g. after T3 heat treatments, show cyclic hardening, cyclic worksoftening is observed for overaged and peak aged microstructures, e.g. after T6 heat treatments. Accordingly, residual stresses should be more stable in T3 states and contribute to an increase in fatigue limit by reducing the crack propagation rate. This is clearly visible when comparing the influence of shot peening on Al2024 (AlCu5Mg2) aluminum specimens for T3- and T6 heat treatments [6.122]. In the case of magnesium alloys, peening intensity must be kept particularly low due to the material's low ductility. Otherwise, fatigue limit will be decreased not only by developing roughness, but also by micro cracks induced in the surface layer. Thus, only small compressive residual stresses are achievable. Nevertheless, increases in fatigue strength can be attained and increased further by subsequent polishing [6.123].

Studies of the fatigue behavior of laser shock treated states which include fatigue strength are still quite scarce, and results are somewhat unsatisfactory. This is shown e.g. by [6.124], who examined X2Cr11 and X6CrNiTi18–10 steels after excimer irradiation and subsequent bending swelling loading and found Wöhler lines hardly changed. By contrast, [6.125] reported increases in fatigue limit from 590 to 1030 MPa – indicated as nominal maximum stresses – for tensile swelling loading of 34NiCrMo6 in a high strength, notched state with a hardness of 54 HRC. For a state of quenched and tempered SAE5155, presumably annealed using a high temperature, [6.126] found increases in 4-point alternating bending strength, which after laser shock treatment was 490 MPa instead of 380 MPa. Additionally, [6.126] showed that greater compressive residual stresses with a slightly reduced depth effect occur when the size of the exposed areas is decreased, and crack initiation and propagation is thus impeded. As a result, fatigue limits are generally increased. According to [6.127], laser shock treatment causes notched specimens of 2024 aluminum alloy in a T351 state to show moderate fatigue limit increases from about 185 to 225 MPa at 10^7 cycles of bending swelling loading. This is more pronounced than the 210 MPa achieved by shot peening. In summary, it can be stated that, compared to most of the other mechanical surface treatment methods (comp. Sect. 3.6.2d) and excepting only shot peened states of smooth surfaces, laser shock treatments create less workhardening in low strength states which show low residual stress stability, and therefore their effect is limited. In the case of medium- and high strength states which show greater residual stress stability, however, the main results are an increase of the number of cycles to crack initiation and a decrease in crack propagation rate, which can increase working life or fatigue limit. Due to the great depth effect of laser shock treatments, their most pronounced effects are found for small loading stress gradients and for states which are already exhibiting, or are particularly prone to, crack initiation.

To date, **isothermal behavior of mechanically surface treated material states at elevated temperatures** has not been thoroughly studied. For TiAl6V4 alloy, [6.102] showed that shot peening has a different influence on crack propagation behavior at room temperature and at 500 °C – as described in Sect. 6.3.4. The reason for this is that residual stresses, workhardening effects and roughness are all essential factors at room temperature, while only workhardening effects and roughness remain effective at 500 °C. The Wöhler lines shown in Fig. 6.97 reflect this. At room temperature, shot peening-induced residual stresses increase the fatigue limit, compared to the electropolished state, while subsequent stress-relief causes roughness to decrease the fatigue limit. At 500 °C, on the other hand, similar decreases in fatigue limit are found after shot peening and after shot peening and subsequent stress-relief, as the effect of roughness is dominant. However, if surface roughness is eliminated by supplementary electropolishing, this will cause the fatigue limit – again, in a similar way for both the shot peened state and the shot peened plus stress-relieved state – to be increased in comparison the state which was only electropolished. This is attributed to the remaining workhardening. Examining TiAl6V4 after a deep rolling treatment, [6.128] found an – albeit small – fatigue limit increase for 450 °C, as well. [6.55] focused on austenitic AISI304 steel to examine the influence of shot peening and deep rolling treatments on fatigue life for push-pull loading with an amplitude of 280 MPa and covering the range from room temperature to 650 °C. According to Fig. 6.98, shot peening causes only small increases of the number of cycles to failure within the entire temperature range. After deep rolling, on the other hand, fatigue life is increased by about one and a half orders of magnitude at room temperature. However, the extent of this increase is quickly diminished by rising temperatures, yet it remains noticeable up to 650 °C.

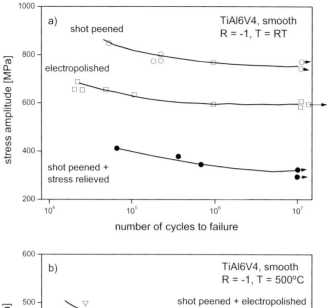

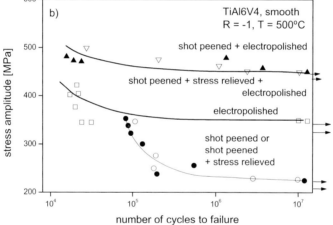

Fig. 6.97: Push-pull-S-N-curves at room temperature (a) and 500 °C (b) for fine lamellar TiAl6V4 in an electropolished state, a shot peened state, a shot peened and stress relieved state and a state electropolished after shot peening or after shot peening and stress relieving [6.102]

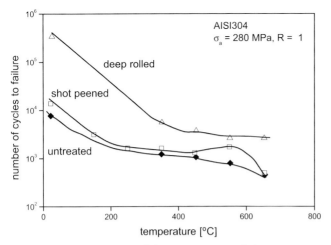

Fig. 6.98: Number of cycles to failure in stress-controlled alternating push-pull tests with a stress amplitude of 280 MPa vs. temperature for AISI304 in an untreated, a shot peened and a deep rolled state [6.55]

The effects of **modified peening treatments on fatigue strength in steels** were studied systematically on the example of alternating bending loading of quenched and tempered AISI4140 steel in different heat treatment states [6.59]. In order to assess the results correctly, the effects of conventional shot peening shall be presented first. According to Fig. 6.99, alternating bending strength of the ground states increases almost linearly with growing material hardness. After shot peening, all heat treatment states show a higher alternating bending strength than the ground state. The value of this increase is similar for low and medium hardness states, while for the quenched state it is higher, not only in absolute terms, but also when related to the value of the ground state. The same alternating bending strength values are plotted in Fig. 6.100, which also provides an assessment of residual stress stability, using cyclic yield strength at the surface as determined in Sect. 6.2.1c. Also shown are the residual stresses after 10^7 cycles, which were determined by experiment for loading in the range of the fatigue limit. For all heat treatment states, there is a good correspondence of the predicted values and the experimental values for alternating bending strength and for cyclic residual stress relaxation, as cyclic yield strength at the surface was determined precisely, by using specimens which had been treated identically. In this way, the shot peening-induced workhardening effect which is pronounced in normalized states and in quenched and tempered states is already incorporated in the values of cyclic yield strength.

The earliest examinations of stress peening to appear in the literature were provided by [6.129]. They resulted in the residual stress Haigh diagram pictured in Fig. 6.101, which shows a linear correlation of alternating bending strength and maximum residual stresses. Residual stress sensitivity shows a value of about 0.7

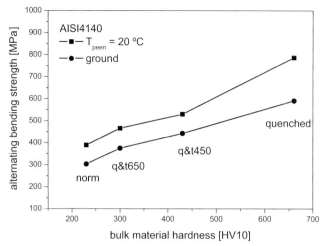

Fig. 6.99: Alternating bending strength ($P = 50$ %) vs. bulk material hardness for AISI4140 in a ground and in a conventionally shot peened state [6.59]

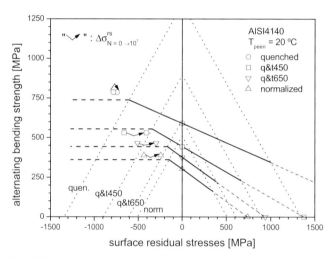

Fig. 6.100: Residual stress-Haigh diagram showing alternating bending strengths of different heat treatment states of AISI4140 after conventional shot peening, including the cyclic yield strengths shown in Sect. 6.2.1c [6.33]

and therefore is much higher than the value expected for quenched and tempered steels. This is due to the fact that additional workhardening effects are included in the plot. Fig. 6.102 shows the correlation of bulk material hardness and alternating bending strength after stress peening, observed at about 40 % of yield strength. Regarding soft material states, stress peening evidently does not offer any significant advantages in comparison with shot peening. According to Fig. 6.28,

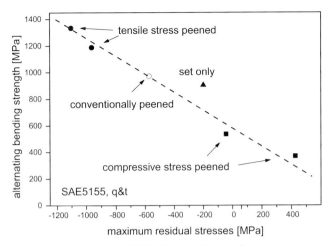

Fig. 6.101: Residual stress-Haigh-diagram showing alternating bending strengths of quenched and tempered SAE5155 after shot peening with different prestrains [6.129]

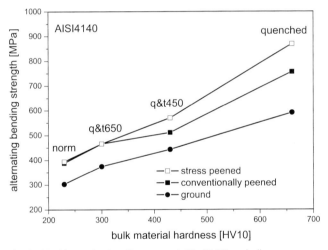

Fig. 6.102: Alternating bending strength ($P = 50\,\%$) vs. bulk material hardness for AISI4140 in a ground state, in a conventionally shot peened state and in a stress peened state [6.33]

residual stress stability is so low that the higher, induced residual stresses will relax quickly. On the other hand, the quenched state and the state quenched and tempered at 450 °C, i.e. higher strength states, show a higher stability of residual stresses, which can result in considerable increases of alternating bending strength. As indicated by the residual stress Haigh diagram in Fig. 6.103, all states except the quenched state show experimental results for alternating bending

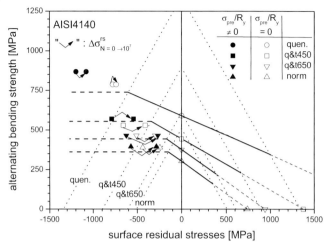

Fig. 6.103: Residual stress-Haigh diagram showing alternating bending strengths of different heat treatment states of AISI4140 after conventional shot peening and stress peening, including the cyclic yield strengths shown in Sect. 6.2.1c [6.33]

strength and residual stress stability that correspond well with the values predicted. The quenched state, however, shows a substantially smaller relaxation of residual stresses than expected, and accordingly, fatigue limit is noticeably higher than expected. The presumable reason for the deviations observed is that the stress peened state shows a higher value of cyclic yield strength at the surface than the conventionally shot peened state on which the Haigh diagram is based.

Compiled in Fig. 6.104 are the values of alternating bending strength found for different heat treatment states of AISI4140 after warm peening at 290 °C. They are slightly increased for the normalized state and for the states quenched and tempered at 650 °C, compared to the respective state after conventional peening. In the case of the state quenched and tempered at 450 °C, on the other hand, warm peening increases alternating bending strength considerably, compared to conventional shot peening. The quenched state shows a small decrease of fatigue limit due to warm peening, which, however, is explained by taking into account the change in hardness which occurs during warm peening. Taking the annealing effect out of the assessment reveals that fatigue limit is in fact increased. The residual stress Haigh diagram in Fig. 6.105 shows that the values of alternating bending strength and residual stress stability which were determined by experiment correspond well with the estimated values, listed in Fig. 6.46 on the basis of cyclic yield strength at the surface for the respective material states. The increases in fatigue limit can be attributed to the increases of cyclic yield strength at the surface and the associated growth in residual stress stability, shown in Fig. 6.29. The dynamic and static strain aging effects discussed in Sect. 3.3.1a are the underlying cause.

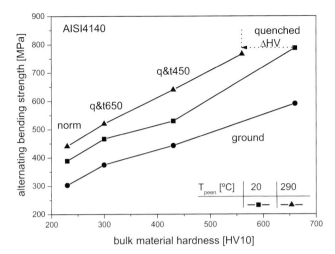

Fig. 6.104: Alternating bending strength ($P = 50\,\%$) vs. bulk material hardness for AISI4140 in a ground state, a conventionally shot peened state and a state warm peened at 290 °C [6.59]

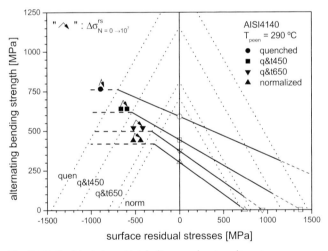

Fig. 6.105: Residual stress-Haigh diagram showing alternating bending strengths of different heat treatment states of AISI4140 after warm peening, including the cyclic yield strengths shown in Sect. 6.2.1c [6.33]

Static strain aging effects alone are also able to increase fatigue limit. Fig. 6.106 shows a residual stress Haigh diagram for AISI4140 quenched and tempered at 450 °C after conventional peening and subsequent annealing treatments. These cause static strain aging effects. Evidently, fatigue limit increases are possible despite the moderate relaxation of residual stresses that comes with annealing. As an example, cyclic yield strength at the surface is plotted for the state annealed at

300 °C for 1 minute. While it shows a clear increase compared to the conventionally peened state, it cannot adequately represent the substantial rise in fatigue limit found in the experiment. Heightened residual stress stability (comp. Sect. 6.2.1) surely is one reason for this increase in fatigue limit. Additionally, the development of a typical fatigue dislocation structure and thus, of fatigue effects, is impeded by Cottrell clouds or finest carbides pinning the dislocations. The latter effect obviously cannot be estimated using a residual stress Haigh diagram, but it is held responsible for the noticeable deviation from the behavior observed in experiments. A point in favor of this view is that the fatigue limit increases grow clearly with the rising values of yield strength induced by the same kind of annealing treatments, as shown in Fig. 6.107. Similar increases in fatigue strength due to conventional peening and subsequent annealing treatments were found for normalized SAE1045 by [6.83] by examining the numbers of cycles at selected stress amplitudes in the range of low cycle fatigue. Strain aging effects and increases in fatigue strength were also found for shot peened and subsequently annealed AISI304 – where the development of ε-carbides and special chromium carbides is essential – and for AZ31 magnesium alloy, where the strain aging effect is attributed to fine $Mg_{17}Al_{12}$ precipitations [6.83].

The influence of **modified shot peening processes in light metals** was also examined, by focusing on the effects of subsequent annealing treatments – as portrayed in Sect. 4.2.4b. Apart from aforementioned strain aging effects, 4 variants of 2 strength-increasing mechanisms were found for AZ31. On the one hand, there is a re-crystallization of the surface layer cold-worked by shot peening which results in finely grained surface layers, or changes of surface layer grain morphology. On the other hand, surface layer hardening may occur "selectively" if the time required for the development of phase nucleation and transformation is reduced in the surface layer to such an extent that hardening is already concluded there before it starts in the core. Alternatively, "preferred" surface layer hardening may occur if the accelerated hardening in the surface is not finished yet when hardening processes begin to appear in the core. Examining the development of finely grained surface layers due to re-crystallization in TiAl8.6 after shot peening, [6.130,6.131] found that 1-minute annealing at 820 °C and a subsequent electropolishing treatment caused an increase in fatigue limit from 270 to 320 MPa at 350 °C, or an increase in working life by a factor of 5. This was attributed to a delayed development of slip bands and an increased duration of crack initiation in the finely grained surface layer, along with a reduced crack propagation rate in the coarsely grained core. [6.131] examined the change of grain morphology in TiAl6Sn2Zr4Mo2 (Ti-6242), which showed a lamellar structure with a high creep resistance. Shot peening was followed by a 1-hour annealing treatment at 850 °C to effect re-crystallization, and by a final electropolishing treatment. The result was a finely grained, equiaxial surface layer which showed a high level of resistance to crack propagation. Correspondingly, an increase in fatigue limit from 400 to 450 MPa was found for 550 °C. Unlike TiAl8.6 alloy, however, this increase is not accompanied by a significant extension of fatigue life in the range of low cycle fatigue. To date, selective surface hardening has been examined mainly for deep

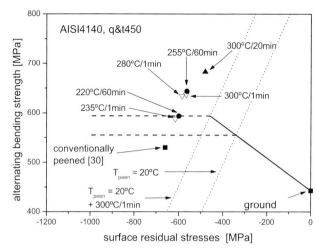

Fig. 6.106: Residual stress-Haigh diagram showing alternating bending strengths for AISI4140 quenched and tempered at 450 °C after conventional shot peening and subsequent annealing under the conditions indicated, including the cyclic yield strengths shown in Sect. 6.2.1c [6.57]

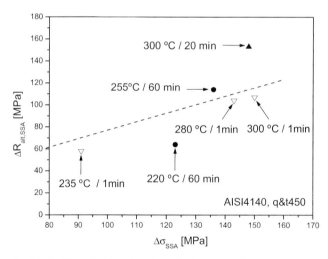

Fig. 6.107: Alternating bending strengths for AISI4140 quenched and tempered at 450 °C after conventional shot peening and subsequent annealing under the conditions indicated vs. yield strength increases determined in static strain aging tests [6.57]

rolled specimens of TiAl3V8Cr6Mo4Zr4 (Ti-38–644, Beta C) [6.132,6.133]. A 2-hour annealing treatment at 440 °C causes aging in the core region only. Therefore, the fictitious notch ground fatigue limit of notched specimens with $k_t = 2.95$

can be increased from about 400 MPa in the electropolished state to 1150 MPa in the state that was deep rolled only, and to 1400 MPa after selective surface hardening. Recent studies of deep rolled notched specimens with $k_t = 2.5$ have shown that a 4-hour annealing treatment at 440 °C can create a fictitious notch ground fatigue limit as high as 1700 MPa [6.134]. Other examinations have shown that selective surface hardening occurs also in shot peened specimens of Timetal 21S [6.135]. [6.132] examined preferred surface hardening for AlCuMg2 (Al2024) aluminum alloy after shot peening, annealing for 12 hours at 190 °C and a final electropolishing treatment. Compared to the electropolished state annealed in the same way, fatigue limit at 10^7 loading cycles was increased from about 120 MPa to 220 MPa. This proves that light metals can show substantial increases in fatigue limit due to re-crystallization- or hardening processes occurring in the surface layer. These are effected by mechanical surface treatments, and annealing, followed by final electropolishing which eliminates the shot peening-induced increase in roughness.

References

6.1 E. Macherauch H. Wohlfahrt: Eigenspannungen und Ermüdung, In: D. Munz (ed.), Ermüdungsverhalten metallischer Werkstoffe, DGM-Informationsgesellschaft Verlag, Oberursel, 1985, pp. 237–283.

6.2 B. Scholtes: Eigenspannungen in mechanisch randschichtverformten Werkstoffzuständen, Ursachen-Ermittlung-Bewertung, DGM-Informationsgesellschaft, Oberursel, 1990.

6.3 H. Holzapfel, V. Schulze, O. Vöhringer, E. Macherauch: Residual stress relaxation an an AISI 4140 steel due to quasistatic and cyclic loading at higher temperatures, Materials Science and Engineering A248 (1998), pp. 9–18.

6.4 V. Schulze: Die Auswirkungen kugelgestrahlter Randschichten auf das quasistatische sowie ein- und zweistufige zyklische Verformungsverhalten von vergütetem 42 CrMo 4, Dissertation, Universität Karlsruhe (TH), 1993.

6.5 J. E. Hoffmann: Der Einfluss fertigungsbedingter Eigenspannungen auf das Biegewechselverhalten von glatten und gekerbten Proben aus Ck45 in verschiedenen Werkstoffzuständen, Dissertation, Universität Karlsruhe (TH), 1984.

6.6 D. Munz: Der Einfluss von Eigenspannungen auf das Dauerschwingverhalten, Härterei Technische Mitteilungen 22 (1967), pp. 52–62.

6.7 H. Wohlfahrt: Einfluss von Eigenspannungen, In: W. Dahl (ed.), Verhalten von Stahl bei schwingender Beanspruchung, 1978, pp. 141–164.

6.8 J. D. Morrow, G. M. Sinclair : Cycle dependent stress relaxation, Symp. Basic Mechanisms of Fatigue, ASTM STP 237, 1959, pp. 83–109.

6.9 A. Glaser: Mittelspannungseinfluss auf das Verformungsverhalten von Ck45 und 42CrMo4 bei spannungs- und nennspannungskontrollierter homogen einachsiger Schwingbeanspruchung, Dissertation, Universität Karlsruhe (TH), 1988.

6.10 L. Bairstow: The elastic limits of iron and steel under cyclic variations of stress, Phil. Trans. Roy. Soc. London 210A (1910), pp. 35–55.

6.11 J. Dubuc, J. R. Vanasse, A. Biron, A. Bazergui: Effect of mean stress and mean strain in low cycle fatigue of A-517 and A-201 steels, J. Eng. Industry, Trans. ASME (1970), pp. 35–52.

6.12 M. Becker: Das Wechselverformungsverhalten von Ck45 und Ck80 im Temperaturbereich 295 K < T < 873 K. Universität Karlsruhe (TH), 1987.

6.13 O. Buxbaum, H. Oppermann, H. G. Köbler, et al.: Bericht Fraunhofer-Institut für Betriebsfestigkeit, Darmstadt, 1983.

6.14 A. Fatemi, R. I. Stephens: Tensile mean stress effects on uniaxial fatigue behavior of 1045HR steel, In: R. O. Ritchie E. A. Starke (eds.), Proc. Fatigue 87, 1987, pp. 537–546.

6.15 H. R. Jhanasale, T. H. Topper: Engineering analysis of inelastic stress response of a structural metal under variable cyclic strains, ASTM STP 519, 1973, pp. 246–260.

6.16 M. R. James: Relaxation of residual stresses – an overview, In: A. Niku-Lari (ed.), Advances in surface treatments; International Guidebook on Residual Stresses, New York, Pergamon, 1987, pp. 349–365.

6.17 J. Bergström: Relaxation of residual stresses during cyclic loading, In: A. Niku-Lari (ed.), Advances in Surface Trestments – Vol.3, Pergamon, New York, 1986, pp. 97–111.

6.18 S. R. Valluri: Some Recent Developments at "Galcit" Concerning a Theory of Metal Fatigue, Acta Metallurgica 11 (1963), pp. 759–775.

6.19 L. F. Impellizzeri: ASTM STP 462, 1970, pp. 40–68.

6.20 S. Kodama: Soc. of Met. Sci. Japan, 1971, pp. 43–47.

6.21 J. M. Potter: ASTM STP 519, 1973, pp. 109–132.

6.22 V. M. Radhakrishwan, C. R. Prasad: Eng.Fract.Mech. 8 (1976), pp. 593–597.

6.23 J. Morrow, A. S. Ross, G. M. Sinclair : Relaxation of residual stresses due to fatigue loading, SA. Transactions 68 (1960), pp. 40–48.

6.24 F. Rotvel: AGAARD CP-118, 1972.

6.25 S. Kodama: Soc. of Mat. Sci Japan, 1972, pp. 111.

6.26 J. F. Flavenot, N. Skalli: Annals of the CIRP 32.1 (2003).

6.27 A. L. Esquivel, K. R. Evans: Boeing Report D6–23377, 1968.

6.28 M. R. James: The relaxation of residual stresses during fatigue, In: E. Kula,

V. Weiss (eds.), Residual Stress and Stress Relaxation, Plenum Publ. Corp., New York, 1982, pp. 297–314.

6.29 J. Lu, J. F. Flavenvot, A. Turbat: In: L. Mordfin (ed.), Mechanical Relaxation of Residual Stresses, ASTM STP 993, 1987, pp. 75.

6.30 J. Lu, J. F. Flavenot: Use of a Finite Element Method for the Prediction of the Shot Peening Residual Stress Relaxation during Fatigue, In: H. Wohlfarth, R. Kopp, O. Vöhringer (eds.), Shot Peening, Proc. Int. Conf. Shot Peening 3, DGM-Informationsgesellschaft, Oberursel, 1987, pp. 391–398.

6.31 J. Zarka, J. Casier: Mech. Today 6 (1981), pp. 93.

6.32 A. Wick: Randschichtzustand und Schwingfestigkeit von 42CrMo4 nach Kugelstrahlen unter Vorspannung und bei erhöhter Temperatur, Dissertation, Universität Karlsruhe (TH), 1999.

6.33 R. Menig: Randschichtzustand, Eigenspannungsstabilität und Schwingfestigkeit von unterschiedlich wärmebehandeltem 42 CrMo 4 nach modifizierten Kugelstrahlbehandlungen, Dissertation, Universität Karlsruhe (TH), 2002.

6.34 H. Wohlfahrt: Auswirkungen mechanischer Oberflächenbehandlungen auf das Dauerschwingverhalten unter Einfluss von Rissbildung und Rissausbreitung, In: H. Wohlfahrt, P. Krull (eds.), Mechanische Oberflächenbehandlungen, Wiley-VCh, Weinheim, 2000, pp. 55–85.

6.35 D. Löhe, K.-H. Lang, O. Vöhringer: Residual stresses and fatigue behavior, In: G. Totten, M. Howes, T. Inoue (eds.), Handbook of residual stress and deformation of steel, AS. International, Metals Park, 2002, pp. 27–53.

6.36 W. Schütz: Über eine Beziehung zwischen der Lebensdauer bei konstanter und veränderlicher Beanspruchungsamplitude und ihre Anwendbarkeit auf die Bemessung von Flugzeugbauteilen, Z.Flugwiss. 15 (1967), pp. 407–419.

6.37 W. P. Evans, R. E. Rickleffs, J. F. Millan: In: Fatigue – An interdisciplinary approach, Proc. 10th Sagamore Conf., Syracuse, 1964, pp. 237.

6.38 J. Goodman: Mechanics Applied to Engineering, Long Man Green & Co., London, 1899.

6.39 B. Syren: Der Einfluss spanender Bearbeitung auf das Biegewechselverformungsverhalten von Ck45 in verschiedenen Wärmebehandlungszuständen, Dissertation, Universität Karlsruhe (TH), 1975.

6.40 K.-H. Kloos, P. K. Braisch: Über die Wirkung einer Randschichtverfestigung auf die Schwingfestigkeit von Proben und Bauteilen, Härterei Technische Mitteilungen 37 (1982), pp. 83–91.

6.41 B. Winderlich: Das Konzept der lokalen Dauerfestigkeit und seine Anwendung auf martensitische Randschichten, insbesondere Laserhärtungsschichten, Materialwissenschaft und Werkstofftechnik 21 (1990), pp. 378–389.

6.42 J. E. Hoffmann, D. Löhe: Einfluss von Makroeigenspannungen auf das Ermüdungsverhalten von glatten und gekerbten Proben aus gehärtetem Stahl Ck45, Härterei Technische Mitteilungen 57 (2002), pp. 79–85.

6.43 T. Fett, D. Munz: Stress intensity factors and weight functions, Computational Mechanics Publications, Southampton, 1997.

6.44 T. Seeger, P. Heuler: Ermittlung und Bewertung örtlicher Beanspruchungen zur Lebensdauerabschätzung schwingbelasteter Bauteile, In: D. Munz (ed.), Ermüdungsverhalten metallischer Werkstoffe, DGM-Informationsgesellschaft, Oberursel, 1984, pp. 213–235.

6.45 A. Bäumel jr.: Experimentelle und numerische Untersuchung der Schwingfestigkeit randschichtverfestigter eigenspannungsbehafteter Bauteile, Dissertation, Inst. f. Stahlbau und Werkstofftechnik, T. Darmstadt, Heft 49, 1991.

6.46 I. Altenberger: Mikrostrukturelle Untersuchungen mechanisch randschichtverfestigter Bereiche schwingend beanspruchter metallischer Werkstoffe, Dissertation, Universität Gesamthochschule Kassel, 2000.

6.47 A. Ebenau: Verhalten von kugelgestrahltem 42 CrMo 4 im normalisierten und vergüteten Zustand unter einachsig homogener und inhomogener Wechselbeanspruchung, Dissertation, Universität Karlsruhe (TH), 1989.

6.48 A. Wick, V. Schulze, O. Vöhringer: Effects of stress- and/or warm peening of AISI 4140 on fatigue life, Steel Research 71 (2000) 8, pp. 316–321.

6.49 P. Starker: Der Größeneinfluss auf das Biegeverhalten von Ck 45 in verschiedenen Bearbeitungs- und Wärmebehandlungszuständen, Dissertation, Universität Karlsruhe (TH), 1981.

6.50 T. Hirsch: Zum Einfluss des Kugelstrahlens auf die Biegeschwingfestigkeit von Titan- und Aluminiumbasislegierungen, Dissertation, Universität Karlsruhe (TH), 1983.

6.51 H. Holzapfel: unveröffentlicht, Universität Karlsruhe (TH), 1994.

6.52 O. Vöhringer, T. Hirsch, E. Macherauch: Relaxation on shot peening induced residual stresses of TiAl6V4 by annealing of mechanical treatment, In: G. Lütjering, U. Zwicker, W. Bunk (eds.), Titanium – Science and Technology, Proc. Int. Conf. Titanium 5, DGM-Informationsgesellschaft, Oberursel, 1984, pp. 2203–2210.

6.53 T. Hirsch, O. Vöhringer, E. Macherauch: Der Einfluss des Kugelstrahlens auf die Biegeschwingfestigkeit von AlCu5Mg2 in verschiedenen Wärmebehandlungszuständen, Härterei Technische Mitteilungen 41 (1986), pp. 166–172.

6.54 W. Zinn, B. Scholtes: Generation and stability of shot peening residual stresses in different aluminium base alloys, In: S. Denis, J.-L. Lebrun, B. Bourniquel, M. Barral, J. F. Flavenot (eds.), Proc. Europ. Conf. Residual Stresses 4, Cluny en Bourgogne, France, 1996, pp. 775–783.

6.55 I. Altenberger, U. Noster, B. Scholtes, R. O. Ritchie: High temperature fatigue of mechanically surface treated materials, In: L. Wagner (ed.), Shot Peening, Wiley-VCh, Weinheim, 2003, pp. 483–489.

6.56 G. Masing: Berechnung von Dehnungs- und Stauchungslinien auf Grund von inneren Spannungen, Wiss. Ver. Siemens Konzern 5 (1926), pp. 135–141.

6.57 R. Menig, V. Schulze, O. Vöhringer: Effects of Static Strain Aging on Residual Stress Stability and Alternating Bending Strength of Shot Peened AISI 4140, Zeitschrift für Metallkunde 93 (2002) 7, pp. 635–640.

6.58 R. Menig, V. Schulze, O. Vöhringer: Kugelstrahlen und anschließendes Auslagern – Steigerung der Eigenspannungsstabilität und der Wechselfestigkeit am Beispiel von 42CrMo4, Härterei Technische Mitteilungen 58 (2003), pp. 127–132.

6.59 R. Menig, V. Schulze, O. Vöhringer: Shot peening at elevated temperatures – Influence of the material state of AISI 4140 on the effect on fatigue strength, In: A. F. Blom (ed.), Proc. Fatigue 2002, Engineering Materials Advisory Services Ltd., West Midlands, 2002, pp. 1257–1266.

6.60 V. Schulze: Warm- und Spannungsstrahlen – Wege zur Erzeugung stabilisierter und erhöhter Druckeigenspannungen, In: Deutscher Verband für Materialforschung und -prüfung e.V., DVM-Tag "Federn im Fahrzeugbau", Berlin, 2002, pp. 45–54.

6.61 R. Menig, V. Schulze, O. Vöhringer: Residual stress relaxation and fatigue strength of AISI 4140 under torsional loading after conventional shot peening, stress peening and warm peening, In: L. Wagner (ed.), Shot Peening, Wiley-VCh, Weinheim, 2003, pp. 316.

6.62 E. Müller: Eigenspannungsabbau an spannungsgestrahlten Torsionsproben unter dynamischer Belastung, Materialwissenschaft und Werkstofftechnik 28 (1997), pp. 549–556.

6.63 J. E. Hoffmann, D. Löhe, E. Macherauch: Influence of Shot Peening on the Bending Fatigue Beha-
viour of Notched Specimens of Ck 45, In: H. Wohlfahrt, R. Kopp, O. Vöhringer (eds.), Shot Peening, Proc. Int. Conf. Shot Peening 3, DGM-Informationsgesellschaft, Oberursel, 1987, pp. 631–638.

6.64 J. Bergström, T. Ericsson: X-Ray Microstructure and Residual Stress Analysis of Shot Peened Surface Layers during Fatigue Loading, In: H. Wohlfahrt, R. Kopp, O. Vöhringer (eds.), Shot Peening, Proc. Int. Conf. Shot Peening 3, DGM-Informationsgesellschaft, Oberursel, 1987, pp. 221–230.

6.65 H. Traiser, K.-H. Kloos: Einfluss der Oberflächenkaltverfestigung auf die Schwingfestigkeit und das Wechselverformungsverhalten der normalisierten Kohlenstoffstähle Ck15 und Ck45 unter Zug-Druck- und Umlaufbiegebeanspruchung, Zeitschrift für Werkstofftechnik 16 (1985), pp. 135–143.

6.66 R. Menig, V. Schulze, O. Vöhringer: Optimized warm peening of the quenched and tempered steel AISI 4140, Materials Science and Engineering A335 (2002), pp. 198–206.

6.67 R. Schreiber: Untersuchungen zum Dauerschwingverhalten des kugelgestrahlten Einsatzstähle 16 MnCr 5 in verschiedenen Wärmebehandlungszuständen, Dissertation, Universität Karlsruhe (TH), 1976.

6.68 U. Ilg: Strukturelle Änderungen in unterschiedlich wärmebehandelten Wälzkörpern aus 100Cr6 und 20MnCr5 bei Wälz- und Wälz-Gleitbeanspruchung, Dissertation, Universität Karlsruhe (TH), 1980.

6.69 H. Holzapfel: Das Abbauverhalten kugelgestrahlter Eigenspannungen bei 42 CrMo 4 in verschiedenen Wärmebehandlungszuständen, Dissertation, Universität Karlsruhe (TH), 1994.

6.70 H.-J. Christ: Wechselverformungsverhalten von Metallen, Springer, 1991.

6.71 C. Laird, Z. Wang, B.-T. Ma, H.-F.Chai: Low Energy Dislocation Structures Poduced by Cyclic Softening, Materials Science and Engineering A113 (1989), pp. 245–257.

6.72 H.-J. Christ, G. Hoffmann, O. Öttinger: History Effects in metals during constant and variable amplitude testing I. Wavy dislocation glide behaviour, Materials Science and Engineering A201 (1995), pp. 1–12.

6.73 O.K. Chopra C. V. B. Gowda: Substructural development during strain cycling of α-iron, Phil. Mag. 30 (1974), pp. 583–591.

6.74 J. Walla: Untersuchungen zur Schädigung in mehrstufig schwingbeanspruchten Proben aus Ck45N und Cu-35%Ni-3,5%Cr, Dissertation, Universität Bremen, 1990.

6.75 E. S. Kayali, A. Plumtree: Stress-Substructure Relationships in Cyclically and Monotonically Deformed Wavy Slip Mode Metals, Metallurgical and Materials Transactions 13A (1982), pp. 1033–1041.

6.76 D. Munz: Ermüdungsverhalten metallischer Werkstoffe, DGM-Informationsgesellschaft, Oberursel, 1984.

6.77 I. Altenberger, B. Scholtes, U. Martin, H. Oettel: Cyclic deformation and near surface microstructures of shot peened or deep rolled austenitic stainless steel AISI 304, Materials Science and Engineering A264 (1999), pp. 1–16.

6.78 G. Kuhn: Zum Zug-Druck-Wechselverformungsverhalten von Kerbproben aus Ck45 mit Bearbeitungseigenspannungen, Dissertation, Universität Karlsruhe (TH), 1991.

6.79 G. Kuhn, J. E. Hoffmann, D. Eifler, B. Scholtes, E. Macherauch: Instability of machining residual stresses in differently heat treated notched parts of SAE 1045 during cyclic deformation, In: H. Fujiwara, T. Abe, K. Tanaka (eds.), Proc. Int. Conf. Residual Stresses 3, Elsevier, London, 1992, pp. 1294–1301.

6.80 A. Ebenau, D. Eifler, O. Vöhringer, E. Macherauch: Influence of shot peening on the cyclic deformation behaviour of the steel 42 CrMo 4 in a normalized state, In: K. Iida (ed.), Proc. Int. Conf. Shot Peening 4, Japan Society of Precision Engineering, Tokyo, 1990, pp. 327–336.

6.81 D. Eifler, D. Löhe, B. Scholtes: Residual Stresses and Fatigue of Metallic Materials, In: V. Hauk, H. Hongardy, E. Macherauch (eds.), Residual stresses, measurement, calculation, evaluation, DGM. Oberursel, 1991, pp. 157–166.

6.82 P. Krull, T. Nitschke-Pagel, H. Wohlfahrt: Influence of shot peening and high pressure water peening on near surface microstructure of 316 Ti stainless steel, In: C. A. Brebbia, J. M. Kenny (eds.), Surface Treatment IV – Computer Methods and Experimental Measurements, Southampton, 1999, pp. 291–300.

6.83 I. Altenberger B. Scholtes: Improvement of fatigue behaviour of mechanically surface treated materials by annealing, Scripta Materialia 42 (1999), pp. 873–881.

6.84 M. Krauss B. Scholtes: Thermal fatigue of shot peened or hrad turned hotwork steel AISI H11, In: L. Wagner (ed.), Shot Peening, Wiley-VCH, Weinheim, 2003, pp. 324–330.

6.85 H. Mughrabi: Mikrokristalline Ursachen der Ermüdungsrissbildung, In: D. Munz (ed.), Ermüdungsverhalten metallischer Werkstoffe, DGM-Informationsgesellschaft, Oberursel, 1984, pp. 7–38.

6.86 T. H. Liu: ASME, 1991.

6.87 H. Nisitani, K. Fujimura: Initiation and growth behaviour of a fatigue crack in shot-peened steel, In: Proc. Int. Conf. Computer Methods and Experimental Measurements for Surface Treatment Effects 3, Oxford, 1997, pp. 13–22.

6.88 R. Herzog: Auswirkungen bearbeitungsbedingter Randschichteigenschaften auf das Schwingungsrisskorrosionsverhalten von Ck45 und X35CrMo17, Dissertation, Universität Braunschweig, 1997.

6.89 Y. Muto, G. H. Fair, B. Noble, R. B. Waterhouse: The effect of residual stresses induced by shot peening on fatigue crack propagation in two high strength aluminum alloys, Fatigue and Fracture of Engineering Materials & Structures 10 (1987), pp. 261–272.

6.90 H. Gray, L. Wagner, G. Lütjering: Influence of Shot Peening induced Surface Roughness, Residual Macrostresses and Dislocation Density on the Elevated Temperature HCF-Properties of Ti-Alloys, In: H. Wohlfahrt, R. Kopp, O. Vöhringer (eds.), Shot Peening, Proc. Int. Conf. Shot Peening 3, DGM-Informationsgesellschaft, Oberursel, 1987, pp. 447–458.

6.91 P. Peyre, P. Merrien, H. P. Lieurade, R. Fabbro: Laser-Induced Shock Waves as Surface Treatment for 7075-T7351 Aluminum Alloy, Surface Engineering 11 (1995), pp. 47–52.

6.92 P. Starker, H. Wohlfahrt, E. Macherauch: Surface crack initiation during fatigue as a result of residual stresses, Fatigue and Fracture of Engineering Materials & Structures 1 (1979), pp. 319–327.

6.93 R. G. Forman, V. E. Kearney, R. M. Engle: Jounal Basic Engineering 89 (1967), pp. 459.

6.94 S. Jaegg: Rissspitzennahe Eigenspannungen und Ermüdungsrissausbreitung des Stahls S690QL1 bei unterschiedlichen Beanspruchungsmoden, Dissertation, Universität Kassel, 1999.

6.95 S. Jaegg B. Scholtes: Crack Propagation and Crack Tip Residual Stresses after Different Loading Histories of Steel S690QL1, In: T. Ericson, M. Oden, A. Anderson (eds.), Proc. Int. Conf. Residual Stresses 5, Linköping University, Linköping, 1997, pp. 1078–1083.

6.96 M. Hilpert: Dauerschwingverhalten von Mangesiumlegierungen: Einfluss von mechanischen Oberflächenbehandlungen und Umgebungsmedien, Fortschritt-Berichte VD. Reihe 5, Nr. 627, Düsseldorf, 2001.

6.97 C. O. Lykins: Laser shock peening vs. shot peening, A damage tolerance investigation, In: J. K. Gregory, H. J. Rack, D. Eylon (eds.), Proc. Symp. Surface Performance of Titanium, Cincinnati, 1997.

6.98 S. D. Thompson, D. W. See, C. D. Lykins, P. G. Sampson: Laser shock peening vs shot peening – a damage tolerance investigation, In:

6.99 W. Cowie, S. Mannava, T. Compton: In: Proc. 1997 USA. Aircraft Structural Integrity Program Conference, San Antonio, 1997.

6.100 J. J. Ruschau, R. John, S. R. Thompson, T .Nicholas: J.Engg.Mat.and Techn. 121 (1999), pp. 321–329.

6.101 A. H. Clauer, C. T. Walters, S. C. Ford: The effects of laser shock processing on the fatigue properties of 2024-T3 aluminum, In: Lasers in Materials Processing, AS. International, Metals Park, 1983, pp. 7.

6.102 L. Wagner: Mechanical surface treatments on titanium alloys: fundamental mechanisms, In: J. K. Gregory, H. J. Rack, D. Eylon (eds.), Proc. Symp. Surface Performance of Titanium, Cincinnati, 1997, pp. 199–215.

6.103 I. Altenberger: Alternative surface treatments: microstructures, residual stresses and fatigue behavior, In: L. Wagner (ed.), Shot Peening, Wiley-VCH, Weinheim, 2003, pp. 421–434.

6.104 K. J. Kang, J. H. Song, Y. Y. Earmme: Fatigue crack growth and closure through a tensile residual stress field under compressive applied loading, Fatigue and Fracture of Engineering Materials & Structures 12 (1989), pp. 363–376.

6.105 P. Starker E. Macherauch: Kugelstrahlen und Schwingfestigkeit, Zeitschrift für Werkstofftechnik 14 (1983), pp. 109–115.

6.106 J. E. Hoffmann, D. Löhe, E. Macherauch: Influence of Machining Residual Stresses on the Bending Fatigue Behaviour of Notched Specimens of Ck 45 in Different Heat Treating States, In: E. Macherauch, V. Hauk (eds.), Residual Stresses in Science and Techology, Proc. Int. Conf. Residual Stresses 1, DGM-Informationsgesellschaft, Oberursel, 1987, pp. 801–808.

6.107 J. E. Hoffmann, D. Eifler, E. Macherauch: Der Einfluss von Bearbeitungseigenspannungen auf das Biegewechselverhalten von Kerbstäben aus Ck 45, In: E. Macherauch, V. Hauk (eds.), Eigenspannungen, DGM-Informationsgesellschaft, Oberursel, 1983, pp. 287–300.

6.108 B. Syren, H. Wohlfahrt, E. Macherauch: Der Einfluss von Bearbeitungseigenspannungen auf das Biegewechselverhalten von Stahl Ck45 im weichgeglühten Zustand, Archiv für das Eisenhüttenwesen 46 (1975), pp. 735–739.

6.109 B. Syren, H. Wohlfahrt, E. Macherauch: The Influence of Residual Stresses and Surface Topography on Bending Fatigue Strength of Machined Ck45 in Different Heat Treatment Conditions, In: Proc. Int. Conf. Mech. Behavior of Materials 2, ASM. Boston, 1976, pp. 807–811.

6.110 R. Schreiber, H. Wohlfahrt, E. Macherauch: Verbesserung des Biegewechselverhaltens eines kugelgestrahlten 16MnCr5 durch Oberflächennachbehandlung, Archiv für das Eisenhüttenwesen 49 (1978), pp. 207–210.

6.111 R. Schreiber, H. Wohlfahrt, E. Macherauch: Der Einfluss des Kugelstrahlens auf das Biegewechselverhalten von blindgehärtetem 16 MnCr 5, Archiv für das Eisenhüttenwesen 48 (1977), pp. 653–657.

6.112 U. Jung, R. Schaal, C. Berger, H.-W. Reinig, H. Traiser: Berechnung der Schwingfestigkeit festgewalzter Kurbelwellen, Materialwissenschaft und Werkstofftechnik 29 (1998) 10, pp. 569–572.

6.113 R. Schaal, U. Jung, B. Kaiser, C. Berger: Berechnung der Dauerfestigkeit festgewalzter Bauteile., In: VDI-Bericht 1472, Werkstoff und Automobilantrieb, Dresden, 1999, pp. 379–389.

6.114 R. Schaal, U. Jung, B. Kaiser, C. Berger: Bewertung der Wirkung von Festwalz-Eigenspannungen auf den Rissfortschritt in Kerben., In: R. Kopp, K. Herfurth (eds.), Werkstoffwoche '98, Band VI, 1998, pp. 875–879.

6.115 Th. Krug, K.-H. Lang, D. Löhe: Influence of the surface layer on the development of fatigue cracks at case har-

dened steels, In: A. F. Blom (ed.), Proc. Fatigue 2002, Engineering Materials Advisory Services Ltd., West Midlands, 2002, pp. 955–962.

6.116 J. K. Gregory, L. Wagner: Property improvement in light metals using shot peening, In: L. Wagner (ed.), Shot Peening, Wiley-VCh, Weinheim, 2003, pp. 349–359.

6.117 A. Drechsler, T. Dörr, L. Wagner: Mechanical surface treatments on Ti-10V-2Fe-3Al for improved fatigue resistance, Materials Science and Engineering A243 (1998), pp. 217–220.

6.118 J. Kiese, J. Zhang, O. Schauerte, L. Wagner: Mechanical Surface Treatments on the high strength a-Titanium alloy KA 120, In: L. Wagner (ed.), Shot Peening, Wiley-VCh, Weinheim, 2003, pp. 399–405.

6.119 J. Lindemann, L. Wagner: Mean stress sensitivity in fatigue of a-, $(a + \geq)$- and $\geq$-titanium alloys, Materials Science and Engineering A 234–236 (1997), pp. 1118–1121.

6.120 U. Holzwarth, J. Kiese, L. Wagner: Verbesserung der Lebensdauer der Implantatlegierung Ti-6Al-7Nb durch Kugelstrahlen und Glattwalzen, In: Mechanische Eigenschaften von Implantatwerkstoffen, DVM, 1998, pp. 75–81.

6.121 J. Lindemann, D. Roth-Fagaraseanu, L. Wagner: Effect of shot peening in fatigue properties of γ-titanium aluminides, In: L. Wagner (ed.), Shot Peening, Wiley-VCh, Weinheim, 2003, pp. 392–398.

6.122 L. Wagner: Mechanical surface treatments on titanium, aluminum and magnesium alloys, Materials Science and Engineering (1999) A263, pp. 210–216.

6.123 M. Hilpert, L. Wagner: Fatigue performance of a shot peened high-strength magnesium alloy, In: C. A. Brebbia, J. M. Kenny (eds.), Surface Treatment IV – Computer Methods and Experimental Measurements, Southampton, 1999, pp. 331–340.

6.124 K. Eisner: Prozesstechnologische Grundlagen zur Schockverfestigung von metallischen Werkstoffen mit einem kommerziellen Excimerlaser,

Dissertation, Universität Erlangen-Nürnberg, 1998.

6.125 T. R. Tucker, A. H. Clauer: Laser processing of materials, MCIC-83–48, Metals and Ceramics Information Center, 1983.

6.126 P. Peyre, L. Berthe, X. Scherpereel, R. Fabbro: J.of Materials Science 33 (1998), pp. 1421–1429.

6.127 P. Peyre et al.: J. of Laser Applications 8 (1996), pp. 135–141.

6.128 U. Noster, I. Altenberger, R. O. Ritchie, B. Scholtes: Isothermal fatigue behavior and residual stress states of mechanically surface treated TiAl6V4: Laser schock peening vs. deep rolling, In: L. Wagner (ed.), Shot Peening, Wiley-VCh, Weinheim, 2003, pp. 447–453.

6.129 R. L. Mattson, J. G. Roberts: The effect of residual stresses induced by strain peening upon fatigue strength, In: G. M. Rassweiler, W. L. Grube (eds.), Symposium internal stresses and fatigue in metals, New York, 1959, pp. 338–357.

6.130 L. Wagner, J. K. Gregory: Thermomechanical surface treatment of titanium alloy, In: Second European AS. Heat Treatment and Surface Engineering Conference, ASM. Metals Park, 1993, pp. 1–24.

6.131 H. Gray, L. Wagner, G. Lütjering: Effect of modified surface layer microstructures through shot peening and subsequent heat treatment on the elevated temperature fatigue behavior of TI alloys, In: H. Wohlfahrt, R. Kopp, O. Vöhringer (eds.), Shot Peening, Proc. Int. Conf. Shot Peening 3, DGM-Informationsgesellschaft, Oberursel, 1987, pp. 467–475.

6.132 J. K. Gregory, C. Müller, L. Wagner: Bevorzugte Randschichtaushärtung: Neue erfahren zur Verbesserung des Dauerschwingverhaltens mechanisch belasteter Bauteile, Metall 47 (1993), pp. 915–919.

6.133 J. K. Gregory, L. Wagner, C. Müller: Selective Surface Aging in the High-Strength Beta Titanium Alloy Beta-C. In: P. Mayr (ed.), Proc. Surface Engineering, DGM-Informationsgesellschaft, Oberursel, 1985, pp. 435–440.

6.134 A. Berg, J. Kiese, L. Wagner: Microstructural gradients in Ti-3Al-8V-6Cr-4Zr-4Mo for excellent HCF strength and toughness, Materials Science and Engineering A243 (1998), pp. 146–149.

6.135 M.-C. Berger, J. K. Gregory: Selective hardening and residual stress relaxation in shot peened Timetal 21s, In: C. A. Brebbia, J. M. Kenny (eds.), Surface Treatment IV – Computer Methods and Experimental Measurements, Southampton, 1999, pp. 341–348.

7

Summary

Mechanical surface treatments allow for inducing specific surface layer states in metallic materials. These states can be utilized to increase fatigue strength. As the historical survey of the development of mechanical surface treatment processes in Sect. 1 shows, the causes of increased fatigue strength were the object of long-lasting and severe dispute. Today it is commonly accepted knowledge that, along with other surface layer characteristics, it is residual stress state, workhardening state, additional microstructural changes and roughness that are crucial for fatigue strength. In particular, it is essential that the surface layer states are sufficiently stable and are not substantially changed by thermal, quasi-static or cyclic loading or combinations thereof. Surface layer state stability in the loading states listed above is the central focus of this study. Beginning with the systematics and the selection of the examined surface treatment processes, this study deals with technological details and with the surface layer states caused by these processes, and then goes on to describe the effects that surface layer states and their stability have on fatigue strength. Special processes aimed at increasing the stability of surface layer states receive particular attention. The processes examined primarily are shot peening, stress peening, warm peening and warm peening at elevated temperatures, deep rolling and laser shock treatment. First, these processes are defined, then selected examples for their application are presented. The concluding discussion of installations, tools and essential influencing parameters serves to outline the technological details of the processes.

The known facts on the **elementary processes** and their relevance to the respective process or material state are described briefly. These elementary processes include the influence of temperature and strain rate on deformation behavior – in the context of thermally activated dislocation movement, viscous damping and shock deformation –, Bauschinger effect and static or dynamic strain aging. The processes are described by means of analytical and numerical model approaches, as well as finite-element approaches. The analytical approaches – such as the theories of Hertz, Tabor and others on shot peening and deep rolling, as well as the shock wave theory for laser shock treatments – serve to provide a fundamental understanding of the processes. Simple numerical models, on the other hand, are particularly suitable for estimating the influence of process parameters on the surface layer states and are utilized industrially for setting parameters. By contrast, finite-element simulations have so far been applied mostly in parameter studies. Only in the context of deep rolling treatments are Finite-element simulations used as a basis for estimating fatigue life.

The **surface layer states** achievable by the individual processes are discussed in detail, focusing on characteristics such as shape, topography, residual stress state, workhardening state, microstructure, phase fractions and texture, thereby demonstrating the influence of process parameters and of the workpiece's material or its state.

- Changes of shape due to mechanical surface treatments occur primarily in thin-walled structures and in processes involving a sufficient degree of plastification, such as deep rolling and laser shock treatment, and are utilized specifically for peen forming, size rolling and roll trueing.
- The impact craters created by shot peening and related processes increase roughness particularly in soft states and at high intensities. Deep rolling and laser shock treatment, on the other hand, reduce roughness, given that treatment intensity is not excessive.
- In all processes examined, the induced residual stress states exhibit compressive residual stresses close to the surface. In laser shock treatments, their maxima are almost always located at the surface. In shot peening and deep rolling, the state of the material clearly influences the residual stress profile, due to surface stretching and Hertzian pressing occurring as competing processes. In these processes, therefore, harder states show maximum values of compressive residual stress below the surface. Changing process parameters – such as the diameter or the velocity of the peening medium in shot peening, rolling force or pressure in deep rolling, power densitiy or the number of exposures in laser shock treatments – primarily effects changes of the depth of the residual stress maximum and the position of zero residual stresses, rather than altering surface residual stresses or their maximum values. Among the modified processes, stress peening causes compressive residual stresses to increase in the direction in which tensile loading stress components are effective during the peening process, while warm peening increases compressive residual stresses slightly. Thus, a combination of both processes creates maximum residual stress values.
- Generally, the increase of workhardening close to the surface is more pronounced in soft states than in hard states. Only high strength states experience a shot peening-induced worksoftening due to dislocations being rearranged into energetically preferable positions. However, values of hardness measured close to the surface may still show increases, as compressive residual stresses increase resistance to penetration by the indentor. Deep rolling and, particularly, laser shock treatment induce less workhardening than shot peening. In laser shock treatment, a growing number of slip systems available within the material increases the workhardening effect. Therefore, workhardening is more pronounced in face-centered cubic material states, in comparison with body-centered cubic and, especially, hexagonal material states. Among the modified peening processes, stress peening causes hardly any changes, compared to conventional peening, while warm peening increases workhardening and hard-

ness, in particular. This is due to strain aging effects impeding the plastic deformation required for testing hardness. Stress peening at elevated temperatures causes similar effects as warm peening, while their depth effect is extended.

- Shot peening causes an increased density of dislocations. In steels, their arrangement is random, due to the obstruction of cross slip. In aluminum alloys, dislocations show cell structures when stacking fault energy is sufficiently high and cross slip is thus abetted. Strain rates are lower in deep rolling than they are in shot peening and therefore facilitate cell formation. Due to shock deformation, laser shock treatment usually results in twin formation and planar dislocation structures, while fine cells are found only when stacking fault energy is very high.
- Shot peening and deep rolling treatments are found to change the phase fractions primarily in steels containing retained austenite – here, retained austenite is transformed to martensite. So far, however, no transformations of retained austenite have been found after laser shock treatments, presumably because the occurring strains are too small. In the case of pure iron, on the other hand, transformations of the body-centered cubic α-structure into the hexagonal, i.e. high-pressure, ε-structure have been observed.
- Additionally, a development of faint fiber- or rolling textures was observed in shot peened steels.

Thermal loading causes changes of the surface layer state which are determined primarily by elementary processes such as dislocation climb – which is controlled by dislocation core- or volume diffusion – and also by processes of re-crystallization and precipitation.

In many cases a quantitative description of the relaxation behavior of macro-residual stresses and characteristics of the workhardening state is achieved by means of an Avrami function, which is applicable to the entire temperature range. The function shows an activation enthalpy which is roughly equal to the enthalpy for self-diffusion. The Norton approach for diffusion-controlled dislocation creep is an alternative way of describing macro-residual stress relaxation. This is achieved by determining the respective correlation between the plastic strain rate and mean residual stress.

- The only statements to be made regarding shape and topography are that the dimensional stability of thin-walled components is impaired and that there should be hardly any changes in topography.
- Residual stress stability is often diminished by growing workhardening states, or degrees of deformation, induced by the surface treatment. Analogously, there is a relaxation of residual stresses which occurs more rapidly in higher strength material states than in low strength states, due to the presence of a greater driving force for dislocation movement. This also causes corresponding changes of the activation enthalpies. The relaxation of residual stresses is more rapid directly at the surface than it is beneath the surface. Internally, however, no further change of residual stress is observable. Thus, the residual stress distributions that remain after thermal loading may be described by an adapted

Avrami function. Transient behavior during heating can also be described by means of the Avrami function. This is carried out by examining isothermal sections on the basis of the so-called stress-transient method.

- The changes of the workhardening state within immediate proximity of the surface are partly accelerated. They can be reduced to values smaller than in the core if an unstable dislocation structure was created by deformation close to the surface. Generally, the relaxation of micro-residual stresses is noticeably delayed, compared to the relaxation of macro-residual stresses, because it requires not only dislocation movement but also dislocation annihilation.
- The microstructure can also be changed by re-crystallization, precipitation hardening and coarsening carbides or sub-grains. Annealing subsequent to mechanical surface treatments can make use of certain microstructural changes for optimizing the surface layer. Steels show coherency stresses caused by static strain aging. They indicate the presence of semi-coherent finest carbides, pinning dislocations while relaxing residual stresses slightly. In titanium and aluminum alloys, on the other hand, annealing can cause processes of re-crystallization or precipitation. These coincide with a substantial, albeit incomplete, relaxation of residual stress, yet they may create fine, or equiaxial instead of lamellar, grains or cause selective or preferred surface hardening. Finally, increases of hardness or flow stress were observed, caused by strain aging effects as well as by developing precipitation.

During **quasi-static loading** of mechanically surface-treated material states, the Bauschinger effect constitutes a significant elementary process, particularly in respect to higher strength material states.

- The description of the deformation behavior and of residual stress relaxation is often carried out by means of a uniaxial surface-core model. Sometimes, a biaxial model is employed. In this type of model, the component or specimen cross section is divided into a surface layer and a core region, both showing constant compressive and tensile residual stresses. A biaxial multi-layer model and a Finite-Element simulation were developed as supplements. The level of precision increases from the surface-core model to Finite-Element models and up to the multi-layer model. The surface-core model is used primarily to describe the start of residual stress relaxation. Thus, it is possible to gain an understanding of fundamental correlations, such as residual stresses changing mathematical signs. The multi-layer model and the Finite-Element model, on the other hand, incorporate the multiaxiality of the stress state. The initial residual stress- and workhardening gradients are observed using a greater degree of local resolution. This allows for more precise descriptions of the dependency of the longitudinal and transversal residual stresses on loading stress. The multi-layer model even makes it possible to describe the deformation behavior of the compound. The deformation behavior of the individual layers is the input required. The efforts made to obtain this data determines how well the relaxation of residual stresses is described.

- Tensile loading completely removes longitudinal stresses if the surface layer shows more workhardening than the bulk material. Additionally, residual stresses show multiple mathematical sign changes if the workhardening rate in the workhardened surface layer is lower than it is in the core area. The material state's ductility is another crucial factor, because generally, residual stress reductions will not be complete when elongations to fracture are small. Compressive loading almost always causes a complete removal of longitudinal residual stresses. Here, ductility is of secondary importance, and the range of loading stresses that can be accessed on an experimental level is limited only by buckling loads. In the surface layer, this complete removal is, at least, an indication of a reduced workhardening rate, and may even indicate worksoftening caused by the Bauschinger effect. An assessment of bending loading must also take into account the supportive effect of those sub-surface regions that are loaded to a lesser degree. An assessment of the effects of torsional loading must additionally take into account the multiaxiality of the loading stress state.

- The surface-core model uses an equation to assess the equivalent stress which results from loading- and residual stresses. This equation also allows for deriving compressive yield strengths at the surface from the loading stress required for residual stress relaxation to set in. While soft material states show workhardening, worksoftening tends to occur when hardness is greater, which can be attributed to an increasing Bauschinger effect. In addition, quenchend steels show evidence of static strain aging. Modified processes, such as warm peening, clearly increase the stability of residual stresses, as compressive yield strength at the surface is increased due to strain aging. Despite being annealed, quenched states show noticeable workhardening when the reduction of hardness is taken into account. Qualitatively and quantitatively similar effects are caused by annealing subsequent to conventional peening treatments. An assessment of the kinetics of the annealing effects shows that strain aging by carbon diffusion is the factor which initially determines the remaining level of residual stresses. The thermal relaxation of residual stress which occurs at longer times, however, is dominated by creep processes.

- Assessing distributions found after tensile loading shows that the workhardening states of the surface layer and the core tend to approximate each other when deformations are great enough. In nearly all cases, the surface values need to be measured using an x-ray line profile analysis in order to permit an assessment of the microstructural processes. The characteristics of the workhardening state show initial minima during compressive and tensile loading, as the dislocation structure is rearranged from an arrangement determined by high-strain rate deformation into an energetically preferable one. During compressive loading, these minima appear at lower strain levels than they do in the case of tensile loading. The reason is that compressive loading effects an immediate onset of plastification at the surface, and plastification is not initially restricted to the core area. A general workhardening occurs when deformations are great enough.

How stable surface layer states created by mechanical surface treatments are during **cyclic loading** is primarily determined by cyclic deformation behavior, and by quasi-static yield strength and cyclic yield strength and their difference to fatigue limit. In steels, rising hardness increases quasi-static and cyclic yield strength more than it increases fatigue limit. Therefore, the stability of the surface layer increases substantially as hardness grows.

- The models describing surface layer changes that occur during fatigue loading necessarily follow the fatigue loading phases. After the first loading cycle – which, apart from the single inverse loading, is determined by the aforementioned phenomena that occur during quasi-static loading – the phase of cyclic deformation without crack initiation is crucial. The residual stress changes which take place in this phase are the more pronounced the greater the number of cycles and the stress amplitude are. There are a number of empirical relations which describe this behavior. The relation describing residual stress relaxation as proportional to the logarithm of the number of cycles and to stress amplitude has proven to be the most useful one. In addition, the changes of the residual stress state can be correlated well to the plastic strain amplitudes which occur e.g. up to half the number of cycles to failure. By contrast, the residual stress relaxation which occurs during the crack propagation phase so far has not been covered by the existing models.

- Examining the start of residual stress relaxation during cyclic loading, on the other hand, allows for estimating cyclic yield strength at the surface and thus, for assessing effects of surface layer workhardening on behaviour during cyclic loading. This may be done in two ways. Firstly, aforementioned approach for a description of residual stress relaxation in the fatigue phase before crack initiation can be extrapolated up to residual stresses that are independent of the number of cycles and therefore, stable. Secondly, it is also possible to determine the stress amplitude which is required for continuous residual stress relaxation between the first and e.g. the 10^{4th} loading cycle. Both methods yield satisfactory descriptions of the resulting cyclic yield strengths at the surface if calculations are based on the residual stress values present after the first cycle. In steels, cyclic yield strengths at the surface show an increase with growing material hardness which is noticeably smaller than the increase of the cyclic yield strength of the bulk material, as the Bauschinger effect becomes increasingly important. Given an otherwise equivalent material state, there is evidence that great residual stresses – such as those present after stress peening – relax more rapidly than smaller ones. Warm peening treatments and conventional peening treatments plus subsequent annealing create similar initial residual stresses, but strain aging effects result in significant increases of cyclic yield strength at the surface and thus, in residual stresses which are more stable.

- The workhardening states close to the surface which have been increased by surface treatment are generally reduced as the number of cycles rises. States exhibiting strain aging effects close to the surface – i.e. warm peened states and states created by conventional peening and subsequent annealing – show workhardening in the surface layer that is particularly stable during cyclic loading or is increased even further. The reason is that due to the small number of mobile dislocations, plastic deformation requires a multiplication of dislocations. The occurrence of typical fatigue dislocation structures is prevented or, at least, delayed.

- Microstructural changes themselves also generally tend to approximate the states of the surface layer and the core, i.e. the microstructure shows typical signs of fatigue, such as cells. Due to workhardening effects, increases of stress amplitudes may occur close to the surface, resulting in smaller cells appearing there. In many cases, however, it is impossible to completely remove the microstructural changes created by pre-deformation.

In simple models, the effects of surface layer states and their stability on **fatigue strength** are divided and additively applied to the individual changes of the surface layer state. Residual stress Haigh diagrams can be used to assess the effects of residual stresses on fatigue strength. These diagrams either assume that residual stresses are stable, or take into account the cyclic relaxation of residual stress which occurs when cyclic yield strength at the surface is exceeded. A more detailed examination, which includes the workhardening effects close to the surface, the topographical changes and the notch- and loading-induced stress gradients, is provided by the concept of the local fatigue limit. If, on the other hand, the ability of crack propagation shows a significant influence, it is possible to use the concept based on linear-elastic fracture mechanics which was also discussed above. The effects of surface layer states on fatigue loading behavior can be examined separately for the individual phases of fatigue.

- Cyclic deformation behavior shows signs of worksoftening to appear sooner, e.g in normalized steels, due to the superposition of loading- and residual stresses, while pronounced workhardening generally reduces the plastic strain amplitudes that occur. However, if mechanical surface treatments are followed by annealing, the subsequent alternating deformation behavior shows that worksoftening occurs later and is reduced. Notched states represent a special case, in which loading without mean stresses can cause a cyclic contraction of the specimen. The reason for this is that residual stress relaxation in notched specimens occurs in a locally concentrated form and the increments of plastic deformation occur primarily during the half cycle of compressive loading. Cyclic stress-strain curves obtained for trepanned specimens after shot peening or deep rolling show noticeable workhardening effects.

- Crack initiation is essentially determined by the size of the plastic strain amplitude and is therefore delayed by the workhardening effects. However, some of the changes in roughness are very important. The smoothing effects of deep rolling and laser shock treatment, at any rate, generally delay the development of crack initiation. Small loading stress gradients and stable residual stresses, such as those found in high strength steels, can shift the position of crack initiation from the surface to a location below the surface.

- Crack propagation is also influenced by the state of the surface layer, particularly by residual stresses. However, this effect can hardly be examined separately, because the areas around the crack tips always develop typical residual stress fields superposing those stemming from surface treatment. Accordingly, mechanical surface treatments have an effect on crack propagation rate. This applies particularly to crack lengths smaller than the depth effect of the mechanical surface treatment. It is primarily the compressive stresses introduced, as well as the workhardening states, which can significantly delay or even stop crack propagation.

- Steels, in particular, have been the subject of systematic studies of the effects which the surface layer states induced by mechanical surface treatments have on fatigue life and fatigue strength. The influence of the different surface layer characteristics results in different stabilities of the surface layer states and, especially, of the residual stresses. This suggests a division into low-, medium- and high strength states. In low strength material states, residual stresses are relaxed significantly. Therefore, they have almost no influence on fatigue limit. Only the workhardening state created by the surface treatment is an essential factor. Medium strength material states, on the other hand, show residual stresses that usually remain stable, at least during loading which does not exceed the fatigue limit. Therefore, the level and distribution of residual stresses significantly influence fatigue limit. Workhardening close to surface and surface roughness can cause significant changes in fatigue limit. Small loading stress gradients, as found e.g. in smooth states, abet crack initiation in the region beneath the surface layer influenced by mechanical surface treatment. In notched states, therefore, significant increases of the fatigue limit are possible, as the position of crack initiation is shifted to the surface. In high strength material states, residual stresses largely remain stable not only during loading in the range of the fatigue limit, but also during loading in the high cycle fatigue range. Therefore, their influence on the fatigue limit is significant when residual stress gradients are sufficiently small or when the stress gradients are sufficiently great, as e.g. in the case of notched states. The fatigue limit of notched material states, in particular, is often determined by cracks which were initiated close to the surface and are unable to propagate in the region of the residual stress maxima. Workhardening close to the surface and surface roughness, on the other hand, are less important, compared to the low- and medium strength states, as both the workhardening state and roughness experience only marginal changes.

- Modified peening treatments, such as stress peening, warm peening or annealing after conventional shot peening, can be utilized to increase fatigue strength significantly. The effects – particularly those of stress peening – are strongest in higher strength steels, due to the residual stress stability required. It is precisely these states for which annealing effects must be taken into account during warm peening and peening treatments with subsequent annealing, as these effects restrict the achievable increases of the fatigue limit.

- Light metals, for the most part, have not been examined in respect to surface layer state stability. Therefore, a simplified assumption is made that surface layer states are more stable in material states that show cyclic workhardening than they are in material states that show cyclic worksoftening. Accordingly, increases of fatigue strength are expected to occur mainly in material states that exhibit cyclic workhardening.

- Annealing after mechanical surface treatment causes re-crystallization effects. These can result in surface layer states which show extremely fine grains or whose grain morphology is altered. Thus, fatigue strength increases are achievable, particularly at elevated temperatures or in combined creep/fatigue loading. Annealing also creates selective or preferred surface layer hardening, which can be utilized to increase fatigue strength, particularly at room temperature.

Acknowledgments

This study was written during my employment as head of the section "Production and Component Behavior" at the *Institut für Werkstoffkunde I. Universität Karlsruhe (TH)*. The section focuses primarily on the study of mechanical surface treatments, resulting surface layer states and their stability during thermal and mechanical loading as well as their effects on component behavior particularly during fatigue loading. This book presents the current state of research on the subject and the essential results of my own work.

The text at hand was originally submitted as a postdoctoral thesis for qualification as an university lecturer. I am greatly indebted to Profs. O. Vöhringer and D. Löhe for giving me the opportunity to carry out my research, for consistently supporting and accompanying my endeavor and offering numerous suggestions which found their way into this book. Furthermore, I am very grateful to Profs. B. Scholtes and J. Fleischer, who kindly provided evaluative reports to the department. My sincere gratitude also goes to Dipl.-Ing. B. Gögdün, Dr.-Ing. H. Holzapfel, Dipl.-Ing. A. Lutz, Dr.-Ing. R. Menig, Dr.-Ing. J. Schwarzer, Dipl.-Ing. F. Theobald and Dr.-Ing. A. Wick. The research they accomplished in the context of their dissertations, seminar papers or theses yielded invaluable experimental results, simulation calculations and model approaches, as well as countless discussions, only the sum of which allowed for the systematic description presented in this study.

Mrs. C. Polixa, Ms. S. Willms, Ms. C. Lauer and Mr. J. Zum Gahr achieved the graphic representation of the results by supplying the schematic diagrams and graphs. I offer my deepest appreciation and thank them for their essential contribution to the manuscript. Lastly, I am especially grateful to Mr. J.K. Schwing for carrying out the complex translation of the German text.

Index

a

ablation   102
absorbing layer   103
athermal mechanisms   104
Avrami function   142

b

back stress models   29
Bauschinger effect   29, 72, 179, 205

c

cell structures   68
compressed-air peening systems   11
concept of the local fatigue limit   3, 257
confined ablation   102
consecutive treatments   100
contact radius   31
Cottrell-atmospheres   2, 82, 83
creep processes   135
critical loading stress amplitude   284, 288
cross slip   105
cyclic deformation   35
cyclic yield strength   253, 288, 324

d

deep rolling   17
deformation-induced martensite   99
density   70
dependence of deformation behavior on
   temperature   27
direct ablation   102
discrete-element calculations   37
dislocation annihilations   163
dislocation core diffusion   135
dislocation density   64
dislocation models   29
dislocation tangles   68
dynamic strain aging   30, 82

e

electron viscosity   28
elevated temperatures   270

f

FEM *see* finite-element
FEM model   211
finest carbides   82
finish rolling   16
finite-element simulation   37, 90, 182, 205

g

Goodman relation   255
grain anisotropy models   29

h

hardening effects   154
Hertzian pressure   28, 33, 53, 89
high strength steels   328
Hugoniot stress   104, 107

i

impact angle   59
injector gravitational peening   11
injector peening systems   11

l

light metals   335
low strength steels   321

m

mass flow rate   57
mean stresses   251
media for shot peening   12
medium strength steels   322
micro-hardness   66
micro-residual stresses   64
microstructure   67, 98, 119
multi-layer model   182, 208

Modern Mechanical Surface Treatment. Volker Schulze
Copyright © 2006 WILEY-VCH Verlag GmbH & Co. KGaA, Weinheim
ISBN: 3-527-31371-0

n

nano-crystalline surface layers   99
Norton approach   139, 143
notched states   276
number of revolutions   57

o

obstruction of plastic deformation   61
Ostwaldt-ripening   82

p

peening pressure   55
phase fractions   70, 99, 121
phonon viscosity   28
plasma   102
plastic strain amplitudes   247
precipitation hardening   2, 137
preferred surface hardening   137
propagation of initiating cracks   259

r

re-crystallization   2, 135, 154
residual stress Haigh-diagrams   255, 257,
    324, 334
residual stress sensitivity   255
retained austenite   70, 99
rotating wheel peening   11

s

selective surface hardening   137
shock waves   102

shot velocity   55
size rolling   17
Snoek Effect   82, 83
stacking fault energy   30, 68, 105
static strain aging   30, 81, 217, 270
strain aging effects   135, 153
stretching   28, 53, 103
surface-core model   180

t

texture   71, 100
thermal activation   104
thermally activated slip   27
time-temperature-precipitation diagrams
    137
torsional prestress   76
torsional swelling loading   276
twin formation   104

u

ultrasonic shot peening   12
uniaxially prestressed   74

v

viscous damping   28, 40
volume diffusion   135

z

Zener-Wert-Avrami equation   137